Recent Advances in Mathematics for Engineering

Mathematical Engineering, Manufacturing, and Management Sciences

Series Editor:
Mangey Ram
*Professor, Department of Mathematics, Computer Science and Engineering,
Graphic Era Deemed to be University, Dehradun, India*

The aim of this new book series is to publish the research studies and articles that bring up the latest development and research applied to mathematics and its applications in the manufacturing and management sciences areas. Mathematical tool and techniques are the strength of engineering sciences. They form the common foundation of all novel disciplines as engineering evolves and develops. The series will include a comprehensive range of applied mathematics and its application in engineering areas, such as optimization techniques, mathematical modeling and simulation, stochastic processes and systems engineering, safety-critical system performance, system safety, system security, high assurance software architecture and design, mathematical modeling in environmental safety sciences, finite element methods, differential equations, and reliability engineering.

Sustainable Procurement in Supply Chain Operations
*Edited by Sachin Mangla, Sunil Luthra, Suresh Jakar, Anil Kumar,
and Nirpendra Rana*

Mathematics Applied to Engineering and Management
Edited by Mangey Ram and S.B. Singh

Mathematics in Engineering Sciences
Novel Theories, Technologies, and Applications
Edited by Mangey Ram

Advances in Management Research
Innovation and Technology
*Edited by Avinash K. Shrivastava, Sudhir Rana, Amiya Kumar Mohapatra,
and Mangey Ram*

Market Assessment with OR Applications
Adarsh Anand, Deepti Aggrawal, and Mohini Agarwal

Recent Advances in Mathematics for Engineering
Edited by Mangey Ram

For more information about this series, please visit:
*https://www.crcpress.com/Mathematical-Engineering-Man-facturing-and-Manage
ment-Sciences/book-series/CRCMEMMS*

Recent Advances in Mathematics for Engineering

Edited by
Mangey Ram

CRC Press
Taylor & Francis Group
Boca Raton London New York

CRC Press is an imprint of the
Taylor & Francis Group, an **informa** business

CRC Press
Taylor & Francis Group
6000 Broken Sound Parkway NW, Suite 300
Boca Raton, FL 33487-2742

First issued in paperback 2021

© 2020 by Taylor & Francis Group, LLC
CRC Press is an imprint of Taylor & Francis Group, an Informa business

No claim to original U.S. Government works

ISBN 13: 978-0-367-19086-6 (hbk)
ISBN 13: 978-1-03-224023-7 (pbk)

DOI: 10.1201/9780429200304

Library of Congress Cataloging-in-Publication Data

Names: Ram, Mangey, editor.
Title: Recent advances in mathematics for engineering / edited by Mangey Ram.
Description: Boca Raton, FL : CRC Press, Taylor & Francis Group, [2020] |
Series: Mathematical engineering, manufacturing, and management sciences |
Includes bibliographical references and index.
Identifiers: LCCN 2019051106 (print) | LCCN 2019051107 (ebook) | ISBN
9780367190866 (hardback : acid-free paper) | ISBN 9780429200304 (ebook)
Subjects: LCSH: Engineering mathematics.
Classification: LCC TA330 .R38 2020 (print) | LCC TA330 (ebook) | DDC
620.001/51—dc23
LC record available at https://lccn.loc.gov/2019051106
LC ebook record available at https://lccn.loc.gov/2019051107

**Visit the Taylor & Francis Web site at
http://www.taylorandfrancis.com**

**and the CRC Press Web site at
http://www.crcpress.com**

Publisher's Note
The publisher has gone to great lengths to ensure the quality of this reprint but points out that some imperfections in the original copies may be apparent.

Contents

Preface

In recent years, mathematics has had an amazing growth in engineering sciences. Everything would be possible with the addition of mathematics to the engineering. *Recent Advances in Mathematics for Engineering* engrosses on a comprehensive range of mathematics applied in various fields of engineering. The topics covered are organized as follows.

Chapter 1 summarizes the current position of surveillance methods in public health and focuses on the inferential part of the surveillance problem including statistical techniques and stochastic modeling for evaluating surveillance systems engineering. Statistical change-point analysis-based methods have several appealing properties compared to the current practice.

Chapter 2 reviews the methods for assessment of earthquake hazards based on statistical models, probability theory, and nonlinear analysis. Also, the steps involved in the assessment of seismic hazard, ground response analysis, and calculation of liquefaction potential have been discussed.

Chapter 3 discusses the use of satellite data in multi-model approach. The main idea is to use the inter-connected models of environment and disaster-forming advances to select the control variables and indicators for satellite observations.

Chapter 4 gives a review of some integral transforms, Parseval–Goldstein-type relationships and its applications to integral transforms, and some well-known differential equations. Laplace transform has been used to generalize Stieltjes transform.

Chapter 5 provides the numerical solution of two kinds of singular integral equations. This chapter proposes the numerical methods to find the approximate solution of Cauchy singular integral equations and hypersingular integral equations.

Chapter 6 discusses an iterative scheme known as generalized minimal residual method (GMRES) to solve linear partial differential equations. This scheme has been used here to solve a transient 2D heat equation with Dirichlet boundary conditions.

Chapter 7 represents the numerical simulation of $(2 + 1)$ dimensional nonlinear sine–Gordon soliton waves. For simulation, hybrid cubic B-spline differential quadrature method has been used.

Chapter 8 presents a qualitative analysis of growth and development processes that involves species distribution and their interplay of spatially distributed populace with diffusion and obtains the stipulations of Hopf and Turing bifurcation in a spatiotemporal region.

Chapter 9 summarizes the contribution and development of Runge–Kutta methods in the field of numerical analysis to solve the ordinary differential equations. The different methods of this category are studied to discuss their applicability and to compare their relative accuracies.

Chapter 10 provides a method to generate all non-dominated points for a tri-objective integer problem. The proposed approach has an advantage: it reduces the number of sub-problems solved and hence the central processing unit (CPU) time.

Chapter 11 investigates a link-weight modification philosophy for solving network optimization problems. The purpose and process of link-weight modification have been examined to understand the philosophy behind these link-weight modification approaches.

Chapter 12 discusses a system for the automatic detection of photorealistic computer-generated images (PRCG). The proposed system utilizes rich models for differentiating PRCG from photographic images (PIM) and achieves close to 99% classification accuracy in many scenarios.

Chapter 13 investigates the flow characteristics of swirling subsonic annular jets based on the swirler vane angle (0°, 25°, and 50°). It has been concluded that the increase in the swirl angle results in the linear increase of swirl number, length of the recirculation zone, and streamwise mass entrainment rate.

Chapter 14 deliberates the reliable tool for solving all types of linear programming problems of real world which would involve industrial and business problems. C language has been used to solve typical problems of linear programming in convergence of simplex method.

Chapter 15 explains the classical inventory control model with reliability influence demand and partially backlogged items. The primary motive of efficient inventory management is to provide a suitable customer service, thereby keeping a low cost of the inventory system. Thus, the aim of this chapter is to minimize the cost of the inventory system.

Chapter 16 discusses a very sensible approach, namely, optimal replenishment policy for non-instantaneous deteriorating items with two storage facilities, with two more major parameters, i.e., multi variate demand and under shortages inflation, which is partially backlogged.

This book will be very useful to the undergraduate and postgraduate students of engineering; engineers; research scientists; and academicians involved in the mathematics and engineering sciences.

Mangey Ram
Graphic Era (Deemed to be University), India

Acknowledgments

The Editor acknowledges CRC press for this opportunity and professional support. My special thanks to Ms. Cindy Renee Carelli, Executive Editor, CRC Press – Taylor & Francis Group for the excellent support she provided me to complete this book. Thanks to Ms. Erin Harris, Editorial Assistant to Ms. Carelli, for her follow-up and aid. Also, I would like to thank all the chapter authors and reviewers for their availability for this work.

Mangey Ram
Graphic Era (Deemed to be University), India

Editor Biography

Dr. Mangey Ram received his Ph.D. degree major in Mathematics and minor in Computer Science from G. B. Pant University of Agriculture and Technology, Pantnagar, India. He has been a faculty member for around eleven years and has taught several core courses in pure and applied mathematics at undergraduate, postgraduate, and doctorate levels. He is currently a professor at Graphic Era (Deemed to be University), Dehradun, India. Before joining the Graphic Era, he was a deputy manager (Probationary Officer) in Syndicate Bank for a short period. He is editor-in-chief of *International Journal of Mathematical, Engineering and Management Sciences*, and the guest editor and member of the editorial board of various journals. He is a regular reviewer for international journals, including IEEE, Elsevier, Springer, Emerald, John Wiley, Taylor & Francis Group, and many other publishers. He has published 150 plus research publications in IEEE, Taylor & Francis Group, Springer, Elsevier, Emerald, World Scientific, and many other national and international journals of repute, and also presented his works at national and international conferences. His fields of research are reliability theory and applied mathematics. He is a senior member of the IEEE and Life Member of Operational Research Society of India, Society for Reliability Engineering, Quality and Operations Management in India, Indian Society of Industrial and Applied Mathematics, Member of International Association of Engineers in Hong Kong, and Emerald Literati Network in the U.K. He has been a member of the organizing committee of a number of international and national conferences, seminars, and workshops. He has been conferred with the "Young Scientist Award" by the Uttarakhand State Council for Science and Technology, Dehradun, in 2009. He has been awarded the "Best Faculty Award" in 2011; "Research Excellence Award" in 2015; and recently "Outstanding Researcher Award" in 2018 for his significant contribution in academics and research at Graphic Era Deemed to be University, Dehradun, India.

Contributors

Deo Datta Aarya
Department of Mathematics
Acharya Narendra Dev College (Delhi
  University)
New Delhi, India

Ali Al-Hasani
Department of Mathematical and
  Geospatial Sciences, School of
  Sciences
RMIT University
Melbourne, Australia
and
Faculty of Sciences, Department of
  Mathematics
Basrah University
Basrah, Iraq

Masar Al-Rabeeah
Department of Mathematical and
  Geospatial Sciences, School of
  Sciences
RMIT University
Melbourne, Australia
and
Faculty of Sciences, Department of
  Mathematics
Basrah University
Basrah, Iraq

Geeta Arora
Department of Mathematics
Lovely Professional University
Phagwara, India

Igor Artemenko
Scientific Centre for Aerospace
  Research of the Earth
National Academy of Sciences of
  Ukraine
Kiev, Ukraine

Anand Chauhan
Department of Mathematics
Graphic Era (Deemed to be University)
Dehradun, India

Neeraj Dhiman
Department of Mathematics
Graphic Era Hill University
Dehradun, India

Phillemon Dikgale
Department of Statistics and Operations
  Research, School of Mathematical
  and Computer Sciences
University of Limpopo, Turf loop
  Campus
Sovenga, South Africa

Santosh Dubey
Department of Physics
University of Petroleum & Energy
  Studies
Dehradun, India

Ramu Dubey
Department of Mathematics
J.C. Bose University of Science and
  Technology
Faridabad, India

Andrew Eberhard
Department of Mathematical and
    Geospatial Sciences, School of
    Sciences
RMIT University
Melbourne, Australia

Anter El-Azab
Material Science & Engineering
Purdue University
West Lafayette, Indiana

Lesia Elistratova
Scientific Centre for Aerospace
    Research of the Earth
National Academy of Sciences of
    Ukraine
Kiev, Ukraine

Sahil Garg
School of Mechanical Engineering
Lovely Professional University
Punjab, India

Isa Sani Garki
College of Science and Technology
Jigawa State Polytechnic
Dutse, Nigeria

Ashwani Jain
Department of Civil Engineering
National Institute of Technology
    Kurukshetra
Kurukshetra, India

Varun Joshi
Department of Mathematics
Lovely Professional University
Phagwara, India

Sharad Joshi
Multimedia Analysis and Security
    (MANAS) Lab, Electrical
    Engineering
Indian Institute of Technology
    Gandhinagar (IITGN)
Gujarat, India

Emmanouil-Nektarios Kalligeris
Laboratory of Statistics and Data
    Analysis
University of the Aegean
Lesbos, Greece

Alex Karagrigoriou
Laboratory of Statistics and Data
    Analysis
University of the Aegean
Lesbos, Greece

Nitin Khanna
Multimedia Analysis and Security
    (MANAS) Lab, Electrical
    Engineering
Indian Institute of Technology
    Gandhinagar (IITGN)
Gujarat, India

Yuriy V. Kostyuchenko
Scientific Centre for Aerospace
    Research of the Earth
National Academy of Sciences of
    Ukraine
Kiev, Ukraine

Mukesh Kumar
Department of Mathematics
Graphic Era (Deemed to be University)
Dehradun, India

Santosh Kumar
Department of Mathematics and
    Statistics
University of Melbourne
and
Department of Mathematical and
    Geospatial Sciences, School of
    Sciences
RMIT University
Melbourne, Australia

Sunil Kumar
Department of Mathematics
Graphic Era Hill University
Dehradun, India

Mohit Lamba
Multimedia Analysis and Security
   (MANAS) Lab, Electrical
   Engineering
Indian Institute of Technology
   Gandhinagar (IITGN)
Gujarat, India

Angeliki Lambrou
Department of Epidemiological
   Surveillance and Intervention
National Public Health Organization
Athens, Greece

'Maseka Lesaoana
Department of Statistics and Operations
   Research, School of Mathematical
   and Computer Sciences
University of Limpopo, Turf loop
   Campus
Sovenga, South Africa

Elias Munapo
School of Economics and Decision
   Sciences
North West University
Mafikeng, South Africa

Philimon Nyamugure
Department of Statistics and Operations
   Research, School of Mathematical
   and Computer Sciences
University of Limpopo, Turf loop
   Campus
Sovenga, South Africa

R.K. Pandey
Department of Mathematics, D.B.S.
   College
Hemvati Nandan Bahuguna Garhwal
   University
Dehradun, India

Mohit Pant
Mechanical Engineering Department
National Institute of
   Technology-Hamirpur
Hamirpur, India

Christina Parpoula
Laboratory of Statistics and Data
   Analysis
University of the Aegean
Lesbos, Greece

Prakhar Pradhan
Multimedia Analysis and Security
   (MANAS) Lab, Electrical
   Engineering
Indian Institute of Technology
   Gandhinagar (IITGN)
Gujarat, India

Nitish Puri
Department of Civil Engineering
Indian Institute of Technology Delhi
New Delhi, India

Neelanjana Rajput
Department of Mathematics, D.B.S.
   College
Hemvati Nandan Bahuguna Garhwal
   University
Dehradun, India

Seema Saini
Department of Mathematics
Graphic Era (Deemed to be) University
Dehradun, India

Gopinath Shanmugaraj
School of Mechanical Engineering
Lovely Professional University
Punjab, India

Vaishali Sharma
Department of Mathematics
BITS Pilani K K Birla Goa Campus
Goa, India

Sanjeev K. Singh
Department of Mathematics
University of Petroleum & Energy
    Studies
Dehradun, India

S. J. Singh
Department of Mathematics
Graphic Era (Deemed to be University)
Dehradun, India

Teekam Singh
Department of Mathematics
Graphic Era (Deemed to be) and
    Graphic Era Hill University
    Dehradun
Dehradun, India

Mohammad Tamsir
Department of Mathematics
Graphic Era (Deemed to be University)
Dehradun, India

Vinay Verma
Multimedia Analysis and Security
    (MANAS) Lab, Electrical
    Engineering
Indian Institute of Technology
    Gandhinagar (IITGN)
Gujarat, India

Osman Yürekli
Department of Mathematics
Ithaca College
Ithaca, New York

Maxim Yuschenko
Scientific Centre for Aerospace
    Research of the Earth
National Academy of Sciences of
    Ukraine
Kiev, Ukraine

1 Statistical Techniques and Stochastic Modeling in Public Health Surveillance Systems Engineering

Emmanouil-Nektarios Kalligeris,
Alex Karagrigoriou, and Christina Parpoula
University of the Aegean

Angeliki Lambrou
National Public Health Organization

CONTENTS

1.1   INTRODUCTION

1.1.1   PRELIMINARIES

Population health is considered to be one of the most valuable commodities that lies at the heart of interest of both the society and the health profession (Starfield et al., 2005). Environmental changes, socioeconomic conditions, and changes in the epidemiology of diseases, along with the burden they cause on humanity, are the main axes that make public health surveillance necessary (see Teutsch and Thacker, 1995; Heath and Smeeth, 1999; Norbury et al., 2011). Public health surveillance can be defined as the "ongoing, systematic collection, analysis, interpretation, and dissemination of data regarding a health-related event that enables public health authorities to reduce morbidity and mortality" (Sosin, 2003). Surveillance provides services for various functions like the estimation of the burden of a disease, the identification of the probability distribution of an illness, the proposal of new research problems, the support and evaluation of prevention and control measures, and finally, facilitating planning (Sosin, 2003). The most significant scope of surveillance is the detection of an outbreak, namely the ability to identify an unusual increase in the disease frequency. However, the syndromic surveillance which is analyzed later is an approach for detecting early enough an outbreak by extending current capabilities.

Along with the rapid advancements in the fields of computing, engineering, mathematics, statistics, and public health, a potentially powerful science of surveillance known as biosurveillance is emerged. The primary challenges in this scientific field are associated with early and accurate outbreak detection to ensure the implementation of effective control measures. The field dealing with these issues and specifically with disease detection is known as biosurveillance (Shmueli and Burkom, 2010; Wagner et al., 2006; Dato et al., 2006). This chapter aims at providing a brief overview of what is being done in the field and how it can be improved with future research considerations.

1.1.2   DEFINITION OF BIOSURVEILLANCE

As indicated earlier, biosurveillance is a continuous process which monitors disease activity. By disease activity, it is meant to encompass not only the emergence and/or manifestation of the disease, but also the preliminary processes involved in the development and/or evolution of the disease. The operational definition of biosurveillance, given by Homeland Security Presidential Directive (HSPD-21, 2014), is that it is "the process of active data-gathering with appropriate analysis and interpretation of biosphere data that might relate to disease activity and threats to human or animal health whether infectious, toxic, metabolic, or otherwise, and regardless of intentional or natural origin in order to achieve early warning of health threats, early detection of health events, and overall situational awareness of disease activity."

It is worth to be noted that biosurveillance, as a science, is relatively young in terms of its origin since the full-thrust of research in this field started only in the early 1990s, along with the emergence of computers and automation. Biosurveillance is a systematic and evolutionary integration of disease and public health surveillance that finds similarities between them in the aspect of systematic data collection and analytical processes for disease detection (Thacker and Berkelman, 1998; Teutsch and Churchill, 2000; Rothman and Greenland, 1998). In this type of processes, case detection is inherent; by case detection, it is meant to be the act of noticing the existence of a single individual with a disease. The evolution, however, is evident on the broader scope of biosurveillance, in the sense that it includes outbreak detection and characterization at a higher rate compared to both disease and public health surveillance. Outbreak detection is a collection of methods and techniques for the identification of an outbreak, whereas outbreak characterization is the mechanism used by researchers to elucidate the outbreak main characteristics. Outbreak detection together with outbreak characterization constitutes the two components of biosurveillance (Wagner et al., 2006). This chapter focuses on a range of statistical and stochastic modeling techniques designed to serve the purposes of early and accurate outbreak detection.

1.1.3 OBJECTIVES OF BIOSURVEILLANCE

As discussed in the previous section, biosurveillance is an evolutionary process which combines disease and public health surveillance. Hence, one of the vital processes in biosurveillance is epidemiological surveillance, defined to be "the process of actively gathering and analyzing data related to human health and disease in a population in order to obtain early warning of human health events, rapid characterization of human disease events, and overall situational awareness of disease activity in the human population" (HSPD-21, 2014). The main goals of epidemiological surveillance are to monitor the distribution and trends of the diseases incidence as well as to design, implement, and evaluate health policies and public health actions after their further processing (Langmuir, 1995), with the ultimate objective to reduce morbidity and mortality, and thus improve population health indicators (Macmahon and Trichopoulos, 1996). Additionally, the present-day definition of epidemiological surveillance includes some key aspects such as ensuring the validity of data, implementing advanced statistical methods for data analysis, as well as deriving scientifically and methodologically adequate conclusions (Fleming and Rotar-Pavlic, 2002). Again, as a result of an evolutionary process, a key difference between traditional epidemiological surveillance and the emerging science of biosurveillance is the development of syndromic surveillance (Fricker, 2013). Sosin (2003) defines syndromic surveillance as the "ongoing, systematic collection, analysis, interpretation, and application of real-time (or near-real-time) indicators of diseases and outbreaks that allow for their detection before public health authorities would otherwise note them." From this definition, it should be noted that the latter process derives from the notion of syndromes, defined as "a set of non-specific pre-diagnosis medical and other information that may indicate the release of a natural disease outbreak or a bioterrorism agent." These syndromes serve as signals for detection of possible high-impact disease-related events (e.g., outbreaks) in the context of biosurveillance processes.

1.1.4 BIOSURVEILLANCE SYSTEMS AND PROCESSES

Biosurveillance is a multidisciplinary science traditionally involving expertise from the fields of epidemiology, medicine, microbiology, veterinary, public health, and health care. Nowadays, as part of the field's evolution, the increased possibility of more powerful biological threats and activities has led this new scientific area to diversify its pool of expertise into a more computer-oriented approach. It has now borrowed expertise from the fields of mathematics, (bio)statistics, computer science, and systems and quality engineering, focusing on the idea of evidenced-based surveillance processes. Moreover, the importance of the latter fields is reflected in the need of conducting biosurveillance at real time and sometimes in forms of big data; hence, a necessity for timely and efficient automation has emerged as pointed out by Wagner et al. (2006).

In the 21st century, the advancements in the field of computing led to newer developments of automated biosurveillance systems. However, such systems, whether automated, manual, or a mixture of both, must still be systematic in terms of its functionality as emphasized by Wagner (2002). A biosurveillance system, as with any engineering system, should be able to meet its functional requirements in order to be considered operational. Such requirements involve specifications of the diseases to be detected and the time frame within which detection occurs (Wagner et al., 2006). It is worth to be noted that the main requirement of such systems is to be able to recognize threat patterns.

1.1.5 OBJECTIVES, GOALS, AND CHALLENGES

According to the above discussion, biosurveillance can be considered as a dynamic activity directly connected to developments and advances in the general area of biosciences. The health community is constantly in search for the early and accurate prediction of the time of an outbreak. As a result, further advances based on various statistical models and methods have been developed in numerous countries in Europe (European Centre for Disease Prevention and Control – ECDC) (Hulth et al., 2010) and the States (Centers for Disease Control and Prevention – CDC). The main challenges in biosurveillance are related to data source, quality control, the monitoring (follow-up), the evaluation of statistical methods for outbreak detection, anomalies and/or outliers in the data, and extreme timeliness of detection.

The combination of a new requirement of timeliness, a high level of applied work building early warning systems and a set of unaddressed research questions, suggests that it may be beneficial to think about the theoretical foundations of this field, what constitutes the relevant existing body of knowledge, and how this scientific field can accommodate the applied work. Towards this end, this chapter aims at the implementation and evaluation of several cutting-edge statistical and stochastic modeling techniques for the automated, accurate, early detection of outbreaks in biosurveillance systems. In particular, this chapter addresses the following issues: (i) a brief literature review for the identification of the mathematical foundations of outbreak detection and the investigation of open relevant research issues; (ii) the development of a variety of novel periodic regression models (with emphasis on the best models for monitoring); (iii) the proposal of guidelines for the implementation and effective use of the change-point analysis mechanism for the detection of epidemics; (iv) the analysis and comparison of the proposed methodologies in terms of several performance metrics, via a retrospective analysis of real epidemiological data.

Conclusively, progress is expected to be facilitated as long as the scientific emphasis will be placed on the valid and very early outbreak detection. This perspective indicates the significance of putting forward the questions to be addressed and identifying valuable prior knowledge and methods from related scientific fields. The range of statistical and stochastic modeling techniques studied in this chapter clearly shows that outbreak detection is an interdisciplinary field where statisticians, informaticians, public health practitioners and experts, engineers, and bioscientists are involved. Progress is also likely provided that outbreak detection is recognized as a discipline with "big data" characteristics. This chapter will bring together the public health community with researchers from Big Data Analysis, Epidemiology and Statistical and Disease Modelling with the ultimate aim of furnishing, for a number of infectious diseases, tools and techniques for situational awareness and outbreak response.

The rest of this chapter is organized as follows. In Section 1.2, the current mathematical foundations of outbreak detection are presented in brief. In Section 1.3, the statistical framework is introduced. In Section 1.4, an empirical comparative study is performed. Finally, in Section 1.5, the main conclusions are presented together with directions and ideas for future extensions and generalizations.

1.2 STATE OF THE ART

In recent years, numerous researchers focus on public health surveillance by considering and implementing advanced statistical tools and models including regression and auto-regressive processes, Bayesian and Markovian processes, and spatiotemporal models, among others. The interested reader is referred to the works of Sonesson and Bock (2003), Farrington and Andrews (2004), Buckeridge et al. (2005), Shmueli and Burkom (2010), and Unkel et al. (2012) for extensive reviews of the relevant literature.

The model proposed by Stroup et al. (1989) although relatively simple it exhibits some robust features due to the fact that the adjustment for seasonal effects is automatic as part of the design instead of the modeling. Although the seasonality is handled automatically, the same is not true for the trend which is not incorporated into the model. The classical fully parametric model for the detection of outbreaks has been proposed by Serfling (1963). The model is based on historical baselines and consists of a linear function of time describing the trend and sine and cosine terms for modeling the seasonality while the errors are assumed to be normally distributed with constant variance. Costagliola et al. (1991, 1994) consider this model for the detection of the onset of influenza epidemics. More recently, Pelat et al. (2007) considered a generalized version of Serfling's model assuming a cubic function of time representing the trend and three trigonometric (sine–cosine) terms for the modeling of the seasonality effect. The ideal model appropriate for both prospective and retrospective surveillance was chosen by the model identification procedure due to Akaike information criterion (AIC).

Regression methods can also be considered as an extension/generalization of Shewhart control charts. Statistical Process Control (SPC) is heavily relying on control charts (Oakland, 2008; Montgomery, 2013) for monitoring the characteristics of a process over time. The implementation of SPC methods to public health surveillance has a very long history. In fact, there is a number of techniques for the detection of outbreaks that are either related to or inspired by SPC methods (Woodall, 2006). Although the Shewhart chart takes into consideration only the last observation, the cumulative sum (CUSUM) chart and the exponentially weighted moving average (EWMA) control chart, proposed by Page (1954), are based on past observations. Adjustments of CUSUM and EWMA methodology for Poisson and binomial data are available in the literature (Lucas, 1985; Gan, 1991; Borror et al. 1998). Extensions, modifications, and variants of these charts were proposed over time to serve the purposes of public health surveillance (Nobre and Stroup, 1994; Rossi et al., 1999; Hutwagner et al., 2003; Rogerson and Yamada, 2004; Burkom et al., 2007; Dong et al., 2008, Höhle and Paul, 2008; Elbert and Burkom, 2009).

Outbreak detection consists of one of the most crucial parts of public health surveillance, and traditionally, its performance is based on the investigation of historic disease records. Having an accurate detection system is of high importance since it helps the authorities to control the spread of outbreak and reduces the mortality rate. There have been several works on outbreak detection methods (Long, 2012; Buehler et al., 2004; Ong et al., 2010; Shmueli and Burkom, 2010). Among them lies change-point analysis which has been proved to be a reliable tool for identifying outbreaks in scientific fields such as medical, climate, public health, speech recognition, and image analysis. There are two kinds of algorithms for change-point detection, namely, offline and online algorithms. For the former, we look back in time to identify the change point having available the entire dataset, while for the latter, the algorithm runs concurrently with the process under monitoring (see for more details Mohsin et al., 2012; Aminikhanghahi and Cook, 2017).

There exist a variety of applications of change-point analysis in disease surveillance data. Kass-Hout et al. (2012) applied various change-point detection methods to the active syndromic surveillance data to detect changes in the incidence of emergency department visits due to daily influenza-like illness (ILI). Monitoring in U.S.A. has a tendency on algorithm procedures like Early Aberration Reporting Systems (EARS) despite their limitation on detecting subtle changes and identifying disease trends. Hence, Kass-Hout et al. (2012) compared a combination of CUSUM method and EARS, and concluded that EARS method in conjunction with change-point analysis is more effective in terms of determining the moving direction in ILI trends between change points. Texier et al. (2016) made use of change-point analysis for evaluating the ability of the method to locate the whole outbreak signal. The use of the kernel change-point model led to satisfactory results for the identification of the starting and ending periods of a disease outbreak in the absence of human resources. In addition, Christensen and Rudemo (1996) studied incidence data by using tests that are modifications of well-known hypothesis tests for retrospective change-point detection. Specifically, they applied the methodology for multiple change points by means of a modification of the forward selection procedure and concluded that the suggested method is an effective tool for exploratory data analysis. Finally, Painter et al. (2012) used both offline and online algorithms for monitoring the quality of aggregate data.

Based on all the above, we conclude that numerous statistical models and methods are available for the early epidemic detection (see Choi (2012), Groseclose and Buckeridge (2017), and references therein). The aforementioned statement raises the query "Which is the most appropriate methodology to use?" As expected, it is not possible to characterize one single methodology as "best" due to the fact that the choice of an ideal method depends on a variety of factors such as the application, the implementation, the purpose, and the context of the analysis.

Some of the factors that may affect any assessment of the relative merits of available methods are (i) the scope and the field application of the public health surveillance system, e.g., the number (from one to a few thousands) of parallel data series to be monitored; (ii) the quality of the data which is related to the method of data collection as well as possible delays between the time of occurrence to the time of reporting; (iii) the spatio-temporal data features which may include the frequency, the trend as well as the seasonality structure, the epidemicity, and finally the time step and spatial resolution; (iv) the nonstationarity and the possible existence of correlations in the distribution of frequency of data; (v) the possible existence of the phenomenon of overdispersion; (vi) the outbreak specific characteristics such as explosive or gradual onset, brief or long duration, severity, or any mixture of the above; (vii) the use for which the system is intended, including the post-signal processing protocols; (viii) the support of the system in terms of processing power and human resources; and (ix) the choice of metrics and measures for performance evaluation.

Therefore, in order to assess the effectiveness of statistical and stochastic modeling techniques for outbreak detection as well as the validity of their results which in turn will result in safe conclusions, the use of appropriately adjusted evaluation criteria is required in order to serve the purposes of public health surveillance. However, in the scientific community, there are no widely accepted evaluation measures for this type of systems (Fricker, 2011, 2013). Consequently, the issue which arises regarding the selection of the optimal statistical methodology for studying the changes of epidemic activity, and thus the early and accurate outbreak detection, together with the selection of the appropriate evaluation criteria of these methods, is a broad, complex, and multifactorial research topic. This thematic area remains underdeveloped to a great extent, in spite of the advances that have been made as pointed out by Buckeridge et al. (2005), Watkins et al. (2006), and Fraker et al. (2008), since it is highly affected by technical approaches associated with SPC and the more empirical epidemiological perspective. In this spirit, this chapter focuses on open epidemiological issues of the interrelated disciplines of (bio)statistics and biosurveillance, and attempts to provide answers to open research issues, such as the fact that existing statistical methods do not apply directly to public health surveillance and that there is a lack of commonly accepted standards for evaluating detection algorithms as well as a lack of common performance evaluation metrics. The description of these open methodological issues, the mathematical formulation of the problem, and the statistical methodology that will be developed for addressing and solving these issues are presented below.

1.3   STATISTICAL FRAMEWORK

Before the statistical reference framework is introduced, it is essential to give a detailed account of the dataset used for analysis purposes and the procedure that was followed. In such a way, we will be able to have a clear aim and action plans on how special cause data points will be analyzed and how the variability of a process (of common and/or special cause) might be interpreted. In addition, it will become possible to comprehend statistical and epidemiological concepts within.

1.3.1   SENTINEL EPIDEMIOLOGICAL SURVEILLANCE SYSTEM

The epidemiological sentinel surveillance system in Greece is in operation since 1999 and is based on voluntary participation of physicians, general practitioners, and pediatricians of Primary Health Care. Such systems often constitute the main source of epidemiological data. In addition, they offer the guidelines for the most effective decision making via a process that includes registration, processing, analysis, and inference. Thus, the frequency of certain diseases and/or syndromes is reported by those health practitioners chosen exclusively for the purpose of making the clinical diagnoses. More specifically, epidemiological data for the number of consultations per syndrome are forwarded weekly to sentinel systems. Based on these data, the National Public Health Organization in Greece provides the estimate of the weekly syndrome cases per 1,000 visits, namely the so-called proportional morbidity reflecting the activity of the syndrome under study (Parpoula et al., 2017).

The reorganization of the Hellenic sentinel system under the Operational Programme "Human Resources Development" of the National Strategic Reference Framework (NSRF) 2007–2013, action "Primary Health Care services (private and public) networking for epidemiological surveillance and control of communicable diseases," took place in the period 2014–2015. The reorganization established that the national priorities of syndromes of great interest include the ILI and gastroenteritis. The fact that both syndromes are traditionally monitored by most European sentinel systems due to their potential for widespread transmission clearly shows that their evolution is of great public health concern. Considering that ILI rate constitutes a potential pandemic risk makes even more important its monitoring. Further note that the surveillance of ILI rate allows not only the study of the seasonality but also the identification of the signaled starting and ending weeks together with the intensity of epidemic waves. On the other hand, the surveillance of gastroenteritis is the key for the determination of epidemic outbreaks.

1.3.2 TWO SEASON INFLUENZA HISTORICAL DATA

The historical data used in this retrospective analysis are the weekly ILI rate data from September 29, 2014 (week 40/2014) to October 2, 2016 (week 39/2016) for the purpose of determining the signaled starting and ending weeks for the past two seasonal influenza outbreaks. Recall that ILI rate is defined as the frequency of influenza-like syndrome cases per 1,000 consultations. The main objectives of the analysis include the prediction of the time interval of influenza outbreak, the estimation of the duration of the time interval, and the early detection of epidemics.

1.3.3 RESEARCH METHODOLOGY

1.3.3.1 The Standard CDC and ECDC Flu Detection Algorithm (Serfling's Model)

The typical approach to ILI rate surveillance (implemented by the ECDC and CDC) is based on Serfling's cyclic regression method by which epidemics are detected and reported when morbidity/mortality exceeds the epidemic threshold. Serfling's model in Serfling (1963) can be described as

$$\mathbf{M11}: X_t = a_0 + a_1 t + \gamma_1 \cos \left(\frac{2\pi t}{m} \right) + \delta_1 \sin \left(\frac{2\pi t}{m} \right) + \varepsilon_t, \qquad (1.1)$$

where X_t is the observed weekly ILI rate, ε_t are the errors terms with mean 0 and variance σ^2, m is the number of observations within 1 year, and model coefficients are estimated by least squares method.

1.3.3.2 An Extended Serfling's Model

Parpoula et al. (2017) developed extended Serfling-type periodic regression models, and through an exhaustive search process, the best fitting model was selected. In particular, four steps were executed: (i) determination of the training period—a retrospective analysis was conducted, making use of all available 2-year historical weekly ILI rate data, in order to estimate the baseline level; (ii) purge of the training period—the 15% highest observations were excluded from the training period so that the baseline level is estimated from truly nonepidemic data (following the suggestions of Pelat et al. (2007)); (iii) estimation of the regression equation—all expressions for X_t were special cases of the model given by

$$\mathbf{M33}: X_t = a_0 + a_1 t + a_2 t^2 + a_3 t^3 + \gamma_1 \cos \left(\frac{2\pi t}{m} \right) + \delta_1 \sin \left(\frac{2\pi t}{m} \right) + \gamma_2 \cos \left(\frac{4\pi t}{m} \right)$$
$$+ \delta_2 \sin \left(\frac{4\pi t}{m} \right) + \gamma_3 \cos \left(\frac{8\pi t}{m} \right) + \delta_3 \sin \left(\frac{8\pi t}{m} \right) + \varepsilon_t. \qquad (1.2)$$

The best fitting model was obtained by an exhaustive search and selection process which was relied on analysis of variance (ANOVA) comparison (significance level was chosen to be 0.05) to select between nested models, and on AIC or on Schwarz's Bayesian information criterion (BIC), to select between non-nested models; (iv) epidemic alert notification—epidemic thresholds were obtained by taking the upper 95th percentile for the prediction distribution, and an epidemic was declared when two weekly successive observations were above the estimated threshold. For a more detailed explanation of the above-mentioned four-step procedure, see Parpoula et al. (2017).

Through this exhaustive search process, the detected best fitting model **M23** is described as follows:

$$\textbf{M23}: X_t = a_0 + a_1 t + a_2 t^2 + \gamma_1 \cos\left(\frac{2\pi t}{m}\right) + \delta_1 \sin\left(\frac{2\pi t}{m}\right) + \gamma_2 \cos\left(\frac{4\pi t}{m}\right)$$
$$+ \delta_2 \sin\left(\frac{4\pi t}{m}\right) + \gamma_3 \cos\left(\frac{8\pi t}{m}\right) + \delta_3 \sin\left(\frac{8\pi t}{m}\right) + \varepsilon_t. \qquad (1.3)$$

Moreover, the aforementioned procedure allowed Parpoula et al. (2017) to correctly identify the epidemics occurred, namely, sw01-ew13/2015, sw01-ew08/2016, where sw and ew denote the start and end weeks of the epidemic period, respectively. Note that the signaled start and end weeks were found to be identical considering either Serfling's model (**M11**) or extended Serfling's model (**M23**).

1.3.3.3   A Mixed Model Including Auto-Regressive Moving Average (ARMA) Terms

The above results motivated Kalligeris et al. (2018) to account for autocorrelation in historical process data and incorporate into the full form of the model (**M33**) described in Equation (1.2), Auto-Regressive Moving Average (ARMA) terms. Covariates related to weather (12 covariates related to wind speed, direction, and temperature) were also included in the model structure and examined for their statistical significance. Following the same four-step exhaustive search process (as previously described), applied to the same 2-year historical weekly ILI rate data, the model chosen as optimal was the simplest one, that is, a mixed model with a linear trend, 12-month seasonality, an ARMA(2,1) process, and the minimum temperature as the only significant random meteorological covariate, described as follows:

$$\textbf{MXM11}: X_t = a_0 + a_1 t + \gamma_1 \cos\left(\frac{2\pi t}{m}\right) + \delta_1 \sin\left(\frac{2\pi t}{m}\right)$$
$$+ \varphi_1 x_{t-1} + \varphi_2 x_{t-2} + \varepsilon_t + \lambda_1 \varepsilon_{t-1} + \omega_1 \text{ mintemp}. \qquad (1.4)$$

In this way, Kalligeris et al. (2018) identified two epidemic periods, namely, sw01-ew12/2015, sw05-ew08/2016.

Recall that the modeling is focused on nonepidemic data after removal of extreme values from the dataset. Both methodologies, based either on extended Serfling's model (**M23**) or on periodic ARMA (PARMA) modeling (**MXM11**), are considered to be sufficient for the modeling of the baseline part of the series, but have a serious drawback. By not considering the extreme values of the dataset, the resulting model is considered unsuitable for predictive purposes. Moreover, these approaches suffer from the absence of scientific justification for excluding nonextreme values to model the baseline distribution. To that end, a variety of ad-hoc rules have been suggested (dividing the time series into typical and non-typical periods) with the most widely used being the removal of the top 15%–25% values from the training period (Pelat et al., 2007). This approach although it has some merits relies on arbitrary pruning (lacking mathematical justification) which constitutes, in that sense, a fundamental obstacle towards the development of an automated surveillance system for influenza.

1.3.3.4 A Mixed Effects Periodic ARMA Model Based on Change-Point Detection

For the aforementioned reasons, Kalligeris et al. (2019) provided a general and computationally fast algorithm which consists of three steps: (i) identification of extreme periods, (ii) modeling of non-extreme periods, and (iii) modeling of extreme periods and estimation accuracy. The goal of the above algorithmic procedure is to simultaneously estimate the baseline level and the extreme periods of the time series via change-point analysis. First, they adapted the segment neighborhood (**SegNeigh**) algorithm (Auger and Lawrence, 1989) applied to time series data. In this way, significant mean shifts were identified at particular, previously unknown time points. Second, for modeling nonepidemic time series data, Kalligeris et al. (2019) excluded from the analysis, based on change-point analysis, the observations leading to an epidemic. Then, using for comparative purposes among candidate models, the AIC and the ANOVA procedure, the algorithm of Kalligeris et al. (2018) was executed for several models with trend, periodicity, ARMA terms, as well as the average minimum weekly temperature (the only covariate identified as significant among a plethora of meteorological variables considered). Finally, the optimal model for baseline influenza morbidity was selected to be the one that includes a quadratic trend, 12- and 6-month seasonal periodicity, ARMA(1,1) terms, and minimum temperature, and is described as follows:

$$\textbf{SegNeigh}: X_t = a_0 + a_1 t + a_2 t^2 + \gamma_1 \cos\left(\frac{2\pi t}{m}\right) + \delta_1 \sin\left(\frac{2\pi t}{m}\right) + \gamma_2 \cos\left(\frac{4\pi t}{m}\right)$$

$$+ \delta_2 \sin\left(\frac{4\pi t}{m}\right) + \varphi_1 x_{t-1} + \varepsilon_t + \lambda_1 \varepsilon_{t-1} + \omega_1 \, \text{mintemp}. \quad (1.5)$$

Finally, Kalligeris et al. (2019) introduced a polynomial approximation of the behavior of the time series in epidemic periods (identified by change-point analysis) and evaluated the estimated ILI rate value of each epidemic time point by the polynomial of each epidemic period. The chosen polynomial describes satisfactorily enough the behavior of the epidemic period. Based on the proposed methodology, the estimation of the epidemic time points (sw01-ew12/2015, sw01-ew08/2016) results by combining the polynomial approximation of the epidemic periods with the baseline model based exclusively on nonepidemic time points. The details of the above technique can be found in Kalligeris et al. (2019).

1.3.3.5 A Distribution-Free Control Charting Technique Based on Change-Point Detection

As discussed earlier, the classical approach used by ECDC and CDC for the ILI rate surveillance is based on the implementation of Serfling's cyclic regression model which requires nonepidemic data for the modeling of the baseline distribution while the observations are treated as being independent and identically distributed.

Towards this end, Parpoula and Karagrigoriou (2020) developed a distribution-free control charting technique based on change-point analysis for detecting changes in location of univariate ILI rate data. The main tool in this methodology is detection of unusual trends, in the sense that the beginning of an unusual trend marks a switch from a control state to an epidemic state. Therefore, it is considered of high importance to timely detect the change point for which an epidemic trend has begun since in such a way the occurrence of a new epidemic could be predicted.

Let x_i represent the ith observation, $i = 1, 2, \ldots, m$, and let us consider the problem of testing the null hypothesis H_0 that the process was in control (IC) against the alternative hypothesis that the process mean experienced an unknown number of step shifts. In such a case, a set of test (control) statistics are needed for detecting $1, 2, \ldots, K$ step shifts. Here, K denotes the maximum number of hypothetical change points. The test statistics $T_k, k = 1, 2, \ldots, K$ were designed for testing H_0 against the alternatives:

$$H_{1,k} \colon E(x_i) = \begin{cases} \mu_0, & \text{if } 0 < i \leq \tau_1 \\ \mu_1, & \text{if } \tau_1 < i \leq \tau_2 \\ \vdots \\ \mu_k, & \text{if } \tau_k < i \leq m, \end{cases} \tag{1.6}$$

where the mean values $\mu_0, \ldots, \mu_k$ and the change points $0 < \tau_1 < \ldots < \tau_k < m$ are assumed unknown. Further, defining $\tau_0 = 0$ and $\tau_{k+1} = m$, it was also assumed that $\tau_r - \tau_{r-1} \geq l_{min}, r = 1, \ldots, k+1$, where l_{min} is a pre-specified constant giving the minimum number of successive observations allowed between two change points. For a sequence of individual observations, the control statistics together with the change points were obtained using a simple forward recursive segmentation and permutation (**RS/P**) approach as described in Capizzi and Masarotto (2013). The different test statistics were standardized and aggregated, obtaining an overall control statistic. Then, given a test statistic, its p-value was calculated, as the proportion of permutations (fixed number, say L) under which the statistic value exceeds or is equal to the statistic computed from the original sample of observations. Choosing an acceptable false alarm probability (FAP), say α, then, for p-value $< \alpha$, the null hypothesis that the process was IC is rejected. This multiple change-point approach is advantageous for epidemiological surveillance purposes for two reasons: (i) there is little to no control over disease incidence, and thus, the distribution of disease incidence is usually nonstationary; (ii) outbreaks are transient, with disease incidence returning to its original state once an outbreak has run its course.

Parpoula and Karagrigoriou (2020) performed the **RS/P** approach for both periods under study executing $L = 100{,}000$ permutations with $K = \max\left(3, \min\left(50, \left[\frac{m}{15}\right]\right)\right)$ and $l_{min} = 5$. The procedure signaled possible changes of the mean for both periods under study (p-value < 0.001 for a change in level), and the extracted signaled epidemics were sw01-ew14/2015 and sw01-ew08/2016.

1.4   COMPARATIVE STUDY

This section deals with the analysis of weekly ILI rate data (provided from the National Public Health Organization) for Greece, between September 29, 2014 (week 40/2014), and October 2, 2016 (week 39/2016).

Parpoula et al. (2019) examined the ability of the **RS/P**, **SegNeigh**, and **MXM11** approaches to detect the true change points compared to the standard and extended CDC and ECDC flu detection algorithm (models **M11** and **M23**). Recall here that the signaled start and end weeks (sw01-ew13/2015, sw01-ew08/2016) were found to be identical considering either Serfling's model (**M11**) or extended Serfling's model (**M23**).

The classic diagnostic test for discriminating between groups (here, epidemic from nonepidemic) is typically evaluated using receiver operating characteristic (ROC) curve analysis (Zweig and Campbell, 1993; Greiner et al., 2000). It is well established that such curves, and the associated statistics/metrics (sensitivity – SENS, specificity – SPEC, accuracy – ACC, area under the ROC curve – AUC) can further be used for evaluating and comparing the performance of different diagnostic tests (see for instance Griner et al., 1981). Thus, Parpoula et al. (2019) estimated these metrics along with their 95% confidence interval (CI) (exact Clopper–Pearson CIs for SENS, SPEC, and ACC, exact binomial CI for each derived AUC) for each method considered (as shown in Table 1.1).

Table 1.1 indicates that **RS/P** and **SegNeigh** approaches (higher ACC, SENS, and AUC values) outperform **MXM11** and seem to detect successfully the true change points compared to the standard approach to influenza surveillance. Both **RS/P** and **SegNeigh** approaches are advantageous since they can be used for the analysis of historical data without the need of identifying typical and non-typical data periods, and single or multiple mean shifts can be detected, hence, opening the way for automated surveillance systems.

1.5   CONCLUDING REMARKS

In this chapter, we studied the implementation of several periodic regression modeling techniques (including standard CDC and ECDC flu detection algorithm) and the mechanism of change-point analysis for the detection of epidemics. Further, we discussed some of the statistical affairs concerning the evaluation and optimal selection among these approaches for the early and accurate outbreak detection. The comparative empirical study of ILI syndrome in Greece for the period 2014–2016

Table 1.1

Performance Metrics for RS/P, SegNeigh, and MXM11 Approaches

	Estimated Value	95% CI
RS/P		
Sensitivity (%)	100.0	83.89–100.0
Specificity (%)	98.81	93.54–99.97
Accuracy (%)	99.05	94.81–99.98
AUC	0.988	0.944–0.999
SegNeigh		
Sensitivity (%)	95.24	76.18–99.88
Specificity (%)	100.0	95.71–100.0
Accuracy (%)	99.05	94.81–99.98
AUC	0.976	0.926–0.996
MXM11		
Sensitivity (%)	76.19	52.83–91.78
Specificity (%)	100.0	95.71–100.0
Accuracy (%)	95.24	89.24–98.44
AUC	0.881	0.803–0.936

provided evidence that **RS/P** and **SegNeigh** change-point analysis-based approaches were found to be superior compared to their competitors in terms of all performance evaluation metrics considered and seem to detect successfully the true change points (**RS/P** and **SegNeigh** change points were compared with those derived after executing the standard CDC and ECDC flu detection algorithm). Thus, the mechanism of change-point detection can be characterized as an excellent retrospective analysis technique.

On the one hand, **RS/P** distribution-free change-point analysis method is able to guarantee a prescribed FAP without any knowledge about the (in-control) underlying distribution. Given these appealing properties, future research will investigate the extension of the **RS/P** approach for estimating the expected baseline level for the time series, associated with a prediction interval. In such a manner, epidemic periods will be obtained along with the estimation of the related morbidity burden. In addition, alert epidemic thresholds could be used for performing online surveillance of ILI syndrome.

On the other hand, **SegNeigh** algorithm in conjunction with periodic-type ARMA time series modeling is capable of describing the behavior of the entire series, extreme and nonextreme parts, without significant loss of accuracy and thus resulting in a strong forecasting capability of the proposed method. The use of change-point detection analysis along with time series modeling techniques seems to provide a useful tool to identify and model outbreaks that may occur in incidence data. Simultaneously, it can be proved useful to the society since it could significantly contribute in the early detection and prevention of any type of extreme/harmful events.

Conclusively, the comparative study provided evidence that statistical methods based on change-point analysis have several appealing properties compared to the current practice for the detection of epidemics. It is worth noting that the purpose of change-point analysis is the accurate identification of epidemic periods via splitting the data into typical and non-typical periods. To that end, a variety of rules have been suggested, as discussed earlier, such as excluding the 15% or 25% highest values from the training period, removing all data above a given threshold, or excluding whole periods known to be epidemic prone. One of the main contributions of this chapter lies on the fact that it presents statistical and stochastic modeling techniques which do not rely on arbitrary pruning and hence is filling up the gap in the relevant literature.

Further, the change-point analysis approach for detecting outbreaks indicates the need of building a flexible model that can adjust in either of the states of the time series data. Thus, as for a future work we will consider Markovian mechanisms such as Markov switching and hidden (semi-) Markov models. Moreover, we will try to implement into those mechanisms, penalized likelihood techniques so that the complexity involved in mixed periodic models, which were discussed throughout this chapter, could be countered.

ACKNOWLEDGMENTS

The authors wish to express their appreciation to the **Hellenic National Meteorological Service** (for the meteorological data) and to the Department of Epidemiological Surveillance and Intervention, **National Public Health Organization** (for the weekly ILI rate data). It should be noted that this work was carried out as part of the activities of LaBSTADA of the University of the Aegean (http://actuarweb.aegean.gr/labstada/index.html).

REFERENCES

Aminikhanghahi, S. and Cook, D.J.: A survey of methods for time series change point detection. *Knowledge and Information Systems* 51, 339–367 (2017).

Auger, I.E. and Lawrence, C.E.: Algorithms for the optimal identification of segment neighborhoods. *Bulletin of Mathematical Biology* 51, 39–54 (1989).

Borror, C.M., Champ, C.W., and Rigdon, S.E.: Poisson EWMA control charts. *Journal of Quality Technology* 30, 352–361 (1998).

Buckeridge, D.L., Burkom, H.S., Campell, M., Hogan, W.R., and Moore, A.: Algorithms for rapid outbreak detection: A research synthesis. *Journal of Biomedical Informatics* 38, 99–113 (2005).

Buehler, J.W., Hopkins, R.S., Overhage, J.M., Sosin, D.M., and Tong, V.: Framework for evaluating public health surveillance systems for early detection of outbreaks: Recommendations from the CDC Working Group. *Recommendations and Reports* 53(RR-1), 1–11 (2004).

Burkom, H.S., Murphy, S.P., and Shmueli, G.: Automated time series forecasting for biosurveillance. *Statistics in Medicine* 26, 4202–4218 (2007).

Capizzi, G. and Masarotto, G.: Phase I distribution-free analysis of univariate data. *Journal of Quality Technology* 45, 273–284 (2013).

Choi, B.C.K.: The past, present, and future of public health surveillance. *Scientifica* 2012, 26 (2012).

Costagliola, D., Flahault, A., Galinec, D., Garnerin, P., Menares, J., and Valleron, A.J.: A routine tool for detection and assessment of epidemics of influenza-like syndromes in France. *American Journal of Public Health* 81, 97–99 (1991).

Costagliola, D., Flahault, A., Galinec, D., Garnerin, P., Menares, J., and Valleron, A.J.: When is the epidemic warning cut-off point exceeded? *European Journal of Epidemiology* 10, 475–476 (1994).

Dato, V., Shephard, R., and Wagner, M.M.: Outbreaks and investigation. In M.M. Wagner, A.W. Moore, and R.M. Aryel (Eds.), *Handbook of Biosurveillance* (pp. 13–26). Burlington: Elsevier Academic Press (2006).

Dong, Y., Hedayat, A.S., and Sinha, B.K.: Surveillance strategies for detecting change point in incidence rate based on exponentially weighted moving average methods. *Journal of the American Statistical Association* 103, 843–853 (2008).

Elbert, Y. and Burkom, H.S.: Development and evaluation of a data-adaptive alerting algorithm for univariate temporal biosurveillance data. *Statistics in Medicine* 28, 3226–3248 (2009).

Farrington, C.P. and Andrews, N.: Outbreak detection: Application to infectious disease surveillance. In R. Brookmeyer and D.F. Stroup (Eds.), *Monitoring the Health of Populations: Statistical Principles and Methods for Public Health Surveillance* (pp. 203–231). Oxford: Oxford University Press (2004).

Fleming, D.M. and Rotar-Pavlic, D.: Information from primary care: Its importance and value. A comparison of information from Slovenia and England and Wales, viewed from the "Health 21" perspective. *European Journal of Public Health* 12, 249–253 (2002).

Christensen, J. and Rudemo, M.: Multiple change-point analysis of disease incidence rates. *Preventive Veterinary Medicine* 26, 53–76 (1996).

Fraker, S.E., Woodall, W.H., and Mousavi, S.: Performance metrics for surveillance schemes. *Quality Engineering* 20, 451–464 (2008).

Fricker, R.D.: *Introduction to Statistical Methods for Biosurveillance: With an Emphasis on Syndromic Surveillance*. Cambridge, CA: Cambridge University Press (2013).

Fricker, R.D.: Some methodological issues in biosurveillance. *Statistics in Medicine* 30, 403–415 (2011).

Gan, F.F.: Monitoring observations generated from a binomial distribution using modified exponentially weighted moving average control charts. *Journal of Statistical Computation and Simulation* 37, 45–60 (1991).

Greiner, M., Pfeiffer, D., and Smith, R.D.: Principles and practical application of the receiver-operating characteristic analysis for diagnostic tests. *Preventive Veterinary Medicine* 45, 23–41 (2000).

Griner, P.F., Mayewski, R.J., Mushlin, A.I., and Greenland, P.: Selection and interpretation of diagnostic tests and procedures. *Annals of Internal Medicine* 94, 555–600 (1981).

Groseclose, S.L. and Buckeridge, D.L.: Public health surveillance systems: Recent advances in their use and evaluation. *Annual Review of Public Health* 38, 57–79 (2017).

Heath, I. and Smeeth, L.: Tackling health inequalities in primary care. *British Medical Journal* 318, 1020–1021 (1999).

Höhle, M. and Paul, M.: Count data regression charts for the monitoring of surveillance time series. *Computational Statistics and Data Analysis* 52, 4357–4368 (2008).

HSPD-21: Homeland Security Presidential Directive. Retrieved from https://fas.org/irp/offdocs/nspd/hspd-21.htm (2014).

Hulth, A., Andrews, N., Ethelberg, S., Dreesman, J., Faensen, D., van Pelt, W., and Schnitzler, J.: Practical usage of computer-supported outbreak detection in five European countries. *Eurosuveillance* 15(36), 1–6 (2010).

Hutwagner, L., Thompson, W.W., Seeman, G.M., and Treadwell, T.: The bioterrorism preparedness and response Early Aberration Reporting System (EARS). *Journal of Urban Health* 80, 89–96 (2003).

Kalligeris, E.-N., Karagrigoriou, A., and Parpoula, C.: On mixed PARMA modeling of epidemiological time series data. *Communications in Statistics-Case Studies: Data Analysis and Applications*, https://doi.org/10.1080/23737484.2019.1644253, to appear (2020).

Kalligeris, E.N., Karagrigoriou, A., and Parpoula, C.: Periodic-type ARMA modeling with covariates for time-series incidence data via changepoint detection. *Statistical Methods in Medical Research*, https://doi.org/10.1177/0962280219871587, to appear (2020).

Kass-Hout, T.A., Xu, Z., McMurray, P., Park, S., Buckeridge, D.L., Brownstein, J.S., Finelli, L., and Groseclose, S.L.: Application of change point analysis to daily influenza-like illness emergency department visits. *Journal of the American Medical Informatics Association* 19, 1075–1081 (2012).

Langmuir, A.D.: The surveillance of communicable diseases of national importance. In C. Buck, A. Liopis, E. Najera, and M. Terris (Eds.), *The Challenge of Epidemiology: Issues and Selected Readings*, 3rd edition (pp. 855–867). Washington, DC: Pan American Health Organization (PAHO) (1995).

Long, Z.A.: Teknik Perlombongan Corak Data Terpencil-kerap (DTK) Bagi Pengesanan wabak Dalam Pegawasan Kesihatan Awam. PhD, Universiti Kebangsaan Malaysia (2012).

Lucas, J.M.: Counted data CUSUM's. *Technometrics* 27, 129–144 (1985).

Macmahon, B. and Trichopoulos, D.: *Epidemiology: Principles and Methods*. Boston, MA: Little, Brown and Co (1996).

Mohsin, M.F.M., Hamdan, A.R., Bakar, A.A.: A review on anomaly detection for outbreak detection. *Proceedings of the International Conference on Information Science and Management (ICoCSIM)*, Indonesia, 22–28 (2012).

Montgomery, D.C.: *Introduction to Statistical Quality Control*, 7th edition. Hoboken, NJ: John Wiley & Sons (2013).

Nobre, F.F. and Stroup, D.F.: A monitoring system to detect changes in public health surveillance data. *Journal of Epidemiology* 23, 408–418 (1994).

Norbury, M., Mercer, S.W., Gillies, J., Furler, J., and Watt G.C.M.: Time to care: Tackling health inequalities through primary care. *Family Practice* 28, 1–3 (2011).

Oakland, J.S.: *Statistical Process Control*, 6th edition. Oxford: Butterworth-Heinemann (2008).

Ong, J.B.S., Chen, M.I.C., Cook, A.R., Lee, H.C., Lee, V.J., Lin, R.T.P., Tambyah, P.A., and Goh, L.G.: Real-time epidemic monitoring and forecasting of H1N1-2009 using influenza-like illness from general practice and family doctor clinics in Singapore. *PLoS One* 5, e10036 (2010).

Page, E.S.: Continuous inspection schemes. *Biometrika* 41, 100–115 (1954).

Painter, I., Eaton, J., and Lober, W.B.: Using change point detection for monitoring the quality of aggregate data. *Statistics in Defense and National Security, Annual Conference Proceedings* 5, 169 (2012).

Parpoula, C., Kalligeris, E.N., and Karagrigoriou, A.: A comparative study of change-point analysis techniques for outbreak detection. *Proceedings of the 21st European Young Statisticians Meeting (EYSM 2019)*, Belgrade, Serbia, accepted (2019).

Parpoula, C., Karagrigoriou, A., and Lambrou, A.: Epidemic intelligence statistical modelling for biosurveillance. In J. Blömer et al. (Eds.), *MACIS 2017, Lecture Notes in Computer Science (LNCS), LNCS 10693* (pp. 349–363). Cham: Springer International Publishing AG (2017).

Parpoula, C., and Karagrigoriou, A.: On change-point analysis-based distribution-free control charts with Phase I applications. In M. V. Koutras, and I. S. Triantafillou (Eds.), Distribution-free Methods for Statistical Process Monitoring and Control, Springer, ISBN: 978-3-030-25080-5, doi:10.1007/978-3-030-25081-2, to appear (2020).

Pelat, C., Boélle, P.Y., Cowling, B.J., Carrat, F., Flahault, A., Ansart, S., and Valleron, A.-J.: Online detection and quantification of epidemics. *BMC Medical Informatics and Decision Making* 7, 29, (2007).

Rogerson, P.A. and Yamada, I.: Monitoring change in spatial patterns of disease: Comparing univariate and multivariate cumulative sum approaches. *Statistics in Medicine* 23, 2195–2214 (2004).

Rossi, G., Lampugnani, L., and Marchi, M.: An approximate CUSUM procedure for surveillance of health events. *Statistics in Medicine* 18, 2111–2122 (1999).

Rothman, J. and Greenland, S.: *Modern Epidemiology*, 2nd edition. Philadelphia, PA: Lippincott Williams and Williams (1998).

Serfling, R.: Methods for current statistical analysis of excess pneumonia-influenza deaths. *Public Health Reports* 78, 494–506 (1963).

Shmueli, G. and Burkom, H.: Statistical challenges facing early outbreak detection in biosurveillance. *Technometrics* 52, 39–51 (2010).

Sonesson, C. and Bock, D.: A review and discussion of prospective statistical surveillance in public health. *Journal of the Royal Statistical Society Series A* 166, 5–21 (2003).

Sosin, D.M.: Syndromic surveillance: The case for skillful investment. *Biosecurity and Bioterrorism: Biodefense Strategy, Practice, and Science* 1, 247–253 (2003).

Starfield, B., Shi, L., and Mackinko, J.: Contribution of primary care to health systems and health. *The Milbank Quarterly* 83, 457–502 (2005).

Stroup, D.F., Williamson, G.D., Herndon, J.L., and Karon, J.M.: Detection of aberrations in the occurrence of notifiable diseases surveillance data. *Statistics in Medicine* 8, 323–329 (1989).

Teutsch, S.M. and Churchill, R.E.: *Principles and Practice of Public Health*. Oxford: Oxford University Press (2000).

Teutsch, S.M. and Thacker, S.B.: Planning a public health surveillance system. *Epidemiological Bulletin* 16, 1–6 (1995).

Texier, G., Farouh, M., Pellegrin, L., Jackson, M.L., Meynard, J.B., Deparis, X., and Chaudet, H.: Outbreak definition by change point analysis: A tool for public health decision? *BMC Medical Informatics and Decision Making* 16, 33 (2016).

Thacker, S. and Berkelman, R.: Public health surveillance in the United States. *Epidemiology Review* 10, 164–190 (1998).

Unkel, S., Farrington, C.P., Garthwaite, P.H., Robertson, C., and Andrews, N.: Statistical methods for the prospective detection of infectious disease outbreaks: A review. *Journal of the Royal Statistical Society: Series A* 175, 49–82 (2012).

Wagner, M.M., Gresham, L.S., and Dato, V.: Case detection, outbreak detection, and outbreak characterization. In M.M. Wagner, A.W. Moore, and R.M. Aryel (Eds.), *Handbook of Biosurveillance* (pp. 27–50). Burlington: Elsevier Academic Press (2006).

Wagner, M.M.: The space race and biodefense: Lessons from NASA about big science and the role of medical informatics. *Journal of the American Medical Information Association* 9, 120–122 (2002).
Watkins, R.E., Eagleson, S., Hall, R.G, Dailey, L., and Plant, A.J.: Approaches to the evaluation of outbreak detection methods. *BMC Public Health* 6, 263 (2006).
Woodall, W.H.: Use of control charts in health care and public health surveillance (with discussion). *Journal of Quality Technology* 38, 88–103 (2006).
Zweig, M.H. and Campbell, G.: Receiver-operating characteristic (ROC) plots: A fundamental evaluation tool in clinical medicine. *Clinical Chemistry* 39, 561–577 (1993).

2 Assessment of Earthquake Hazard Based on Statistical Models, Probability Theory, and Nonlinear Analysis

Nitish Puri
Indian Institute of Technology Delhi

Ashwani Jain
National Institute of Technology Kurukshetra

CONTENTS

2.1 INTRODUCTION

Earthquake is an unstoppable force of nature, and its occurrence cannot be predicted. However, sound design of earthquake-resistant structures, planning of rescue arrangements and implementation of mitigation measures, can greatly reduce the vulnerability to earthquakes. Therefore, in plain areas of high seismicity, assessment of earthquake-induced hazards, i.e., wave amplification and soil liquefaction, is crucial.

The chapter reviews methods for assessment of earthquake hazards based on statistical models, probability theory, and nonlinear analysis. This includes description of steps involved in assessment of earthquake hazard, ground response investigation, and calculation of liquefaction potential. The results can be formulated as earthquake hazard maps based on peak ground acceleration (PGA), spectral acceleration (Sa), response spectra, amplification factors for PGA and Sa, and liquefaction susceptibility map. These results could be useful for engineers and town planners to obtain site-specific design parameters for calculation of seismic loading for earthquake resistant design of structures and identification of potentially liquefiable areas for planning of mitigation measures.

2.2 EARTHQUAKE HAZARDS

The world has withstood several great earthquakes documented from the times of 1652 BC Xia China Earthquake to the recent on 2019 Peru Earthquake that occurred on May 26. The people have lost their lives, households, and relations in the earthquakes. Outdated structural design provisions, lax building by-laws, broad zoning of earthquake hazard, and ignorance towards state-of-the-art practices in earthquake-resistant design are some of the reasons behind the disasters that happened during these earthquakes at many places. Many investigators have raised these issues and suggested detailed assessment of earthquake hazard in line with local seismotectonic setting for all important projects.

Some of the earthquake-induced hazards have been explained in upcoming subsections.

2.2.1 STRONG GROUND MOTIONS

Earthquake is an event which can inflict severe damage to the infrastructure of a region and take it back to a few decades. The recent example is the Canterbury earthquakes of 2010 and 2011 that caused heavy damages in the Christchurch city. The estimated cost to rebuild is around 20% of total gross domestic product (GDP) of New Zealand, i.e., NZ$40 billion approximately (Potter et al. 2015). The Christchurch city is located near several active tectonic features developed due to the clash of Australian plate and Pacific plate, and these features can generate large-magnitude earthquakes in the area. It was the main reason behind such a big earthquake in Christchurch and the subsequent mass destruction. The New Zealand construction standards require a structure with a 50-year design period to withstand loads estimated for a 475-year earthquake event (MacRae et al. 2011). However, ground motions during 2011 Christchurch earthquake considerably exceeded even 2475-year design motion (Kaiser et al. 2012). Unfortunately, no structure was built to take that high seismic loading which led to severe damage to structures (Figure 2.1). Some slope failure events were also observed (Figure 2.2). The city also experienced hazards of liquefaction, which include significant ground movements, undermining of foundations, destruction of infrastructure, and gushing out of more than 200,000 ton of silt, making it the worst liquefaction event ever recorded anywhere in a modern city (Figure 2.3). Hence, it is necessary for the areas located in the vicinity of tectonically active sources to be ready with proper mitigation measures and rescue arrangements for earthquake-induced hazards.

The mere estimation of seismic hazard would not help in full preparedness against earthquake-induced hazards. Amplification of seismic waves and liquefaction of soils could result due to earthquake in plain areas.

2.2.2 SEISMIC WAVE AMPLIFICATION

On surface of the Earth, an earthquake manifests itself as shaking of ground and sometimes its displacement. Earthquake engineers are primarily concerned with the

Figure 2.1 Destruction of the Pyne Gould building in Christchurch city. (Source: en.wikipedia.org.)

Figure 2.2 A major landslide near Kaikoura (South Island). (Source: en.wikipedia.org.)

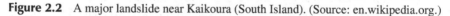

ground motions, which are strong enough to be felt during an earthquake. Earthquake waves radiate from the point of rupture to travel rapidly towards surface and can produce minor to severe shaking lasting from seconds to minutes. Attributes of earthquake waves are modified as they propagate through different soil profiles as shown in Figure 2.4.

Earthquake waves travel long distance in rock strata and a few meter in soil before reaching the ground surface. The soil acts as a filter and has great influence on the magnitude and time of shaking at location under consideration. This phenomenon is called "local site effects." The local soil conditions significantly affect amplitude, frequency, and duration of ground motions during an earthquake. The degree of influence is controlled by thickness and properties of the soil layers, topography of the site, and attributes of the ground motion itself.

Figure 2.3 Aerial view of liquefaction-affected areas in Christchurch city. (Source: en.wikipedia.org.)

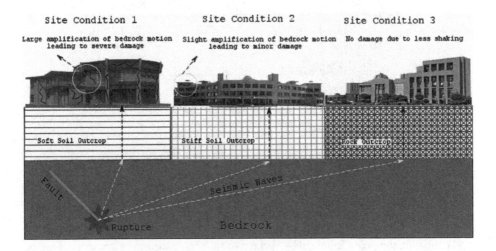

Figure 2.4 Site conditions and their possible effect on structures.

The soft soil sites intensify long-period bedrock motions to greater degree in comparison with stiff soil sites, and opposite trend is generally seen for low-period motions as reported from theoretical ground response analysis (GRA) performed by Kramer (2013) (Figure 2.5). Also, the GRAs performed by Idriss (1990) show that at low to moderate PGA levels of bedrock motion (<0.4 g), the amplification in soft soils is likely to be on the larger side than observed in rocks. However, at high PGA levels of bedrock motion, the low shear modulus and the nonlinearity of soft soils generally intercept amplifications as high as those observed for rocks. The relationship for PGA for soft soil sites and rock sites determined by Idriss (1990) is shown in Figure 2.6.

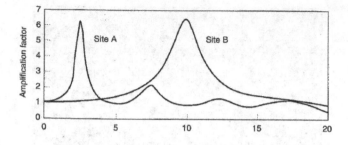

Figure 2.5 Amplification factors for soft site A and stiff site B. (After Kramer 2013.)

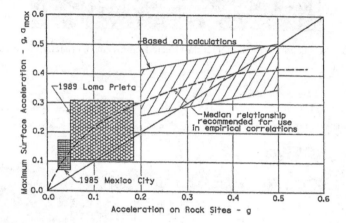

Figure 2.6 Relationship of PGA on soft soil sites and rock sites. (After Idriss 1990.)

2.2.3 LIQUEFACTION HAZARD

Liquefaction is a phenomenon observed during an earthquake due to which the effective stress is reduced to zero, and the sand–water mixture acts as a viscous material, and consolidation starts followed by surface settlement resulting in denser packing of sand particles. It is demonstrated in the form of sand ejecta and soil spouts at the surface created due to seepage of water, or in some circumstances by the development of quicksand. Structures may settle substantially into the surface or tilt excessively, and lightweight constructions and foundations may get displaced laterally causing structural failures and the conventional measure of reinforcing the upper part of the structure in such a situation is entirely useless.

2.3 PROBABILISTIC SEISMIC HAZARD ANALYSIS

2.3.1 ASSESSMENT

Earthquake hazard manifests itself in severe ground shaking or subsidence that can severely damage the structures build under and over the surface. The main aspects in seismic hazard assessment for site are magnitude, source-site distance, recurrence interval, and time-span of ground shaking. Assessment of seismic hazard is the initial step towards mitigation. The purpose of assessment is to estimate potential damage and loss due to possible earthquakes by estimating the probable ground quaking at a site. Hence, the purpose of maintaining desired level of serviceability of structures to withstand a given level of ground shaking can be achieved. Macrozonation and microzonation are two scales at which these kinds of studies are carried out to develop earthquake hazard maps which are useful for the design and construction of earthquake-resistant structures, planning of land use, planning of emergency rescue and relief, and estimation of probable economic loss.

2.3.2 METHODS FOR SEISMIC HAZARD ANALYSIS

A probabilistic seismic hazard analysis (PSHA) or a deterministic seismic hazard analysis (DSHA) approach is generally adopted for seismic hazard assessment. In DSHA, a specific earthquake scenario is presumed based on earthquake data and tectonics of the seismic study region, and hazard is calculated based on attenuation characteristics of the region. DSHA provides the worst-case scenario earthquake that can occur in the region, and the hazard characteristics are obtained for the largest expected earthquake magnitude supposed to occur at the smallest distance from the location under consideration. This is done without taking into account the possibility of occurrence of the earthquake for a definite exposure period during the design life of the structure. It is used extensively for the design of nuclear power stations, large bridges, big dams, facilities for harmful waste containment, and as an upper limit for PSHA. On the other hand, PSHA rectifies several problems inherent in its deterministic precursor, namely, lack of quantification of uncertainties in magnitude, location of an earthquake, and probability of its occurrence. The detailed procedure for probabilistic approach was originally given by Cornell (1968) and explained in detail by Baker (2008). It quantitatively represents the association between seismogenic sources, related ground motion parameters, and respective chances of occurrence. It also computes the likelihood of exceedance of a definite intensity of earthquake ground motion at a particular location, which is represented as function of recurrence interval and fault displacement. PSHA has now become a necessity for earthquake resistant design across the globe due to its capability to accommodate uncertainties. The detailed process for the calculation of seismic hazard using both approaches has been explained below.

2.3.2.1 Deterministic Seismic Hazard Analysis (DSHA)

DSHA is a simple analysis, often carried out at an initial stage for the assessment of earthquake hazard for a specific location or an area. It was used widely in the initial stages in geotechnical earthquake engineering, nowadays, used for worst-case situation. For this, all the probable sources associated with seismic activities must be recognized and their potential for generating earthquakes assessed. A typical DSHA consists of four stages as described by Kramer (2013):

1. Identification and characterization of all the tectonic sources with the potential to generate strong ground motion at a location.
2. Estimation of source to site distances (R).
3. Determination of a controlling earthquake.
4. Description of hazard with reference to a ground motion parameter (Y), usually expressed in terms of PGA and Sa.

The DSHA procedure has been shown schematically in Figure 2.7. The steps in DSHA are very simple to carry out and do not require any specialized computer program or software.

DSHA seems to be a simple procedure as expressed in these four steps but involves many subjective decisions, particularly regarding estimation of maximum earthquake magnitude. It requires collective knowledge and judgment

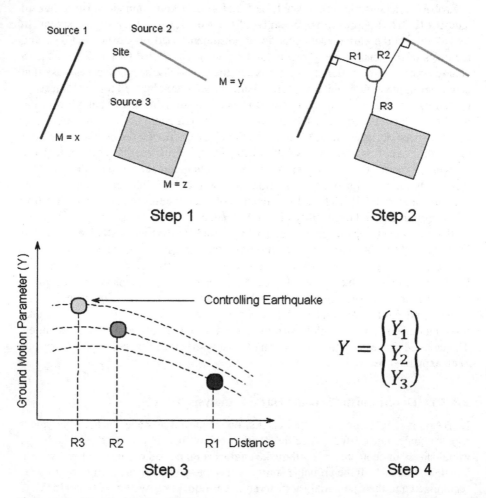

Figure 2.7 Steps involved in DSHA.

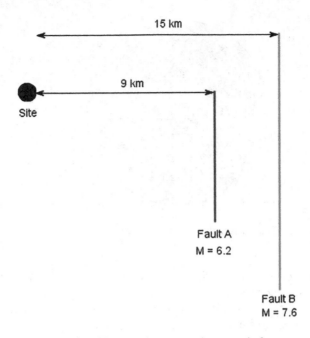

Figure 2.8   Case 1: Study area with more than one active tectonic feature.

of seismologists, geophysicists, engineers, hazard analysts, economists, and administrative officers. Even when it worked out from the consensus amongst the above, there is never a single "worst-case" event when there is more than one active tectonic feature in the study area (Figure 2.8). In such a case, a smaller magnitude event nearby can produce a larger spectral acceleration amplitude at shorter period, while a larger magnitude event can produce a larger amplitude at longer period (Baker 2008).

Much greater challenge arises in DSHA when there are lot of epicenters, but the faults are not obvious, or when the faults are diffused together. These sources are then together considered as an area source likely to generate earthquakes at any location (Figure 2.9). In such a situation, the worst-case earthquake is with the largest possible magnitude for location right underneath place under consideration at the ground surface, i.e., with zero distance. Clearly, it is a worst-case earthquake, no matter how improbable is its chance of occurrence. Hence, in DSHA, uncertainty in size, location, as well as occurrence of an earthquake event cannot be accounted for.

2.3.2.2   Probabilistic Seismic Hazard Analysis (PSHA)

PSHA offers a framework in which the uncertainties in seismic occurrences can be identified, allocated values, and combined in a logical manner to provide a clearer representation of seismic hazard. PSHA, broadly, can be carried out in five stages (Baker 2008):

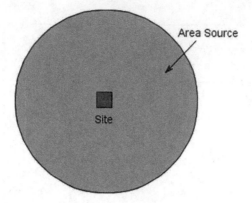

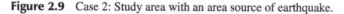

Figure 2.9 Case 2: Study area with an area source of earthquake.

1. Identification and characterization of tectonic sources capable of generating harmful seismic waves.
2. Calculation of recurrence periods at which various magnitudes of earthquake in a region are likely to occur.
3. Calculation of source to site distances corresponding to possible seismic events and their distribution.
4. Distribution of ground motion usually as PGA and Sa for particular earthquake magnitude and distance.
5. Integration of uncertainties in various attributes of an earthquake ground motion using total probability theorem.
The PSHA procedure is shown schematically in Figure 2.10.

PSHA can also accommodate the uncertainty in the choice of a model by using logic tree approach that allocates weighting factor for probability of a model to be accurate. Decision of assigning weights to models is quite subjective as it is difficult to rate models one over another and hence requires an expert opinion. The outcome of these computations would be a complete allocation of ground motion intensities to each site with their respective probabilities of exceedance. These outcomes can be used to find a ground-shaking intensity with an acceptable minor chance of exceedance instead of a "worst-case" scenario.

The accuracy of PSHA relies on the precision with which uncertainties in magnitude, place, and occurrence of an earthquake are characterized. A wise engineering judgment is necessary for the interpretation of PSHA results. The procedure for PSHA is quite complex as compared to DSHA. It is difficult to use it in an area where statistically significant earthquake catalogue is not available. In addition, PSHA cannot be used for areas where tectonic sources are not properly delineated. The method was initially evolved by Cornell in 1968. The first computer form EQRISK was developed by McGuire in 1976, and it is referred to as Cornell-McGuire method. The computer form was further modified as FRISK in the year 1978. Till now, many programs have been developed for carrying out PSHA, e.g., FRISK 88M

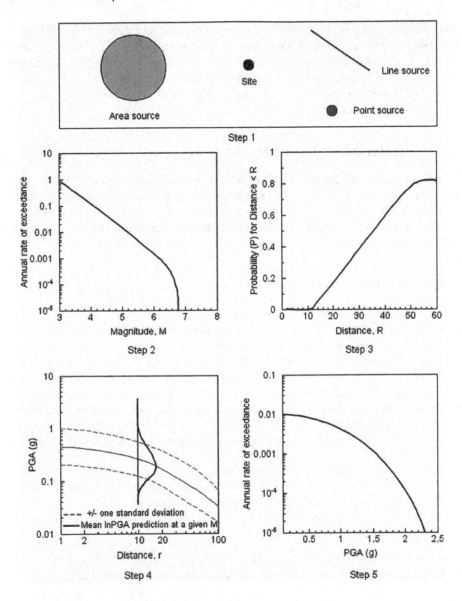

Figure 2.10 Steps involved in PSHA.

(McGuire 2001), OpenSHA (Field et al. 2003), OPENQUAKE (Pagani et al. 2014), and R-CRISIS (Ordaz and Salgado-Gálvez 2017).

2.3.3 DEVELOPMENT OF A COMPREHENSIVE EARTHQUAKE CATALOGUE

Comprehensive earthquake catalogue is a prerequisite for seismic hazard estimation. Reliable seismic hazard assessment of a region strongly depends upon the data statistics of the events. There are two types of database accessible for analysis: pre-instrumental and instrumental. Instrumental data can be collected from various seismological agencies. Non-instrumental measurements of pre-instrumental events are of immense importance and are required for compiling earthquake catalogue. The catalogue is then homogenized, cautiously inspected to eliminate matching events, de-clustered, and tested for completeness.

2.3.4 CATALOGUE HOMOGENIZATION

The seismic data are generally collected from various repositories. A wide range of magnitudes are reported by these repositories in terms of moment (M_w), body wave (m_b), surface wave (M_s), local (M_L), duration (M_d), etc. The M_w is extensively used as it does not saturate. All the magnitudes are changed to M_w, by empirical relationships taken from the work done by various investigators.

2.3.5 DE-CLUSTERING OF CATALOGUE

An accurate estimation of earthquake hazard requires a Poisson model, which has a random set of earthquake occurrences. The instrumental catalogues include a lot of dependent events like foreshocks and aftershocks with the mainshock, leading to a major deviation from the Poisson model. It gives false information about the seismicity giving wrong assessment of seismic hazard. De-clustering is the process of elimination of dependent events from the catalogue. This gives an approximate Poisson or random dataset to achieve more reliable and accurate calculation of the recurrence intervals of earthquakes. This can be done by taking help of time and space windows as suggested in the work done by Gardner and Knopoff (1974).

1. Considering all the events in a chronological order, if an ith event (smaller magnitude event) is in the window of a preceding larger shock that has not been removed, then ith event is removed from the catalogue.
2. If a larger shock is in the window of ith event, then ith event is deleted; otherwise, the ith event is retained.

The other methods can also be used, e.g., given by Savage (1972), Reasenberg (1985), Davis and Frohlich (1991a,b), and Molchan and Dmitrieva (1992).

2.3.6 CHECK FOR COMPLETENESS

Instrumental recording of earthquakes started in 1900 outside India and in 1960 in India. It is very important to compile a catalogue comprising of sufficient number of seismic events covering a reasonable span of time. Use of an inadequate catalogue will lead to incorrect calculations of earthquake parameters. Therefore, catalogue is scrutinized for completeness. For this purpose, catalogue is divided into several magnitude classes, and completeness periods are calculated by cumulative visual inspection (CUVI) technique (Tinti and Mulargia 1985) and Stepp (1972) technique.

The CUVI method assesses the span of the time for which magnitude class is complete. It is an easy criterion for testing completeness of the data. In this method, catalogue is divided into several magnitude classes, e.g. $4 \leq M_w < 5, 5 \leq M_w < 6,$ $6 \leq M_w < 7, 7 \leq M_w < 8$ and $8 \leq M_w < 9$, with a class taken as point process in time. The total events in a year are plotted versus time of occurrence for a magnitude range. The completeness is estimated from a year with sharp rise in plot. Figure 2.11 depicts completeness analysis using CUVI method. It is observed that for the range of the earthquake events of magnitude between 4.0 and 4.9 M_w, there is a big gap before year 1962 as no record is available. Beyond the year 1962, there is a steep rise in the plot indicating that the catalogue is complete for this range of magnitude for a duration between the years 1962 and 2015.

In Stepp (1972) method, earthquake data is also grouped into several magnitude classes and each magnitude class is considered as a point process in time. The average number of events per year in each magnitude range is determined for a time interval of 10 years. The mean rate (λ) for this sample is calculated by the following equation:

$$\lambda = \frac{1}{n} \sum_{i=1}^{n} x_i \tag{2.1}$$

The variance (σ_λ^2) and standard deviation (σ_λ) are calculated by the following equations:

$$\sigma_\lambda^2 = \frac{\lambda}{T} \tag{2.2}$$

$$\sigma_\lambda = \sqrt{\frac{\lambda}{T}} \tag{2.3}$$

where T is the sample length.

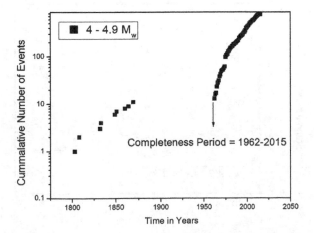

Figure 2.11 Completeness analysis using CUVI technique.

If λ is to be a constant, then σ_λ would vary as $\sqrt{\frac{1}{T}}$. The standard deviation of the mean rate of the magnitude intervals is plotted as a function of sample length along with nearly tangent lines with slope $\sqrt{\frac{1}{T}}$. The deviation of plot from the tangent line indicates completeness of magnitude range in terms of the time duration. Figure 2.12 gives completeness check using the Stepp (1972) method.

It is observed that for the range of earthquake events of magnitude between 4.0 and 4.9 M_w, standard deviation of the estimate of the mean deviates from the tangent line of the plot beyond 1963–2015. This shows that the catalogue is complete for this range of magnitude for a period between the years 1963 and 2015. The completeness for greater earthquake magnitudes cannot be verified, as recurrence interval may be greater and therefore, catalogue is taken as complete for whole period.

2.3.7 SEISMOGENIC SOURCE CHARACTERIZATION

It comprises preparation of a tectonic map, identification of tectonic features having potential of causing significant ground motions, calculation of highest perceived magnitude (M_{obs}), estimating fault length and focal depth, and calculating maximum magnitude (M_{max}).

2.3.8 SEISMICITY PARAMETERS

These can be calculated by linear least squares regression analysis by exponential distribution of magnitudes with the following equation.

$$\lambda_m = 10^{a-bM_w} = \exp(\alpha - \beta M_w) \tag{2.4}$$

where λ_m = mean annual rate of exceedance, a = coefficient, ath power of ten gives mean yearly number of earthquakes of magnitude ≥ 0, $\alpha = 2.303a$, b = coefficient

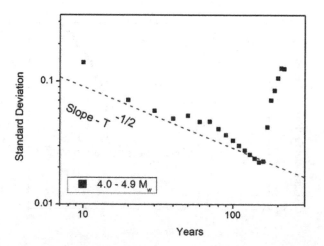

Figure 2.12 Completeness analysis using the Stepp (1972) technique.

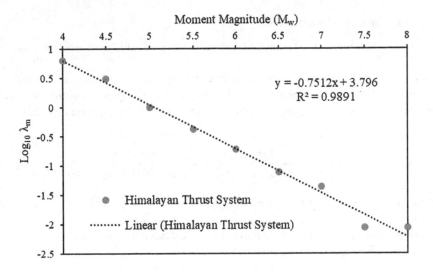

Figure 2.13 Seismicity parameters for Himalayan Thrust System.

for possibility of large and small earthquakes and $\beta = 2.303b$. Figure 2.13 shows a typical plot for Himalayan Thrust System. The reciprocal λ_m is taken as return period (T_R) of earthquake exceeding a magnitude.

2.3.9 GROUND MOTION PREDICTION EQUATION (GMPE)

The earthquake-resistant design of a structure is based on an approximation of probable strong ground motion. The ground motion model is generally developed based on strong motion characteristics of the region; plate boundary, subduction and intraplate, and accelerogram records of different magnitude of earthquakes at various epicentral distances. A regression analysis is then carried out to get the mean value of PGA with minimum variance, and subsequently, site coefficients are estimated at different periods. However, a sufficient number of strong ground motion records are rarely available for direct estimation. Therefore, selection can be made from globally available GMPEs (Douglas 2019).

2.3.10 FORMULATION OF PSHA

The accuracy of a PSHA demands vigilant consideration to the problems of characterization of seismogenic sources and prediction of ground motion parameters and probability calculations. Characterization of seismogenic source requires consideration to spatial characteristics of the source, distribution of earthquakes within the source, distribution of earthquake size for the source, and distribution of earthquakes with time. These characteristics, however, involves some level of uncertainty. The hazard is generally calculated for recurrence intervals in years with percent probability of exceedance.

2.3.10.1   Spatial Uncertainty

Uncertainty in source-site distance may be given as probability distribution function (PDF).

Figure 2.14 illustrates the spatial uncertainty associated with an area source of earthquake. This can be done by assuming that there is an equal possibility of occurrence of an earthquake anywhere within the area source by using Equation 2.5.

$$F_R(r) = P(R \le r) = \frac{\text{Area of circle with radius } r}{\text{Area of circle with a radius of 100 km}} = \frac{r^2}{10,000} \quad (2.5)$$

where $F_R(r)$ is the cumulative distribution function (CDF) and r between 0 and 100 km.

The following equation gives a more complete description as it accounts for other ranges as well.

$$F_R(r) = \begin{cases} 0 & \text{if } r < 0 \\ \frac{r^2}{10,000} & \text{if } 0 \le r < 100 \\ 1 & \text{if } r \ge 100 \end{cases} \quad (2.6)$$

The PDF for a distance can be calculated as follows:

$$f_R(r) = \frac{d}{dr} F_R(r) = \begin{cases} \frac{r}{5,000} & \text{if } 0 \le r < 100 \\ 0 & \text{otherwise} \end{cases} \quad (2.7)$$

2.3.10.2   Size Uncertainty

The uncertainty in size of an earthquake can be described by PDF of Gutenberg–Richter law (Gutenberg and Richter 1954). For a known maximum and minimum magnitude for a source, CDF is given by the following equation

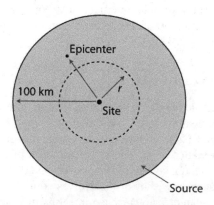

Figure 2.14   Illustration of spatial uncertainty of an area source.

$$F_m(m) = \frac{1 - 10^{-b(m - m_{\min})}}{1 - 10^{-b(m_{\min} - m_{\max})}} m_{\min} < m < m_{\max} \tag{2.8}$$

and the PDF as

$$f_m(m) = \frac{b \ln(10) \, 10^{-b(m - m_{\min})}}{1 - 10^{-b(m_{\max} - m_{\min})}} m_{\min} < m < m_{\max} \tag{2.9}$$

2.3.10.3 Temporal Uncertainty

The Poisson model can be combined with Gutenberg–Richter recurrence law to predict the probability of at least one exceedance in a period of t years by the following expression:

$$P[N \geq 1] = 1 - e^{-\lambda_m t} \tag{2.10}$$

where λ_m = mean annual rate of exceedance that can be calculated using seismicity parameters.

2.3.10.4 Uncertainty in GMPE

This can be calculated by the following equation:

$$P(\text{PGA} > x | m, r) = 1 - \Phi\left(\frac{\ln x - \overline{\ln \text{PGA}}}{\sigma_{\ln \text{PGA}}}\right) \tag{2.11}$$

where $\Phi()$ is the standard normal CDF.

2.3.10.5 Hazard Calculation Using Total Probability Theorem

This can be determined by the following equation:

$$\lambda(\text{PGA} > x) = \sum_{i=1}^{n_{\text{sources}}} \lambda(M_i > m_{\min}) \int_{m_{\min}}^{m_{\max}} \int_0^{r_{\max}} (\text{PGA} > x | m, r) f_{M_i}(m) f_{R_i}(r) \, dr \, dm \tag{2.12}$$

where n_{sources} is the number of sources considered, and M_i/R_i denotes the magnitude/distance distributions for source i.

2.3.10.6 Disaggregation

Disaggregation analysis helps in segregating a magnitude and/or distance combination that gives the largest contribution towards the ground motion. Disaggregation is useful in the analysis for which value of magnitude and/or distance corresponding to the most likely earthquake is essential, e.g., GRA and liquefaction analysis. The possibility of having earthquake of magnitude m is calculated by the following equation:

$$P(M = m | \text{PGA} > x) = \frac{\lambda(\text{PGA} > x, M = m)}{\lambda(\text{PGA} > x)} \tag{2.13}$$

The denominator of above equation is exactly the same as the Equation 2.12. The numerator can be calculated by omitting the integration over M in Equation 2.12 and is given by

$$\lambda\,(\text{PGA} > x, M = m) = \sum_{i=1}^{n_{\text{sources}}} \lambda\,(M_i > m_{\text{min}}) \int_0^{r_{\text{max}}} P\,(\text{PGA} > x | m, r)$$
$$\times f_{M_i}\,(m)\, f_{R_i}\,(r)\, dr dm \qquad (2.14)$$

2.4   GROUND RESPONSE ANALYSIS (GRA)

The most crucial as well as frequently required investigation in soil dynamics is the estimation of ground response during earthquakes. GRA is often carried out for the assessment of seismic wave amplification/de-amplification, development of response spectra, determination of peak accelerations and corresponding resonant frequencies, evaluation of deformation characteristics to assess liquefaction hazard, and calculation of earthquake loads to be imposed on embankment dams.

There are a number of approaches available to estimate the degree of wave amplification, e.g., linear, equivalent linear, and nonlinear analysis offering varying dimensionality (1-D, 2-D, and 3-D). Schnabel et al. (1972) approximated the nonlinear hysteretic behavior of soils by an equivalent linear model and developed SHAKE program. The computer program is now widely used for 1-D equivalent linear GRA. The soil shows nonlinear behavior; therefore, shear modulus of soil would vary constantly during cyclic loading. The incapability of equivalent linear model to represent true variation in soil stiffness that actually occurs during cyclic loading has been highlighted by many investigators, e.g., Finn et al. (1978), Arslan and Siyahi (2006), Hosseini and Pajouh (2012), and Kramer (2013). The nonlinear models have been found to better represent the response of soils to earthquake ground motions (Hosseini and Pajouh 2012). Also, for the sites with deep soft soils or sites where strong earthquakes are expected, the use of equivalent linear model is not considered as good practice (Hashash et al. 2010). Hence, nonlinear approach is preferred by most of the investigators. The standard nonlinear models that are being popularly used for analysis have been developed by Ramberg and Osgood (1943), Matasovic and Vucetic (1993), and Hashash and Park (2001). Generally, amplification of seismic waves is computed by 1-D model that assumes that the horizontal shear waves originating from bedrock propagate in a vertical direction through several layers of the soil profile. For the rare cases, dealing with 2-D or 3-D problems, constitutive models proposed by Mroz (1967), Momen and Ghaboussi (1982), Dafalias (1986), Kabilamany and Ishihara (1990), Gutierrez et al. (1993), and Cubrinovski and Ishihara (1998) are generally preferred.

2.4.1   METHODS FOR NONLINEAR GROUND RESPONSE ANALYSIS

The present DSHA and PSHA methodologies account for the earthquake hazard for rock sites only, and the effect of wave amplification is rarely considered in ground motion models. Therefore, estimation of ground motion amplification for soil sites is an important task for accurate assessment of seismic hazard in earthquake-prone areas. The development of site-specific ground motion characteristics involves analysis of both seismic hazard and seismic wave amplification. Over the years, nonlinear method has greatly evolved to give a more precise characterization of nonlinear behavior of the soil (Stewart and Kwok 2008). The description of various approaches offering different levels of dimensionalities for carrying out nonlinear GRA is as follows.

2.4.1.1 One-Dimensional Approach

When a fault ruptures, body waves are produced, which radiate in various directions in the continuum. These waves are reflected and refracted at the interface of various geological materials. The shear wave velocity of top layer is usually lesser than lower layer, and therefore, the incident waves are reflected more vertically. As the waves arrive at the surface, they become nearly vertical due to several refractions (Figure 2.15), giving 1-D wave propagation. One-dimensional approach assumes that the geological margins are parallel and that the response of ground is mainly produced by horizontal seismic shear waves traveling vertically from bedrock. Also, strata along with bedrock are supposed to extend infinitely in the horizontal. 1-D GRA is based on these assumptions. The methods developed based on above assumptions have been generally observed to predict response quite close to the observed values for various events. The problems in which a single dimension is unable to provide an appropriate assessment of ground response are usually solved by adopting 2-D and 3-D approaches.

2.4.1.2 Two-Dimensional Approach

The 1-D GRA predicts accurately the response for level or slightly sloping sites with horizontal material boundaries and hence is extensively used in most of the ground response problems. For some specific problems, the assumptions of 1-D wave propagation are not applicable; for example, for steep or uneven ground surfaces, sites with large structures, buried structures, or walls and tunnels or pipelines, all require 2-D GRA. The cases where one dimension is significantly greater are solved as 2-D problem. Figure 2.16 shows some cases of 2-D ground response problems.

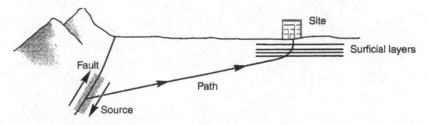

Figure 2.15 Multiple refractions produce practically a vertical wave near the ground surface. (After Kramer 2013.)

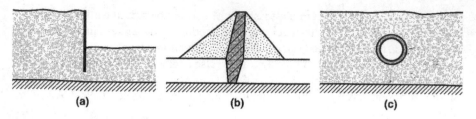

Figure 2.16 Typical problems analyzed by 2-D approach: (a) cantilever retaining wall, (b) terrain dam, and (c) pipeline. (After Kramer 2013.)

Ground response problems involving 2-D and 3-D are most generally resolved using a finite-element analysis or shear beam approach.

2.4.1.3 Three-Dimensional Approach

The situations where soil conditions vary in 3-D and boundaries change in 3-D, e.g., earthfill dam in a narrow valley, soil–structure interaction problems, and where response of one building may affect response of another, a 3-D approach is more appropriate (Figure 2.17).

Three-dimensional GRA is carried out just like 2-D analysis. Programs based on finite-element methods (FEM) offering both equivalent linear and nonlinear approach are available. Several 3-D analyses are available to solve soil–structure interaction problems such as direct method and multistep method. For approximating the response of terrain dam in narrow valleys, shear beam analysis can be used.

2.4.2 PROCEDURE OF GRA

A typical GRA requires a standard set of data and input. The procedure involves of the following steps (Yoshida 2015): (i) collection of data, (ii) modeling of data in computer programs, (iii) executing the program, and (iv) interpretation of results.

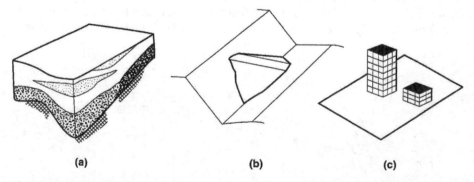

Figure 2.17 Situations requiring 3-D GRA: (a) soil conditions change in 3-D, (b) problem boundaries change in 3-D, and (c) problem of soil–structure interaction. (After Kramer 2013.)

The inputs are classified into following:

i. Geological or geotechnical vertical profiles
ii. Engineering properties of soil layers
iii. Recorded/scaled/artificial input ground motions
iv. Parameters to regulate the working of the software.

More state-of-the-art analyses need additional soil properties, e.g., permeability (K), dry and saturated density of soil; model, i.e., shear modulus degradation curves, damping ratio curves, and curve-fitting parameters; and hysteretic and viscous damping formulation. These input parameters may have different effects on the output of the GRA. Generally, the characteristics of input ground motions significantly affect the results. The low strain shear modulus and nonlinear behavior at high strains largely govern ground response represented by the response spectrum (Roblee et al. 1996).

Several computer programs are available for carrying out GRA. All these programs estimate tangent shear modulus which depicts true nonlinear behavior of soil and solve the dynamic equation of motion in time domain by a step-by-step time integral scheme. However, they may simulate the nonlinear behavior and material damping of soil differently. The widely used nonlinear computer programs are D-MOD2000 (Matasović and Ordóñez 2007), DEEPSOIL (Hashash et al. 2016), OpenSees (Mazzoni et al. 2006), SUMDES (Li et al. 1992), TESS (Pyke 2000), etc. All can compute 1-D nonlinear ground response, while OpenSees and SUMDES can simulate 2-D and 3-D shaking.

The D-MOD2000 and DEEPSOIL programs simulate the stiffness and damping of soil with nonlinear hysteretic springs connected to lumped masses. The equation of motion is solved in time domain using dynamic response scheme developed by Lee and Finn (1978). Viscous dashpots are used for additional viscous damping. The nonlinear soil behavior is simulated by coupling, (i) a backbone curve which can be curve-fitted to match G/G_{max}-γ curves and D-γ input by the user, and (ii) the extended Masing rules that govern unload–reload behavior and develop hysteretic damping. The initial backbone curve in D-MOD2000 is defined by modified Konder–Zelasko (MKZ) constitutive model (Matasovic and Vucetic 1993, 1995), whereas in DEEPSOIL, the extended MKZ model is used for this purpose. Both codes offer the use of simplified as well as full Rayleigh damping formulations, which match a target damping ratio at one or two frequencies. Additionally, DEEPSOIL also offers extended Rayleigh damping for four matching frequencies, which can be useful for deep profiles (Park and Hashash 2004).

OpenSees and SUMDES programs utilize nonlinear finite-element analyses that can solve the multi-directional earthquake shaking problem with full coupling of wave propagation and pore water pressure generation and dissipation effects. Both have total stress as well as effective stress analysis capabilities. SUMDES utilizes a bounding surface hypo-plasticity model (Wang et al. 1990), whereas OpenSees utilizes a multi-surface plasticity model (Yang et al. 2003). Both programs utilize Rayleigh damping. SUMDES utilizes simplified Rayleigh damping with the matching frequency set at 1 Hz, and OpenSees utilizes full Rayleigh damping.

TESS is a 1-D nonlinear ground response assessment program that uses an explicit finite difference method to solve the dynamic equation of motion. In line with D-MOD2000 and DEEPSOIL, the backbone curve is fit using coefficients, but instead of Masing rules, the Cundall–Pyke hypothesis is used to model unloading and reloading behavior (Pyke 1979). The viscous damping formulation is not included in this program. However, the program introduces a low strain damping for which a parameter VT is used to quantify the rate of loading effect on shear stress.

On comparison of approaches available for performing GRA, it is useful to adopt 1-D nonlinear ground response approach to estimate the effect of local site conditions on ground motions considering the following facts (Govindaraju et al. 2004):

a. This approach is believed to provide conservative results, and many commercial software applications with different constitutive models are available. Moreover, the approach is time-tested, and most structures designed by this approach have survived earthquakes.
b. When a fault ruptures, body waves are produced that radiate in various directions and get reflected and refracted at the boundaries of different geological materials. The shear wave velocity in top soil is usually lesser than the soil below, and therefore, the incident waves are refracted more vertically.
c. Earthquake ground motions in vertical direction are not as significant from the viewpoint of structural design as horizontal earthquake ground motions.
d. Soil properties generally change much rapidly in the vertical path than in the horizontal path.

2.4.3 GEOTECHNICAL SITE CHARACTERIZATION

2.4.3.1 Site Class

The sites can be classified on the basis of provisions of NEHRP (FEMA 450 2003) using average standard penetration test (SPT) value of the profile. The existing NEHRP provisions organize the strata into the classes for average SPT value (N_{30}) of the profile for 30 m depth as per following:

$$N_{30} = \frac{\sum_{i=1}^{n} d_i}{\sum_{i=1}^{n} \frac{d_i}{N_i}} \tag{2.15}$$

where N_i is SPT value for the ith stratum, d_i is thickness of the stratum.

2.4.3.2 Bedrock Definition

The location of bedrock needs to be defined as it sets a boundary condition. The boundary condition implies that behavior of strata below the bedrock does not have any impact on the result of GRA. This also means that an earthquake motion at the engineering bedrock can be assumed as an incident wave, and it cannot be affected by the behavior of strata below. The wave can reflect back into the bedrock, and the boundary condition can also consider radiation damping. An engineering bedrock following such a boundary condition is called as an elastic bedrock.

2.4.4 ESTIMATION OF DYNAMIC SOIL PROPERTIES

The dynamic soil characteristics essential for modeling the cyclic behavior of soils are (i) maximum shear modulus or low strain shear modulus (G_{max}), (ii) modulus degradation curves (G/G_{max}-γ), and (iii) damping ratio curves (D-γ).

2.4.4.1 Low Strain Shear Modulus (G_{max})

Characterization of the stiffness of an element of soil requires consideration of both maximum shear modulus (G_{max}), and the way the modulus ratio G/G_{max} varies with cyclic strain amplitude and other parameters. Hence, G_{max} plays a fundamental role in the estimation of the ground response parameters in seismic microzonation studies. The field value of G_{max} is generally computed using shear wave velocity by the following equation:

$$G_{max} = \rho V_s^2 \qquad (2.16)$$

where ρ = density of soil layer and V_s = shear wave velocity. The value of shear modulus based on shear wave velocity is very reliable, as most of the geophysical tests conducted to determine V_s induce shear strain $<3 \times 10^{-4}$ %.

2.4.4.2 Standard G/G_{max}-γ and D-γ Curves

The shear modulus of clays degrades much slowly than that of sands. The plasticity of clays has profound influence on the shape of shear modulus degradation curves, and it was first reported by Zen et al. (1978). Many other investigators (e.g., Sun et al. 1988, Vucetic and Dobry 1991) have also reported considerable influence of the plasticity index in comparison with void ratio, on the shape of G/G_{max}-γ curve. On the other hand, for the soils of low plasticity, effective confining stress influences the degradation behavior of shear modulus (Iwasaki et al. 1982). The damping behavior is also influenced by plasticity characteristics as observed by Kokusho et al. (1982). The damping ratio decreases with the increase in plasticity index for the same cyclic shear strain amplitude. However, damping behavior of the low plastic soils depends upon effective confining pressure. The damping behavior of gravels is quite similar to that of sands (Seed et al. 1986).

The G/G_{max}-γ and D-γ curves are extremely important for GRA as they influence extent of attenuation of seismic waves in a deposit. Many investigators, e.g., Seed and Idriss (1970), Seed et al. (1986), Sun et al. (1988), Vucetic and Dobry (1991), Darendeli (2001), Menq (2003), and Vardanega and Bolton (2011), have developed standard curves for different types of soils. Often due to time and practical constraints, ground response studies are carried out using the standard curves, as developing the site-specific curves for soils is a tedious process which requires advanced dynamic soil tests.

2.4.5 SELECTION OF INPUT EARTHQUAKE MOTION

The last step in wave amplification analysis is generating or getting an acceleration time history, which is compatible with the maximum dynamic loading expected at the site of interest. This time history is then used, as an input motion assuming it to be originating from the engineering bedrock as an incident wave. As per the recommendations of Pacific Earthquake Engineering Research Center (PEER), a rock outcropping motion should be applied without any modification, for an elastic base, for time-domain analyses, or a within motion be used with no change with rigid bedrock (Stewart and Kwok 2008).

Modern seismic codes, e.g., UBC 1997 and IBC 2000, motivate the use of real records, at the same time allowing the design engineer to supplement these with simulated motions where sufficient suitable real records are not available (Bommer and Acevedo 2004). Suitable acceleration time histories can be selected on the basis of PGA value, magnitude of controlling earthquake, and source to site distance and site class.

2.4.6 NONLINEAR GROUND RESPONSE

A comprehensive GRA requires an accurate characterization of sites, reliable estimates of various dynamic soil properties, and an input motion representing characteristics of the expected ground motion. Nonlinear analysis can be carried out using DEEPSOIL v.6.1 (Hashash et al. 2016). The pressure-dependent hyperbolic model (MKZ) relates shear modulus and damping ratio of the soil layers to shear strains developed during earthquakes. For each layer of the soil deposit, reference strain (γ_r), stress–strain curve parameter-β, stress–strain curve parameter-s, pressure-dependent parameter-b, reference stress (σ_{ref}), and pressure-dependent parameter-d need to be defined. A curve-fitting procedure, MRDF-UIUC, is then adopted for each layer to find the above parameters. For sandy soils, effective vertical stress is required for defining the variation of shear modulus with shear strain at a particular depth from modulus reduction curves, whereas for clayey soils, effective vertical stress and plasticity index are required. The hysteretic behavior is governed by Masing and extended Masing criteria. The number of iterations in the software is kept 15.

2.4.6.1 Formulation of Ground Response

All the sites are assumed to have horizontal soil layers, which extend infinitely. The soil profiles have been modeled as a series of lumped masses connected by springs and dashpots making a multiple degree of freedom system (MDOF) as shown in Figure 2.18.

The following incremental dynamic equation of motion is solved to carry out the nonlinear dynamic analysis of the soil column.

$$M\Delta\ddot{u} + C\Delta\dot{u} + K\Delta u = -M\Delta\ddot{u}_g \tag{2.17}$$

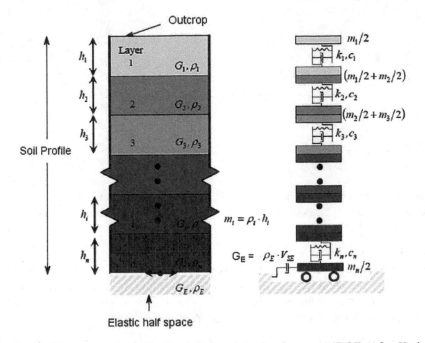

Figure 2.18 Representation of horizontally layered soil column as MDOF. (After Hashash et al. 2010.)

where the coefficients M, C, and K represent mass, viscous damping, and stiffness, respectively, and $\ddot{u}$, $\dot{u}$, u, and $\ddot{u}_g$ represent acceleration, velocity, displacement, and exciting acceleration at the elastic base, respectively.

The soil response is estimated from a constitutive model that defines the cyclic behavior of soil. For modeling the hysteretic behavior, most widely used software uses the variation of hyperbolic model, to represent the backbone curve of the soil, with the extended unload–reload Masing rules (Masing 1926). The loading and unloading equations of MKZ model (Matasovic 1993), further modified by Hashash and Park (2001) used in DEEPSOIL software, are, respectively, as follows:

$$\tau = \frac{\gamma G_{\max}}{1 + \beta \left(\frac{\gamma}{\gamma_r} \right)^S} \tag{2.18}$$

$$\tau = \frac{2 G_{\max} \left(\frac{\gamma - \gamma_{\text{rev}}}{2} \right)}{1 + \beta \left(\frac{\gamma - \gamma_{\text{rev}}}{2 \gamma_r} \right)^S} + \tau_{\text{rev}} \tag{2.19}$$

where τ = shear strength, $G_{\max}$ = low strain shear modulus, γ = shear strain, γ_r = reference shear strain, τ_{rev} = shear stress at reversal, γ_{rev} = shear strain at reversal, and β and S = model fitting parameters.

Loading and unloading (cyclic loading) are introduced by extended Masing rules, which are as follows (Kramer 2013) (Figure 2.19):

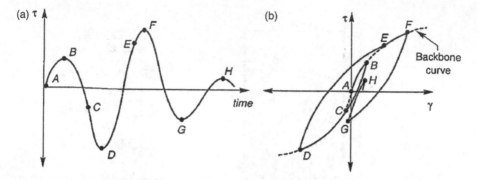

Figure 2.19 Extended Masing rules (after Kramer 2013). (a) Variation of shear stress with time and (b) resulting stress–strain behavior.

1. The backbone curve is used for initial loading.
2. Stress–strain plot tracks a path given by Equation 2.19, as stress reversal occurs at a point B. This means that the unloading, reloading curves would have the same shape as the backbone curve, and the origin is shifted to load reversal point. The path is twice expanded.

The above rules (Masing 1926) are insufficient for describing the soil response. Hence, following additional guidelines are required:

1. If unloading or reloading curve exceeds maximum previous strain and intersects backbone curve, it follows backbone curve until next stress reversal.
2. The stress–strain curve follows the stress–strain curve of previous cycle, if unloading or reloading curve of the present cycle intersects unloading or reloading curve of previous cycle.

The modification in MKZ model allows the effect of confining pressure on secant shear modulus of soil. The coupling of confining pressure and shear stress is introduced by making reference shear strain (γ_r) dependent on effective stress by using the following equation:

$$\gamma_r = a \left(\frac{\sigma_v'}{\sigma_{\text{ref}}} \right)^b \tag{2.20}$$

where a and b are curve-fitting parameters, σ_v' = vertical effective stress, and σ_{ref} = reference shear stress of 0.18 MPa.

Low strain damping (ξ) is induced separately by the following equation:

$$\xi = \frac{c}{(\sigma_v')^d} \tag{2.21}$$

where c and d are curve-fitting parameters.

It is observed that overestimation of damping at large strain can result, when the hysteretic damping (ξ_{Masing}) is calculated using unload–reload cycles as per using

Masing rules based on modulus reduction curves. This overestimation can be avoided by multiplying ξ_{Masing} with $F(\gamma_m)$ which is as follows:

$$F\left(\gamma_m\right) = p_1 - p_2 \left(1 - \frac{G_{\gamma_m}}{G_{\max}}\right)^{p_3} \tag{2.22}$$

where G_{γ_m} = shear modulus at maximum strain, and p_1, p_2, and p_3 are fitting parameters. The reduction factor modifies the reloading cycle, and the expression is as follows:

$$\tau = F\left(\gamma_m\right) \left[\frac{2G_{\max}\left(\frac{\gamma - \gamma_{\text{rev}}}{2}\right)}{1 + \beta \left(\frac{\gamma - \gamma_{\text{rev}}}{2\gamma_r}\right)^S} - \frac{G_{\max}\left(\gamma - \gamma_{\text{rev}}\right)}{1 + \beta \left(\frac{\gamma_m}{\gamma_r}\right)^S} \right] + \frac{G_{\max}\left(\gamma - \gamma_{\text{rev}}\right)}{1 + \beta \left(\frac{\gamma_m}{\gamma_r}\right)^S} + \tau_{\text{rev}} \tag{2.23}$$

where γ_m = maximum shear strain. The β method (Newmark 1959) is then used to get response of the soil column.

2.5 LIQUEFACTION POTENTIAL

Based on the type of data available, liquefaction hazard mapping can be carried out by different methods, e.g., susceptibility mapping based on geological and geomorphological characteristics and assessment based on geotechnical criteria using deterministic or probabilistic approach. The simplified procedure based on SPT value is described next.

2.5.1 SIMPLIFIED PROCEDURE BASED ON SPT

Two approaches, deterministic and probabilistic, are available for the assessment of factor of safety against liquefaction. A number of generalized assumptions and approximations regarding dominant sources of uncertainty and their distributions across the case history database have been taken for developing the probabilistic model. This is due to the reason that the majority of liquefaction case histories lack sufficient information for quantifying various sources of uncertainty in magnitudes or distributions (Boulanger and Idriss 2012). Therefore, the deterministic approach is generally used for the assessment of liquefaction potential. The simplified procedure is summarized as follows:

1. Using compositional criteria, it is ascertained, whether or not, the soil would liquefy during an earthquake.
2. The observations for the position of ground water table must be available from the geotechnical data to ensure that liquefaction analysis is done on submerged soil. The possibility of rise in water table at a future date must also be accounted for in the liquefaction analysis.
3. The cyclic stress induced by the earthquake normalized for the effective overburden pressure (cyclic stress ratio or CSR) is calculated.

4. The cyclic strength of the soil, assessed using field test data, normalized for the effective overburden pressure (cyclic resistance ratio or CRR) is calculated.
5. Factor of safety (FS) = CRR/CSR is calculated. The higher FS would imply that the soil is more resistant to liquefaction.
6. Based on the factor of safety (FS) values, liquefaction potential index (LPI) is calculated using FS values of all the liquefiable layers of a soil column.

For liquefaction susceptibility mapping, PGA values obtained from PSHA for rock sites are revised appropriately using results of GRA. The corresponding expected earthquake magnitudes (M_w) are determined from disaggregation of PSHA results. The procedure for the assessment of liquefaction potential of a soil deposit, as given by Idriss and Boulanger (2006) using SPT N-value, is described in detail next.

2.5.1.1 Cyclic Stress Ratio (CSR)

The value of CSR is adjusted for equivalent uniform shear stress induced by earthquake ground motions having a moment magnitude of 7.5 and equivalent overburden pressure of 1 atmosphere by following equation:

$$(CSR)_{M=7.5,\sigma=1} = 0.65 \left(\frac{\sigma_{vo}}{\sigma'_{vo}} \right) (a_{max}) (r_d) \left(\frac{1}{MSF} \right) \left(\frac{1}{K_\sigma} \right) \tag{2.24}$$

where, σ_{vo} = total overburden stress, σ'_{vo} = effective overburden stress, a_{max} = PGA corresponding to 475-year return period, r_d = stress reduction coefficient, MSF = magnitude scaling factor, and K_σ = overburden correction factor.

Stress reduction coefficient (r_d) accounts for the flexibility and dynamic response of the soil and represents the variation of shear stress amplitude with depth.

$$r_d = \exp(\alpha + \beta M) \tag{2.25}$$

$$\alpha = -1.012 - 1.126 \sin \left(\frac{z}{11.73} + 5.133 \right) \tag{2.26}$$

$$\beta = 0.106 + 0.118 \sin \left(\frac{z}{11.28} + 5.142 \right) \tag{2.27}$$

These equations are applicable for depth $z \leq 34$ m.
For $z > 34$ m.

$$r_d = 0.12 \exp(0.22M) \tag{2.28}$$

Magnitude scaling factor (MSF) is used to adjust the CSR induced by an earthquake magnitude (M).
MSF ≤ 1.8 for $M_w \leq 5.4$ and is given as follows:

$$MSF = 6.9 \exp \left(\frac{-M}{4} \right) - 0.058 \leq 1.8 \tag{2.29}$$

For an equivalent overburden pressure of one atmosphere, correction factor (K_σ) is used to adjust CSR values and computed as follows:

$$K_\sigma = 1 - C_\sigma \ln\left(\sigma'_{vo}/P_a\right) \leq 1.0 \tag{2.30}$$

$$C_\sigma = \frac{1}{18.9 - 2.55\sqrt{(N_1)_{60}}} \tag{2.31}$$

where P_a is the reference pressure.

2.5.1.2 Cyclic Resistance Ratio (CRR)

$$CRR = (CRR_{\sigma=1,\alpha=0}) K_\sigma K_\alpha \tag{2.32}$$

where K_σ = overburden correction factor and K_α = static shear stress correction factor.

The SPT N-values need to be normalized to an equivalent effective vertical overburden pressure σ'_{vo} of about 101 kPa to obtain blow count values that are more uniquely dependent on relative density (D_R) rather than on the overburden pressure coming from the above soil layers. The corrected blow count can be expressed as follows:

$$(N_1)_{60} = C_N (N)_{60} \tag{2.33}$$

$$C_N = \left(\frac{P_a}{\sigma'_{vo}}\right)^\alpha \leq 1.7 \tag{2.34}$$

$$\alpha = 0.784 - 0.0768\sqrt{(N_1)_{60}} \tag{2.35}$$

where $N_1 = C_N (N_m)$, N_m = SPT value at field, C_N = overburden correction factor to normalize SPT value, and N_{60} = SPT value after correction to an equivalent 60% hammer efficiency. The value of $(N_1)_{60}$ is limited to 46. The calculation of C_N is iterative as both C_N and $(N_1)_{60}$ depend on each other. The expression for N_{60} is as follows:

$$N_{60} = N_m C_R C_S C_B E_m / 0.60 \tag{2.36}$$

where C_R = rod length correction, C_S = sampling method correction, C_B = borehole diameter correction, and E_m = hammer efficiency (0.45). The value of correction factors for N_{60} can be adopted from Youd et al. (2001).

Fine content (FC) correction has to be applied to $(N_1)_{60}$, if FC > 5%, to convert it into equivalent clean sand value. The description of correction factor is as follows, for FC $\leq$ 35:

$$(N_1)_{60CS} = (N_1)_{60} + \Delta(N_1)_{60} \tag{2.37}$$

where

$$\Delta(N_1)_{60} = \exp\left\{1.63 + \frac{9.7}{FC + 0.1} - \left(\frac{15.7}{FC + 0.1}\right)^2\right\} \tag{2.38}$$

These $(N_1)_{60cs}$ values are further used to compute CRR by using the following formulation:

$$CRR_{\sigma=1,\alpha=0} = \exp\left\{ \frac{(N_1)60cs}{14.1} + \left(\frac{(N_1)60cs}{126}\right)^2 - \left(\frac{(N_1)60cs}{23.6}\right)^3 \right.$$
$$\left. + \left(\frac{(N_1)60cs}{25.4}\right)^4 - 2.8 \right\} \qquad (2.39)$$

However, the layers with FC > 35% are considered non-liquefiable.

2.5.1.3   Factor of Safety (FS) and Liquefaction Potential Index (LPI)

The FS shows the potential of a given layer of soil against liquefaction.

The factor of safety (FS) against liquefaction is determined as follows:

$$FS = \frac{CRR}{CSR} \qquad (2.40)$$

On the other hand, LPI quantifies the severity of liquefaction at a given location for down to a depth of 20 m (Iwasaki et al. 1978, Luna and Frost 1998). It is computed by taking integration of one minus the factors of safety (FS) against liquefaction, for liquefiable layers, along the entire depth of soil column below the ground surface, at a specific location. The LPI value is considered zero for a layer with FS ≥ 1. A weighting function has also been added to give more weight to the layers closer to the ground surface. The LPI is calculated using the following expression:

$$LPI = \sum_{i=1}^{n} w_i F_i H_i \qquad (2.41)$$

and

$$F_i = 1 - FS \qquad (2.42)$$

where w_i is the weighting factor = $10-0.5z_i$, and z_i is the depth of ith layer (m).

The liquefaction severity level with respect to LPI is given in Table 2.1.

Table 2.1
Severity of Liquefaction Based on LPI

LPI	Severity of Liquefaction
LPI = 0	Little to none
0 < LPI < 5	Minor
5 < LPI < 15	Moderate
LPI > 15	Major

Source: After Luna and Frost (1998).

REFERENCES

Arslan H. and Siyahi B. (2006). A comparative study on linear and nonlinear site response analysis. *Environmental Geology* 50:1193–1200.

Baker J.W. (2008). An introduction to probabilistic seismic hazard analysis. White Paper, Version 1.3, Stanford University, Stanford, CA.

Bommer J.J. and Acevedo A.B. (2004). The use of real earthquake accelerograms as input to dynamic analysis. *Journal of Earthquake Engineering* 8(1):43–91.

Boulanger R.W. and Idriss I.M. (2012). Probabilistic standard penetration test based liquefaction triggering procedure. *Journal of Geotechnical and Geoenvironmental Engineering* 138(10):1185–1195.

Cornell C.A. (1968). Engineering seismic risk analysis. *Bulletin of Seismological Society of America* 58:1583–1606.

Cubrinovski M. and Ishihara K. (1998). Modelling of sand behaviour based on state concept. *Soils and Foundations* 38(3):115–127.

Dafalias Y.F. (1986). Bounding surface plasticity. I: Mathematical foundation and hypoplasticity. *Journal of Engineering Mechanics* 112(9):966–987.

Darendeli M.B. (2001). Development of a new family of normalized modulus reduction and material damping curves. *PhD Dissertation*, University of Texas, Austin, TX.

Davis S.D. and Frohlich C. (1991a). Single-link cluster analysis, synthetic earthquake catalogues, and aftershock distribution. *Geophysical Journal International* 104:289–306.

Davis S.D. and Frohlich C. (1991b). Single-link cluster analysis and earthquake aftershocks; decay laws and regional variations. *Journal of Geophysical Research* 96:6335–6350.

Douglas J. (2019). Ground motion prediction equations 1964–2014. University of Strathclyde, Glasgow, United Kingdom.

FEMA 450. (2003). NEHRP recommended provisions for seismic regulations for new buildings and other structures. Building Seismic Safety Council of the National Institute of Building Sciences, Washington, DC.

Field E.H., Jordan T.H., and Cornell C.A. (2003). OpenSHA: A developing community-modeling environment for seismic hazard analysis. *Seismological Research Letters* 74:406–419.

Finn W.D.L., Martin G.R., and Lee M.K.W. (1978). Comparison of dynamic analysis of saturated sand. *Proceedings of the ASCE Geotechnical Engineering Division Specialty Conference*, Pasadena, CA, pp. 472–491, June 19–21, 1978.

Gardner J.K. and Knopoff L. (1974). Is the sequence of earthquakes in southern California with aftershocks removed, poissonian? *Bulletin of Seismological Society of America* 64(5):1363–1367.

Govindaraju L., Ramana G.V., Hanumantharao C., and Sitharam T.G. (2004). Site specific ground response analysis. *Current Science* 87(10):1354–1362.

Gutenberg B. and Richter C.F. (1954). *Seismicity of the Earth and Associated Phenomena.* Princeton University Press, Princeton, NJ.

Gutierrez M., Ishihara K., and Towhata I. (1993). Model for the deformation of sand during rotation of principal stress directions. *Soils and Foundations* 33(3):105–117.

Hashash Y.M.A. and D. Park. (2001). Non-linear one-dimensional seismic ground motion propagation in the Mississippi embayment. *Engineering Geology* 62(1–3):185–206.

Hashash Y.M.A., Park D., Tsai C.C., Philips C., and Groholski D.R. (2016). DEEPSOIL: 1-D wave propagation analysis program for geotechnical site response analysis of deep soil deposits, version 6.1. Tutorial and User Manual, University of Illinois at Urbana-Campaign.

Hashash Y.M.A., Phillips C., and Groholski D.R. (2010). Recent advances in non-linear site response analysis. *International Conference on Recent Advances in Geotechnical Earthquake Engineering and Soil Dynamics*, San Diego, CA, May 24–29.

Hosseini S.M.M.M. and Pajouh M.A. (2012). Comparative study on the equivalent linear and the fully nonlinear site response analysis approaches. *Arabian Journal of Geosciences* 5:587–597.

Idriss I.M. (1990). Response of soft soil sites during earthquakes. *Proceedings of H. Bolton Seed Memorial Symposium*, Vancouver, Canada, vol. 2, pp. 273–289, May 01.

Idriss I.M. and Boulanger R.W. (2006). Semi-empirical procedures for evaluating liquefaction potential during earthquakes. *Soil Dynamics and Earthquake Engineering* 26:115–130.

Iwasaki T., Tatsuoka F., and Takagi Y. (1982). Shear modulus of sands under torsional shear loading. *Soils and Foundations* 18(1):39–56.

Iwasaki T., Tatsuoka F., Tokida K., and Yasuda S. (1978). A practical method for assessing soil liquefaction potential based on case studies at various site in Japan. *Proceedings of 5th Japan Earthquake Engineering Symposium* 2:641–648.

Kabilamany K. and Ishihara K. (1990). Stress dilatancy and hardning laws for rigid granular model of sand. *Soil Dynamics and Earthquake Engineering* 9(2):66–77.

Kaiser A., Holden C., Beavan J., Beetham D., Benites R., Celentano A., Collett D., Cousins J., Cubrinovski M., Dellow G., Denys P., Fielding E., Fry B., Gerstenberger M., Langridge R., Massey C., Motagh M., Pondard N., McVerry G., Ristau J., Stirling M., Thomas J., Uma S.R., and Zhao J. (2012). The M_w 6.2 Christchurch earthquake of February 2011: Preliminary report. *New Zealand Journal of Geology and Geophysics* 55(1):67–90.

Kokusho T., Yoshida Y., and Esashi Y. (1982). Dynamic properties of soft clay for wide strain range. *Soils and Foundations* 22(4):1–18.

Kramer, S.L. (2013). *Geotechnical Earthquake Engineering*. Pearson, Singapore, South Asia.

Lee M.K.W. and Finn W.D.L. (1978). *DESRA-2: Dynamic Effective Stress Response Analysis of Soil Deposits with Energy Transmitting Boundary Including Assessment of Liquefaction Potential*. Soil Mechanics Series 36, Department of Civil Engineering, University of British Columbia, Vancouver, Canada.

Li X.S., Wang Z.L., and Shen C.K. (1992). *SUMDES: A Nonlinear Procedure for Response Analysis of Horizontally-Layered Sites Subjected to Multi-Directional Earthquake Loading*. Department of Civil Engineering, University of California, Davis, CA.

Luna R. and Frost J.D. (1998). Spatial liquefaction analysis system. *Journal of Computing in Civil Engineering* 12(1):48–56.

MacRae G., Clifton C., and Megget L. (2011). Review of NZ building codes of practice. Report to the Royal Commission of inquiry into the building failure caused by the Christchurch earthquakes, ENG.ACA.0016.1, 57pp.

Masing G. (1926). Eigenspannungeu und verfertigung beim Messing. *Proceedings of the 2nd International Congress on Applied Mechanics*, Zurich, Swisse (in German), pp. 332–335.

Matasovic N. (1993). Seismic response of composite horizontally-layered soil deposits. *PhD Dissertation*, University of California, Oakland, CA.

Matasović N. and Ordóñez G.A. (2007). D-MOD2000: A computer program package for seismic response analysis for horizontally layered soil deposits, earthfill dams, solid waste landfills. GeoMotions, LLC.

Matasovic N. and Vucetic M. (1993). Cyclic characterization of liquefiable sands. *Journal of Geotechnical Engineering* 119(11):1805–1822.

Matasović N. and Vucetic M. (1995). Generalized cyclic degradation-pore pressure generation model for clays. *Journal of Geotechnical Engineering* 121(1): 33–42.

Mazzoni S., McKenna F., Scott M.H., and Fenves G.L. (2006). *Open System for Earthquake Engineering Simulation (OpenSees): Command Language Manual.* Pacific Earthquake Engineering Center, University of California, Berkeley, CA.

McGuire R.K. (2001). *FRISK-88M: User's Manual (Version 1.80).* Risk Engineering Inc., Boulder, CO.

Menq FY. (2003). Dynamic properties of sandy and gravelly soils. *PhD Dissertation*, University of Texas, Austin, TX.

Molchan G.M. and Dmitrieva O.E. (1992). Aftershock identification: Methods and new approaches. *Geophysical Journal International* 109(3):501–516.

Momen H. and Ghaboussi J. (1982). Stress dilatancy and normalized work for sands. *Proceedings of IUTAM Conference on Deformation and Failure of Granular Materials*, Delft, Netherlands, p. 10.

Mroz Z. (1967). On the description of anisotropic workhardening. *Journal of the Mechanics and Physics of Solids* 15(3):163–175.

Newmark N.M. (1959). A method of computation for structural dynamics. *Journal of the Engineering Mechanics Division* 3:67–94.

Ordaz M. and Salgado-Gálvez M.A. (2017). R-CRISIS validation and verification document. Technical Report. Mexico City, Mexico.

Pagani M., Monelli D., Weatherill G., Danciu L., Crowley H., Silva V., Henshaw P., Butler L., Nastasi M., Panzeri L., Simionato M., and Vigano D. (2014). OpenQuake engine: An open hazard (and risk) software for the global earthquake model. *Seismological Research Letters* 85(3):692–702.

Park D. and Hashash Y.M.A. (2004). Soil damping formulation in nonlinear time domain site response analysis. *Journal of Earthquake Engineering* 8(2):249–274.

Potter S.H., Becker J.S., Johnston D.M., and Rossiter K.P. (2015). An overview of the impacts of the 2010-2011 Canterbury earthquakes. *International Journal of Disaster Risk Reduction* 14:6–14.

Pyke R.M. (1979). Nonlinear soil models for irregular cyclic loadings. *Journal Geotechnical Engineering* 105(GT6):715–726.

Pyke R.M. (2000). TESS: A computer program for nonlinear ground response analyses. TAGA Engineering Systems & Software, Lafayette, CA.

Ramberg W. and Osgood W.R. (1943). Description of stress: Strain curves by three parameters. National advisory committee for Aeronautics, Report No. NACA-TN-902, Washington, DC.

Reasenberg P. (1985). Second-order moment of central California seismicity, 1969–1982. *Journal of Geophysical Research* 90:5479–5496.

Roblee C.J., Silva W.J., Toro G.R., and Abrahamson N.A. (1996). Variability in site-specific seismic ground motion design predictions. ASCE Geotechnical Special Publication Number 58, Uncertainty in the Geologic Environment: From Theory to Practice 2.

Savage W. (1972). Microearthquake clustering near Fair View Peak, Nevada and in the Nevada seismic zone. *Journal of Geophysical Research* 77:7049–7056.

Schnabel P.B., Lysmer J., and Seed H.B. (1972). SHAKE: A computer program for earthquake response analysis of horizontally layered sites. Report No. EERC72-12, University of California, Berkeley, CA.

Seed H.B. and Idriss I.M. (1970). Soil moduli and damping factors for dynamic response analyses. Report No. EERC-70/10, Earthquake Engineering Research Center, University of California, Berkeley, CA.

Seed H.B., Wong R., Idriss I.M., and Tokimatsu K. (1986). Moduli and damping factors for dynamic analyses of cohesionless soils. *Journal of Geotechnical Engineering* 112(11):1016–1032.

Stepp J.C. (1972). Analysis of the completeness of the earthquake sample in the Puget Sound area and its effects on statistical estimates of earthquakes hazard. *Proceedings of International Conference on Microzonation for Safer Construction Research and Application*, Seattle, Washington, DC.

Stewart J.P. and Kwok A.O.L. (2008). Nonlinear seismic ground response analysis: Code usage protocols and verification against vertical array data. *Proceedings of Geotechnical Earthquake Engineering and Soil Dynamics IV Congress*, Sacramento, CA.

Sun J., Golesorkhi R., and Seed H.B. (1988). Dynamic moduli and damping ratios for cohesive soils. Report EERC 88-15, Earthquake Engineering Research Center 1988, University of California, Berkeley, CA.

Tinti S. and Mulargia F. (1985). Completeness analysis of a seismic catalog. *Annales Geophysicae* 3(3):407–414.

Vardanega P.J. and Bolton M.D. (2011). Strength mobilization in clays and silts. *Canadian Geotechnical Journal* 48:1485–1503.

Vucetic M. and Dobry R. (1991). Effect of soil plasticity on cyclic response. *Journal of the Geotechnical Engineering Division, ASCE* 111(1):89–107.

Wang Z.L., Dafalias Y.F., and Shen C.K. (1990). Bounding surface hypoplasticity model for sand. *Journal of Engineering Mechanics* 116(5):983–1001.

Yang Z., Elgamal A., and Parra E. (2003). Computational model for cyclic mobility and associated shear deformation. *Journal of Geotechnical and Geoenvironmental Engineering* 129(12):1119–1127.

Yoshida N. (2015). *Seismic Ground Response Analysis*. Geotechnical Geological and Earthquake Engineering Series 36, Springer, Berlin, Germany.

Youd T.L., Idriss I.M., Andrus R.D., Arango I., Castro G., Christian J.T., Dobry R., Finn W.D.L., Harder L.F., Hynes M.E., Ishihara K., Koester J.P., Liao S.S.C., Marcuson W.F., Martin G.R., Mitchell J.K., Moriwaki Y., Power M.S., Robertson P.K., Seed R.B., and Stokoe K.H. (2001). Liquefaction resistance of soils: Summary report from the 1996 NCEER and 1998 NCEER/NSF workshops on evaluation of liquefaction resistance of soils. *Journal of Geotechnical and Geoenvironmental Engineering* 127(10):817–833.

Zen K., Umehara Y., and Hamada K. (1978). Laboratory tests and in-situ survey on vibratory shear modulus of clayey soils with different plasticities. *Proceedings of Fifth Japan Earthquake Engineering Symposium*, Tokyo, pp. 721–728.

3 Multi-Model Approach in the Risk Assessment Tasks with Satellite Data Utilization

Yuriy V. Kostyuchenko, Maxim Yuschenko, Lesia Elistratova, and Igor Artemenko
Scientific Centre for Aerospace Research of the Earth, National Academy of Sciences of Ukraine

CONTENTS

3.1   INTRODUCTION: ON THE METHODOLOGY OF SATELLITE DATA UTILIZATION IN MULTI-MODELING APPROACH FOR SOCIO-ECOLOGICAL RISKS ASSESSMENT TASKS – A PROBLEM FORMULATION

A multi-modeling approach for socio-ecological risks assessment is a quite new research field. There is lack of formalizations, decisions, and approaches in this area. Also the utilization of the data of satellite observations in framework of the multi-modeling approach is enough new. Developed methods and algorithms are required in this field. Besides, new complex challenges and interlinked threats in field of ecological and socio-ecological security require new approaches to decision-making. So current and future graduates, professionals, practitioners, and researchers still need more books on these issues.

The proposed multi-modeling approach has methodological advantages: it is wider than that used in the listed books, i.e., multi-level and multi-objects models. Second, the proposed approach includes the utilization of the remote sensing data, and it is illustrated by the number of important examples in field of socio-ecology. So the proposed chapter has a number of theoretical and practical advantages.

For at least past 40 years, since its origin, remote sensing demonstrates a striking evolution (Elachi and Van Zyl, 2006; Lillesand, Kiefer, and Chipman, 2014), first of all in field of methodology of remote sensing, which developing, basing on the models of signal formation, models of individual indicators, and natural processes studying (Elachi and Van Zyl, 2006).

Development of natural sciences and ecology is also rapid, which leads to more comprehensive understanding of ecosystems structure and functioning: now, it is possible to understand and model complex interactions between the processes and phenomena, to simulate the feedbacks in multi-agent environment, model the integrated dynamics of the processes, and predict the behavior of multi-component systems (Suter, 2016). In this analytic context, the satellite observation technology is a source of information about behavior of the variables in these complex, interlinked models; remote sensing should be tooled not only for monitoring, but also for predictions and forecasts.

So, it requires a further development of remote sensing methodology, in particular, development of new methods of processing and interpretation of satellite data, harmonized with models and measurements, in the context of new foundation to select the sets of interrelated indicators, based on the environmental models (Campbell and Wynne, 2011). Further, application of satellite data-based indicators will also require other approaches to environmental risk assessment: this kind of risk assessment approaches should not be based on assessments of deviations of observed values from the mean and will be focused on optimal decision-making in a complex multi-component and multi-physics environment (Ermoliev, Makowski, and Marti, 2012). Therefore, this chapter is aimed to the formal task definition of the utilization of satellite data in the multi-model approach for socio-ecological risk assessment tasks.

3.2 ON THE METHODOLOGY OF MODELING: SELECTION OF VARIABLES TO ASSESSING RISKS

3.2.1 DATA UTILIZATION APPROACH TO VARIABLES SELECTION

The problem of the models' application to selection of the optimal set of remote sensing indicators in risk assessment tasks should be considered (Kostyuchenko et al., 2015; Kostyuchenko, 2018).

Let the set of a priori assumptions observed or measured values described the state of the system studied is captured in a vector x $\left(x = \left(x^1, x^2, \ldots, x^s\right)\right)$ at the initial stage of the process of forecast generation.

Hereafter, the model recalculates these values in a group of core hydrological, bio-physical, and climatological series $\left(\text{with } F(x, \varepsilon, y) = 0, (x, \varepsilon) \rightarrow y\right)$, collected in a vector y, and, based on the information from the pair (x, y), calculates values for a list of parameters, grouped into what referred as the vector of satellite indicator-based models: $z = \left(z^1, z^2, \ldots, z^s\right)$ $\left(\text{with } z^s = g^s(x, y), (x, y) \rightarrow z^s\right)$. As the result, the parameters summarized in triplet (x, y, z) will be obtained. This combined vector is an input data for the modeling of socio-economic, socio-ecological, and risk parameters as shown in Figure 3.1.

In the group of S developed satellite data-based models, labeled $s \in \{1, 2, \ldots, s\}$, each equation is such that the endogenous variables z^s can be obtained as an explicit mapping of the core variables $z^s = g^s(x, y)$ (Kostyuchenko, 2018). So, developed satellite model may be presented as time series of (x, y, z), which will determine the behavior of z_t^s, with a residual term "μ_t^s":

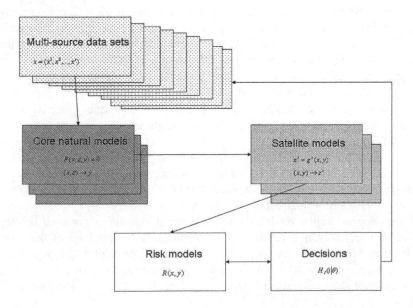

Figure 3.1   Multi-model approach to socio-ecological risks assessment.

$$z_t^s = f\left(x_t, x_{t-1}, \ldots, x_{t-L}, y_t, y_{t-1}, \ldots, y_{t-L}, z_{t-T}^s, \mu_t^s\right). \tag{3.1}$$

As we can see this relationship is unidirectional, this simple time-series satellite data-based model formally allows no interactions with other satellite variables nor any feedback between z_t^s and the core assumptions in (x,y) (Campbell and Wynne, 2011; Kostyuchenko, 2018), so we can use any methodology utilizing as separate as well-interlinked indicators.

Further, it can utilize any traditional time-series models such as autoregressive moving average models to find the most usable model of the data-generating process for a given risk metric Z_t:

$$Z_t = c + \sum_{l=0}^{N} \beta_l X_{t-l} + \sum_{l=0}^{P} \rho_l Y_{t-l} + \sum_{l=1}^{L} \partial_l Z_{t-l} + \sum_{k=0}^{K} \theta_k \varepsilon_{t-k}, \tag{3.2}$$

where Z_t is a satellite variable, X_t is a row vector of initial exogenous core variables, Y_t is a row vector of the layer of core parameters series, and ε_t is the value of the stochastic error term. The parameters c, β, ρ, ∂ are unknown and should be estimated.

Herein, it should be noted that including autoregressive terms into the model may result in a muted impact of core drivers on a target variable (Engle and Russell, 1998). Therefore, depending on risk metric Z_t and on type of supplementary variable may be applied different forms of Equation (3.2); for example, for the analysis of climate-related risk, an approach based on copulas utilization may be used (Kostyuchenko et al., 2013).

According to modern developments of Earth sciences, the approach proposed is based on the analysis of variables, which reflect a combination of ecology, climatology, hydro-geology, hydrology, and geostatistics as consideration of the statistical properties of the estimated model (Kostyuchenko, 2015). So a variable selection to identify which core drivers best explain the dynamic behavior of the studied socio-ecological risk variable as a key aspect of satellite model development is based on the models both physical and statistical (Kostyuchenko et al., 2013). Separately, it should be stated that models built using data-mining techniques (such as machine learning and neural network) may fit the existing data well, but more likely to fail in a changing external environment because they lack theoretical underpinnings. So the best analytical and prediction models should employ a combination of statistical rigors with physical principles, combine geo-ecological models with statistical optimization (see Figure 3.2) (Kostyuchenko, 2018). Additional benefit of models built this way is easiness of interpretation.

To develop a satellite model, the set of optimal exogenous potential drivers X_t should be selected from a set of Y_t in Equation (3.2). On the base of the selected and estimated final model, the conditional dynamic forecasts of Z_t can be generated, taking into account the estimated sets of final parameter and the forecasts of the core variables from the initial stage (Figure 3.2). The final step is to validate the final model.

The optimal drivers can be selected according to the following procedure. Initial set of potential drivers are identified based on relevant theory and ensuring with

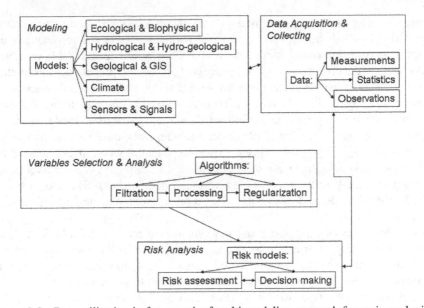

Figure 3.2 Data utilization in framework of multi-modeling approach for socio-ecological risk assessment.

calibration measurements; this ensures that the most robust and predictive model available from the tested variables will be obtained. The selected drivers should be significant at a conventional level and have the analyzed parameters of distribution; to obtain a required distribution, a regularization procedure should be applied; the final models selected by the search procedure are reviewed for consistency with initial assumptions (Kostyuchenko, 2018).

Therefore, formally, there is a problem with the selection of variables for each model type $(x, \varepsilon) \rightarrow y$, the search of the relevant type of formal relationship $(x, y) \rightarrow z^s$ between the physical and observable variable parameter, and the development of the total distribution for each type of risk investigated Equation (3.2). The problem of regularization of initial distributions of variables should also be separately considered (Kostyuchenko et al., 2015). Besides, after we obtain distributions of parameters that determine the state of the system, we need to estimate the distribution of risk and make the management decisions.

3.2.2 FORMAL MODELS OF RISK ASSESSMENT AND DECISION SUPPORT

Utilizing the obtained collection of indicators, the methodology of risk assessment, based on the optimal decisions, can be proposed (Ermoliev, Makowski, and Marti, 2012).

A key assumption in the framework of task of risk assessment and risk management was formulated as the reducing or non-increasing of losses should be used as quantitative characteristics. In real cases, variables that affect the characteristics of the management and/or decision-making system may be controlled or unmanageable. The controlled variables are used as the parameters of decision-making under the influence of information (input data) to the behavior of unmanageable variables.

Analysis of the effectiveness of full process of collecting, processing, interpreting information about the system studied, decision-making, and of system's response to decisions may be considered as the "information-response" formalization. This type of formalization was presented as follows (Schlaifer and Raiffa, 1961): $I(x,y,z)$ – stochastic information obtained from direct measurements, observations, and model forecasts; and $H_I(i|\theta)$ – probability distribution function, where θ is a state of the studied natural system. In the majority of cases, the state of the system cannot be determined accurately, so the appropriate probability distribution $p(\theta)$ and distribution $H_I(i|\theta)$ should be defined, which describes a priori incompleteness of information available.

Management decision-making and implementation may be formalized as a response to incoming information as decision function $d(I)$. Proposed approach assumes that in the case of certain specific strategy of decision-making in conditions of constant or slow changing state of natural systems, θ, the losses will be defined as function $l(d(I), \theta)$. Thus, for the decision function d, the expected losses (of risks) associated with the dangerous processes, and connected with the management decisions, can be described as

$$R(I,d) = R[H_I(\cdot),d),d] = \iint l(d(i),\theta)dH_I(i|\theta)p(\theta)d\theta. \qquad (3.3)$$

These risks can be minimized by optimal decision function d^*, entitled as Bayes decision function, which are determined by information I:

$$R(I(z^s),d^*) = \min_{d(i)} \iint l(d(i),\theta(x,y))dH_I(i|\theta)p(\theta)d\theta. \qquad (3.4)$$

Therefore, the risk can be presented in the form of a simple functional of decision function.

In the proposed approach, the minimization of losses requires an intention to completeness of information about the studied system, i.e., determination of θ states, for each of which can be defined a solution $-a^{\sim}$, which builds up the set of possible decisions A. Let's consider the realization of the set of data i^*, which optimizes the decision function d^* and minimizes appropriate risk and therefore nominally makes information (I) formally completed (I^*). This set of data can be presented as the information obtained from direct ground measurements, remote observations, archive statistics, and model forecasts.

Completeness formally means that there is a single state of studied natural object or system that meets all of the set of data—a separate realization of information I, or from a formal viewpoint, information is nominally full, $I^* \equiv I$, if there exists a function $\phi(i)$, that $H_I(i|\theta) \neq 0$, $\theta = \phi(i)$. Therefore, it means that we have to develop a set of models $(x,\varepsilon) \to y$ that have operated the set of parameters $(x,y) \to z^s$, which can be controlled by certain technological tools within the framework of a sustainable methodology of measurement. In our case, these requirements correspond to the data and methods of satellite observation of the Earth's surface z_t^s. In the described case, an optimal decision function may be presented as follows:

$$d^*(i^*) = b, \ l(b, \phi(i^*)) = \min_a l(a, \phi(i^*)). \tag{3.5}$$

Then, using the before-presented form of optimal decision function and with respect to the nominally complete information about the studied system, the risk may be defined as

$$R(I^*, d^*) = \int \min_{a \in A} l(a, \theta) p(\theta) d\theta. \tag{3.6}$$

In every separate case, depending on the task, data availability and properties of their distributions, as the optimal decision function in this approach, may use stochastic (Kopachevsky et al., 2016), Bayesian (Kostyuchenko et al., 2019), or fuzzy operators (Kostyuchenko et al., 2017).

Besides, models aimed to analyze the behavior of the distribution of $H_I(i|\theta)$ and $p(\theta)$, and to determine the realization of i^* of set I should be described . Complex analytical models should be aimed to calculate a unique set of parameters that will be obtained from determined observation systems, using defined tools of processing and interpretation of data.

Equation (3.6) allows to estimate the distributions of risk of disasters and also to develop a basement for a system of risk management decision-making.

3.3 GENERALIZED STOCHASTIC MODEL OF HYDROLOGICAL THREATS

3.3.1 ANALYSIS OF KEY PROCESSES FORMING FLOOD EMERGENCY

Determination of the parameters of the flood hazard is based on the system model of the river bed, which includes the catchment basin and channel runoff. The basic equations of the model can be represented in the following general form.

Formation of the channel flow is directly determined by the characteristics of the catchment basin. The model of the drainage basin with a variable supply area can be presented in general form as (Laurenson, 1964)

$$q(t) = \left[A_1(t) k \frac{dH}{dx} \right] + [A_2(t) R(t)] + [A_3 P(t)], \tag{3.7}$$

where $A_1(t)$ is the square of total saturation zone, A_2 is the horizontal projection of the total saturation zone within the basin, A_3 is impenetrable for water area, $R(t)$ is the precipitation intensity, k is the filtration coefficient for full saturation, and H is the hydraulic pressure.

Such a representation allows us to correctly formulate the forecasting of dangerous natural phenomena, in particular, floods and induced phenomena (such as landslides) in order to determine the characteristics of a basin catchment. To clarify the presented equations, consider some detailed individual processes. A separate but very important case of the considered catchment process is the sloping runoff.

In the general case on the slope under the conditions of minimal influence of inertial forces, the equation of the kinematic wave can be used to describe the surface runoff of a single small slope:

$$\frac{\partial h}{\partial t} + \frac{\partial q}{\partial x} = R - f;$$
$$q = \alpha h^{\beta};$$
$$h(0,t) = h(x,0) = 0$$

(3.8)

Here, x – spatial coordinates, t – time, h – depth of the surface runoff stream, q – volume of flowing water, R – precipitation intensity, f – filtration intensity, α and β – empiric coefficients. For the laminar flow $\alpha = 8gi/k\gamma$, $\beta = 3$, and for turbulent $\alpha = i^{0.5}/ni$ $\beta = 5/3$, where g – gravitational constant, i – slope angle, γ – coefficient of kinematic viscosity, and k i n – coefficients of roughness.

In addition, in the context of the formation of the channel flow, the model of the subsurface runoff on the water-tight surface should be considered. Such a process may be responsible for the transfer of certain, in some cases a significant, amount of water and for the formation of an extreme channel flow, i.e., for the formation of flood:

$$m\frac{\partial h_g}{\partial t} + \frac{\partial q_g}{\partial x} = f,$$
$$q_g = k_f h_g \left(i_g - \frac{\partial h_g}{\partial x} \right),$$
$$q_g(0,t) = 0, h_g(L_s,t) = h_r, h_g(x,0) = h_g^0(x).$$

(3.9)

where m – soil porosity, h_g – the distance between the surface of groundwater depression and the runoff surface (subsurface runoff power), q_g – volume of subsurface runoff, i_g – runoff angle, k_f – coefficient of horizontal filtration, L_s – length of slope, h_r – river water level, and h_g^0 – the initial power of subsurface runoff.

The difference between the entry of water onto the surface of the catchment and the replenishment of the water reserve of the surface layer of the soil can be described using a water-absorbing layer y, which is given in a simple formula:

$$y = \begin{cases} 0, & w < w_{min}, \\ (I - \partial w)\Delta t, & w_{min} \leq w \leq w_{max}, \\ I, & w > w_{max}, \end{cases}$$

(3.10)

Here, I – water extraction layer on the surface of the catchment (depending on the intensity of precipitation and/or snow melting), w – surface soil layer moisture, δw – replenishment of the water reserve of the surface layer of soil during the time Δt; w_{min} and w_{max} – minimum and maximum value of soil moisture.

Technically, drainage can be divided into two components. The, first, so-called "fast" consists of surface and subsurface drains, and "slow" is a ground runoff. The next step is to recalculate the runoff into a hydrographic network, which is described in the following equation:

$$Q(t) = \frac{1}{\tau} \int_{t-\tau}^{t} q(\eta) d\eta,$$

(3.11)

where Q – water flow in the river channel, q – the flow of water falling into the channel from the catchment area, and τ – flood wave parameter.

Thus, the task is to determine the parameters of the in- and out-flow of water to the basin and the parameters of the basin (terrain, location of infrastructure, and population, etc.), which will allow to assess the risks of flooding and to determine the methods of losses reducing. However, the correct calculation of the parameters of income and expenditure of water in the area of the basin is the most uncertain. Even the most simple basic process models show the need to operate spatially distributed data and take into account the inherent errors of the definition of individual variables. To demonstrate this, the general view of the models of individual processes, in particular, soil moisture and infiltration processes, should be considered.

The following diffusion equations have been successfully used for the description of the moisture transfer in the aeration zone of the soil (under saturated and unsaturated groundwater) for many years (Parker, 2014):

$$\frac{\partial w}{\partial t} = \frac{\partial}{\partial z}\left[D(w)\frac{\partial w}{\partial z} - K(w)\right],$$

$$D(w) = K(w)\frac{\partial \Psi}{\partial z}, \tag{3.12}$$

$$w(z,0) = w_0(z), w(H_s,t) = \text{const},$$

$$w(0,t) = m, R \geq f$$

Here, z – vertical coordinate oriented to the down, w – volumetric soil moisture, D – coefficient of soil moisture diffusion, K – hydraulic conductivity of soil, Ψ – soil moisture potential, and H_s – lower boundary of the calculated layer.

In the simplest case, the simplified formulas can be used:

$$K = k_f M^\mu,$$

$$\Psi = \Psi_0 M^\nu, \tag{3.13}$$

$$M = \frac{w - w_0}{m - w_0}$$

where k_f – vertical filtration coefficient, w_0 – the initial value of soil moisture, and μ and ν – calculation parameters. The given model describes the movement of the front of full saturation of soil pores with water.

A general description of the processes of water intake with precipitation, for instance, due to the melting of snow, compared to rainfall, should be proposed.

Based on the general equation of conservation of mass of water in the simulated area (e.g., $dM/dt = f_{in} - f_{out}$, where M is the water mass, and f_{in}, f_{out} are the water input and output respectively), it can be written as

$$\frac{dX}{dt} = R - h_m, \tag{3.14}$$

where X – water equivalent of snow cover, R – precipitation intensity, and h_m – intensity of water return of the snow.

To calculate the dynamics of snow deposits and water return, the various solutions of this balance equation can be used, for example, based on the daily averaging:

$$h_m = \begin{cases} [z_n - \gamma_n(1-z_n)]X - [z_{n-1} - \gamma_{n-1}(1-z_{n-1})]X \\ 0, z_n \le z_0 \end{cases}, \qquad (3.15)$$

Here, z_n and z_{n-1} – relative amounts of melting snow for given and previous intervals of time, z_0 – the value of the relative reduction of snow, at which water return begins, and γ_n – current humidity of snow.

To estimate the relative intensity of snow melting, the following equation can be used:

$$z_n = \frac{\sum\limits_{i=1}^{n} h_{ci}}{X}, \qquad (3.16)$$

where $h_{ci} = aT_i$ – layer of snow melting during the time interval i, a – average coefficient of melting, T_i – air temperature at the time i, and n – calculation step number from the beginning of the snow melting.

Relative reduction of snow reserves z_0 at which the return of water begins can be determined by the following equation:

$$z_0 = \begin{cases} 0.34\gamma_{max} + 0.059, \gamma_{max} \le 0.28 \\ 0.25\gamma_{max} + 0.083, \gamma_{max} > 0.28 \end{cases}, \qquad (3.17)$$

where γ_{max} – maximum water content of snow ($\gamma_{max} = \exp(-4\rho) - 0.04$), ρ – snow density before melting. Current water content of snow γ_n at each time interval can be estimated as

$$\gamma_n = \begin{cases} (\gamma_{max} - 0.06)\exp(-4z_n) + 0.06 \\ 0, \gamma_m < 0.063 \end{cases}, \qquad (3.18)$$

Layer of water supply to the catchment surface I during the time Δt can be determined by taking into account the distribution of snow cover, which varies during the melting period from 100% to 0%:

$$I = (1-f)h_m + R. \qquad (3.19)$$

Here, R – layer of liquid precipitation over time Δt, and f – relative square of the basin, which has been released from the snow in steps since the beginning of melting:

$$f = \sum_{i-1}^{n} \{\Delta z_i \exp[\alpha_0 \ln\alpha_0 - \ln\Gamma(\alpha_0) + (\alpha_0 - 1)\ln z_i - \alpha_0 z_0]\}. \qquad (3.20)$$

Here, α_0 – parameter, which is equal to $1/C_v^2$, C_v – is the coefficient of variation of distribution on the area of basin of water reserves in the snow cover, $\Delta z_i = z_i - z_{i-1}$ – change in the relative loss of snow in time Δt, and Γ – the gamma function.

With the monthly averaging, the other model can be applied:

$$\frac{dX}{dt} = r_1 - r_2,$$

$$r_1 = \begin{cases} R, T < 0, \\ 0, T \geq 0, \end{cases}$$

$$r_2 = \begin{cases} 0, T < 0, \\ X, T \geq 0 \end{cases}$$

(3.21)

where r_1 – input from the precipitation, r_2 – water output from the snow cover, R – precipitation layer, and T – mean air temperature.

Describing the formation of a runoff, we can define the key natural processes that determine the characteristics of the water catchment and form the flood potential of the territories.

3.3.2 DETAILED MODELS OF MOISTURE AND SOIL WATER CONTENT

Method of hydrological risk assessment using satellite data should be based on the model of dangerous processes. Hazardous dynamics (floods and swamping) might be described as the variation of moisture and water content in the accumulation zone s_1 and in the discharge zone s_2, according to Kostyuchenko et al. (2019):

$$ds_1 = A(s_1, s_2)\, dt + B(s_1, s_2)\, dW_t$$

(3.22)

$$ds_2 = C(s_1, s_2)\, dt + D(s_1, s_2)\, dW_t$$

(3.23)

Here, dW_t – is the Wiener increment: $\langle dW_t \rangle = 0, \langle dW_t dW_{t'} \rangle = 1$ if $t = t'$, and $dW_t = 0$ for all any cases. This increment is used for the description of long-term fluctuations of evapotranspiration and precipitation, or describes the parameter:

$$\alpha = LE_p/2wu,$$

(3.24)

where L – the total size of studied area, E_p – evapotranspiration, w – average volume of atmospheric humidity, and u – average wind speed.

Functions $A(s_i)$, $B(s_i)$, $C(s_i)$ i $D(s_i)$ can be determined as (Rodriguez-Iturbe et al., 1991)

$$A(s_1, s_2) = \frac{P_a}{nz_r}\left\{1 + \langle\alpha\rangle\left[(1-f_g)s_1^c + f_g s_2^c\right]\right\}(1 - \varepsilon s_1^r) - \frac{E_p s_1^c - k_s s_1^b}{nz_r},$$

(3.25)

$$B(s_1, s_2) = \frac{P_a}{nz_r}\left[(1-f_g)s_1^c + f_g s_2^c\right](1 - \varepsilon s_1^r)\sigma,$$

(3.26)

$$C(s_1, s_2) = \frac{P_a}{nz_r}\left\{1 + \langle\alpha\rangle\left[(1-f_g)s_1^c + f_g s_2^c\right]\right\}(1 - \varepsilon s_2^r) - \frac{E_p s_2^c - Q_2(t)}{nz_r},$$

(3.27)

$$D(s_1, s_2) = \frac{P_a}{nz_r}\left[(1-f_g)s_1^c + f_g s_2^c\right](1 - \varepsilon s_2^r)\sigma.$$

(3.28)

Here, P_a – precipitation; f_g – partition of territory, covered by discharge zone; ε, r, c, – empirical coefficients; n – porosity; and z_r – thickness of the active root layer.

The flows of underground water Q_i may be described as (Entekhabi et al., 1992; Kostyuchenko et al., 2019)

$$Q_1(s_1) = k_s s_1^b, \tag{3.29}$$

$$Q_2(t) = \left(\frac{1-f_g}{f_g}\right)\frac{k_s}{J}\int_{-\infty}^{t} s_1^b(t-\tau)e^{-\tau/J}d\tau, \tag{3.30}$$

where k_s – permeability with the total saturation, t, τ – time, b – empirical coefficient, and J – average groundwater delay, described as $J = S_y l^2/\pi^2 T$, where S_y – debit of saturated zone, T – average permeability, and l – average distance between discharge zones (drainage density). Parameters J and f_g describe the hydro-geological (parameters of water table) and geo-morphological (location and size of discharge zones) features of the surface.

Combined Penman–Monteith equation could be used to calculate average daily evapotranspiration on the level on the top of vegetation (Kostyuchenko et al., 2019):

$$E_p = \left\{\frac{[f(A)+1][R_n-G]\Delta}{[\sigma f(A)+1]C_p\rho} + [f(A)+1]\frac{\rho_2^* - \rho_2}{r_a}\right\}\left\{\frac{r_a+r_x}{r_a} + \frac{[f(A)+1]L_v\Delta}{[\sigma f(A)+1]C_p\rho}\right\}^{-1} \tag{3.31}$$

Here, Δ – derivative of saturated vapor pressure; C_p – specific heat of air at the constant pressure; L_v – latent heat of transformation of water into vapor; σ – the ratio of the area of convection to the area of evapotranspiration; r_a – resistance from the atmosphere to vapor motion from the vegetation surface; R_n – amount of solar heat entering to the surface of evapotranspiration; G – the amount of energy that goes from vegetation to the soil for a certain time; r_x – resistance from the surface of the evaporator to the exit of water vapor; ρ – air density calculated with average pressure and actual humidity; ρ_2^* – vapor density with full saturation at daily average temperature; ρ_2 – actual vapor density in the atmosphere over the vegetation cover; $f(A)$ – effective area of vegetation per unit of the total area of the investigated territory.

Thus, the task of control of hydrological and hydro-geological risks (in particular, flooding and swamping) using satellite data may be reduced to determination of the methodology for analyzing the set of surface indicators corresponding to the variable of Equations (3.22)–(3.31), and thus to control of changes in indicators of the reaction of local ecosystems to changes in the water and heat balance according to certain types of terrestrial cover.

3.4 SATELLITE MODELS: SPECTRAL RESPONSE MODELS

3.4.1 SPECTRAL MODEL OF SURFACE RESPONSE TO THE HEAT AND WATER STRESS

The surface reflection spectra, detected using satellite sensors, are forming by integrating energy input from large areas. This essentially distinguishes them from the spectra obtained by ground measurements and should be taken into account the comparative analysis. In some narrow spectral bands, these effects may not be essential.

However, when the bandwidth is considerable, or when the composition of spectral bands (spectral indices) should be analyzed, the spatial and temporal variability of the energy fluxes from the surface should be taken into account.

The energy balance of surface site R_n may be described as follow (Gupta et al., 1999):

$$R_n\left(T_{\text{aero}}, \mathbf{P}, \mathbf{W}\right) = H\left(T_{\text{aero}}, \mathbf{P}, \mathbf{W}\right) + LH\left(T_{\text{aero}}, \mathbf{P}, \mathbf{W}, \Theta\right) + G\left(T_{\text{aero}}, \mathbf{P}, \mathbf{W}, \mathbf{T}, \Theta\right)$$
(3.32)

Here, T_{aero} – aerodynamic surface temperature, LH – latent heat (the residual heat energy that came with radiation after absorption by vegetation and soil); parametric description of soils and vegetation ($\mathbf{P}$), precipitation and solar radiation ($\mathbf{W}$), thermal ($\mathbf{T}$), and hydrological (Θ) parameters of surface.

Variations of thermal and hydrological parameters may be described as (Castelli et al., 1999)

$$\frac{d\mathbf{T}}{dt} = f\left(T_{\text{aero}}, \Theta, \mathbf{T}, \mathbf{P}, \mathbf{W}\right),$$
(3.33)

$$\frac{d\Theta}{dt} = f\left(LH, \Theta, \mathbf{T}, \mathbf{P}, \mathbf{W}\right).$$
(3.34)

Based on these general equations and taking into account evaporation and evapotranspiration, Kostyuchenko et al. (2019) proposed a general equation for the description of energy flux from the surface site:

$$g_T = f_v g_v + f_s g_s,$$
(3.35)

where f_v and f_s – sites of studied area covered by vegetation or with bare soils. In the most common case, with no ground verification of hydro-geological and geo-morphological parameters, Kostyuchenko et al. (2019) proposed $f_s = (1 - f_v)$. For more accurate calculation of f_v and f_s, the satellite-based indices may be used (Kostyuchenko et al., 2019):

$$f_v = 1 - \left(\frac{\text{NDVI} - \text{NDVI}_{\min}}{\text{NDVI}_{\max} - \text{NDVI}_{\min}}\right)^p,$$
(3.36)

$$f_s = g\left(\frac{\text{NDWI} - \text{NDWI}_{\min}}{\text{NDWI}_{\max} - \text{NDWI}_{\min}}\right)^q.$$
(3.37)

In these equations, algorithm values of Normalized Difference Vegetation Index (NDVI) and Normalized Difference Water Index (NDWI) indices were calculated according to Jackson et al. (1983) and Gao (1995) in the framework of the observation interval; g, p, q – empirical coefficients.

There are two ways to consider a variability of the energy flux: (i) construction of a special algorithm for spectral indices calculation, which takes into account features of energy balance of the surface, and (ii) application of statistical procedures to ground measurement data analysis aimed to data regularization toward satellite observations. Besides, all data should be temporally regularized.

The task of qualitative and quantitative definition and determination of the spectral inter-calibration procedure requires the detailed analysis and processing of huge massive of lab, ground, and remote data. Our overview will be limited by analysis of two spectral indices: NDVI and NDWI, which correspond to our task.

Classic equations for calculating these indices, based on lab experiments, were proposed in the study of Jackson et al. (1983) for NDVI and in the study of Gao (1995) for NDWI:

$$NDVI^{lab} = \left(\frac{r_{800} - r_{680}}{r_{800} + r_{680}} \right),$$

$$(3.38)$$

$$NDWI^{lab} = \left(\frac{r_{857} - r_{1241}}{r_{857} + r_{1241}} \right).$$

$$(3.39)$$

Here, r_λ – reflectance in corresponding band λ, nm.

Based on the balance Equations (3.32)–(3.35) and taking into account usual bands of satellite sensors, Kostyuchenko et al. (2010 proposed algorithms for the calculation of these indices using specific sensors:

$$NDVI^{MSS} = \left(\left[\int_{700}^{800} I d\lambda - \int_{600}^{700} I d\lambda \right] \Big/ \left[\int_{700}^{800} I d\lambda + \int_{600}^{700} I d\lambda \right] \right) \Big/ g. \quad (3.40)$$

For data from MSS sensor of Lansat USGS satellite and for data from TM and ETM sensors of Lansat satellite,

$$NDVI^{ETM} = \left(\left[\int_{760}^{900} I d\lambda - \int_{630}^{690} I d\lambda \right] \Big/ \left[\int_{760}^{900} I d\lambda + \int_{630}^{690} I d\lambda \right] \right) \Big/ g, \quad (3.41)$$

$$NDWI^{ETM} = \left(\left[\int_{760}^{900} I d\lambda - \int_{1550}^{1750} I d\lambda \right] \Big/ \left[\int_{760}^{900} I d\lambda + \int_{1550}^{1750} I d\lambda \right] \right) \Big/ g. \quad (3.42)$$

Application of reduced spectral intervals equations allows to obtain distributions of spectral indices reflecting both the specificity of the used sensors and the spatial variations of the energy balance of the earth's surface. Thus, more correct comparison of the results of satellite observations and ground spectrometric measurements may be provided.

Further regularization may be provided in different ways. In the framework of the task solving, the relatively simple way may be proposed. It is based on the determination of distribution of studied parameters over the whole area $f_{x,y}$ toward the distribution on studied sites f_m (Kostyuchenko, 2015):

$$f_{x,y} = \sum_{m=1}^{n} w_{x,y} \left(\tilde{f}_m \right) f_m,$$

$$(3.43)$$

where $w_{x,y} \left(\tilde{f}_m \right)$ – weighting coefficient, determined as the minimum:

$$\min \left\{ \sum_{m=1}^{n} \sum_{f_m \in F} w_{x,y} \left(\tilde{f}_m \right) \left(1 - \frac{f_m}{\tilde{f}_m} \right)^2 \right\}.$$

$$(3.44)$$

In this equation, m – number of measurement points, n – number of measurement series, f_m – distribution of measurement results, F – set of measurement data, and $\tilde{f}_m$ – average distribution of measured parameters.

Application of the regularization procedure allows to obtain distributions of the measured parameters, which with controlled reliability correspond to spatial and temporal parameters of satellite data. Comparison of two regular datasets from ground measurements and satellite observations may be carried out with the approach (Acarreta and Stammes, 2005) according to the following equation:

$$\overline{R} = \int R(\vec{r})d\vec{r}. \tag{3.45}$$

Here, $\vec{r}$ – two-dimensional vector of the site coordinates and R – measured spectral distribution ($\overline{R}$ – with worse spatial resolution).

Thus, ground measurement and satellite observation data can be correctly compared.

3.4.2 SPECTRAL RESPONSE TO THE SNOW MELTING: THE STOCHASTIC APPROACH

In accordance with the models (3.14)–(3.21), we can give a set of calculation equations that will determine the important parameters of snow cover for risk management tasks. For example, the total weight of snow and the water content (water equivalent) are dynamic values, which depend on the density of the snow cover.

We will determine the dynamic density of the snow cover during the observation interval as a stochastic value:

$$\rho_{\text{snow}} = \frac{\sum_i (\rho_i \Delta z_i + \rho_0 \Delta z_0)}{\Delta z_{\text{snow}}}, \tag{3.46}$$

where ρ_i, Δz_i, – the density and capacity of the snow cover recorded during the observation period i; and ρ_0, Δz_0 – density and snow cover capacity at the beginning of the observation period. At the same time, during the interval of observations i, the dynamics of density is determined by a complex of meteorological indicators:

$$\rho_i = \rho_i^{\text{min}} + \kappa \frac{1 - Q_p}{f_l^{\text{max}}}, \tag{3.47}$$

where ρ_i^{min} – initial precipitation; κ – empiric coefficient (usually equal 181); Q_p – the thermal precipitation functional, which depends on temperature, pressure, and humidity; f_l^{max} – maximum content of water in precipitation (in the forecasting calculations is taken at the level of 0.5).

The set of empirical coefficients is determining by the complex of ground measurements; the change in local meteorological indicators is determined according to meteorological measurements, and the distributions of spatial indicators can be determined by the remote sensing data.

Water equivalent of the deposited snow W_{ij} is assessed by taking into account current and forecasting meteorological parameters, as well as the spatial heterogeneity of snow cover. In the simplest case, it may be presented as

$$W_{ij} = \frac{\rho_w H_s}{c_s} \sum_{ij} \frac{d_{ij}^{cc}}{\left(0°C - T_{ij}^s\right)}, \tag{3.48}$$

where ρ_w – water density, T_{ij}^s – snow temperature, H_s – latent heat of snow cover (80 cal/g), c_s – thermal conductivity of the snow (0.5 cal/g °C), and d_{ij}^{cc} – power of solid component.

To calculate the catchment from the whole territory, an integral form of the water equivalent equation can be used:

$$W_{\text{tot}} = \left[\frac{1 - A_0}{A_i - A_0} \left(W_i - W_0 \right) \right] + W_0. \tag{3.49}$$

In this equation, W_{tot} – total water content in the snow cover (snow cover water equivalent); W_0 – snow cover water equivalent at the beginning of melting; W_i – snow cover water equivalent at the observation moment i; A_0 – square of the snow cover at the beginning of melting; A_i – current square of cover (at the moment of observation i). In this way, we obtain a stochastic algorithm for estimating snow cover parameters based on observation data, including satellite.

Thus, the key variables will be the area of snow cover and characteristics that determine the water equivalent of snow at the moment of observation.

The normalized index of snow cover can be proposed in the form (Salomonson and Appel, 2004):

$$\text{NDSI} = \frac{R_{[0.55-0.65]} - R_{[0.75-0.85]}}{R_{[0.55-0.65]} + R_{[0.75-0.85]}} \approx \frac{R_{\text{VIS}} - R_{\text{NIR}}}{R_{\text{VIS}} + R_{\text{NIR}}}, \tag{3.50}$$

where R – reflectance in the defined spectral band.

According to Salomonson and Appel (2004), it is possible to define a limit value of NDSI index, which corresponds to the presence of snow cover. For open areas, it is 0.32–0.36, and for areas covered with dense vegetation, it is 0.28–0.31. So if pixel x_{ij} has NDSI ≥ 0.3, we classify it to snow-covered cluster A_i with high reliability. The ratio of the NDSI value and the snow cover capacity needed to estimate a water equivalent can be determined from field measurements at the local level.

3.5 SATELLITE DATA FOR ASSESSMENT OF HYDROLOGICAL CLIMATE-RELATED RISKS

3.5.1 LAND COVERS CLASSIFICATION APPROACH

In accordance with the model solutions, the application of satellite observations has two main purposes: first, to provide a reliable classification of terrestrial covers in the risk assessment sites, and second, to collect a sufficient statistics of spectral observations of varied classes of terrestrial covers (Kostyuchenko, 2019).

For the most parts of Earth, in particular, for the essentially anthropogenic changed landscapes of Northern Eurasia, sufficient experience of solving of these problems has been accumulated (Kopachevsky et al., 2016). At the same time, in view of drastic changes of precipitation patterns and land-use practices induced by social transformations, the hydrological hazards are still dangerous, and their risks still require a more accurate assessment in the context of regional climate, and environmental and social change.

As the Equations (3.25)–(3.37) show, the key monitoring parameters are the square and density of the plant cover and the hydrographic network distribution, and key controlled variables are humidity, moisture, photochemical processes intensity, and temperature. So, surface cover classification should be directed to the detection of water objects and key local plant classes. With respect to water balance parameters, plants covers are divided into forests, shrubs, grasslands (natural meadows and farmland), wetlands, and peat bogs with herbaceous vegetation. Water objects are divided into lakes, rivers, abandoned channels, natural river tributaries, artificial water channels, and artificial reservoirs. Besides, from the viewpoint of risk assessment, i.e., estimation of the possible losses, it is important to identify the elements of the infrastructure: buildings, roads, bridges, dams, etc.

Based on the parameters of radiometric, spectral, spatial, and temporal resolution, the required set of satellite data should be collected. Usual set of data for problem-oriented classification and analysis of spectral characteristics of terrestrial covers may include the data from USGS Landsat satellites, from EOS Terra MODIS, acquired during the vegetation and the snow periods, and individual scenes from high-resolution satellites such as Ikonos, QuickBird, or GeoEye1, acquired during the vegetation period for model calibration and field measurements validation.

For change and anomaly detection, a basic period, when variables demonstrated relative predictable behavior, should be determined. The basic period for determined variables is usually determined based on the data availability. For Northern Eurasia, this period is 1986–2009. The cartographic data since 1972 and 1986 and field verification data since 2007 can be used for data and model calibration and land covers classification. This set of data allows to calculate relatively reliable assessments using current observations in the framework of methodology Equations (3.1)–(3.2) and (3.3)–(3.6).

To process the satellite observation data, the hybrid classification procedure with Bayesian maximum-likelihood classifier can be successfully applied (Blackburn, 1998). After obtaining of land cover classification, further analysis of the data should be directed to study individual spectral characteristics of certain types of terrestrial cover, taking into account the existing tendencies of changes local parameters, in particular climatic (Kostyuchenko, 2019).

3.5.2 SPECTRAL DATA CALIBRATION USING IN-FIELD SPECTROMETRY MEASUREMENTS

The basic idea of the proposed method of hydrological risk assessment is to analyze a set of indicators of ecosystems stresses generated by variations of water and heat dynamics. Ecosystems reaction to water and heat stresses can be detected through analysis of vegetation spectral response on the set of spectral indices (Blackburn, 1998; Dobrowski, 2005; Choudhury, 2001; Verma, 1993; Penuelas, 1995).

Analysis of a wide range of spectral indices allows to determine the tendencies of the change, which indicates the presence of water and temperature stress. Number of existing spectral indices may be used in form adapted to task, reduced to sensor and satellite.

For example, the analysis of long-term water, temperature, and radiation stresses that would result a noticeable reaction of regional ecosystems can be accomplished by analyzing the changes in vegetation photosynthetic activity, estimated by Photochemical Reflectance Index (PRI) for the studied region. PRI index usually calculates for TM sensor of Landsat satellite according to algorithm (Gamon et al., 1997) in the form (Kostyuchenko et al., 2019):

$$\mathrm{PRI^{TM}} = \left(\left[\int_{0.45}^{0.52} Id\lambda - \int_{0.52}^{0.60} Id\lambda \right] \Big/ \left[\int_{0.45}^{0.52} Id\lambda + \int_{0.52}^{0.60} Id\lambda \right] \right) \Big/ g, \quad (3.51)$$

And the adapted form for ETM sensor:

$$\mathrm{PRI^{ETM}} = \left(\left[\int_{0.45}^{0.52} Id\lambda - \int_{0.52}^{0.60} Id\lambda \right] \Big/ \left[\int_{0.45}^{0.52} Id\lambda + \int_{0.52}^{0.60} Id\lambda \right] \right) \Big/ g. \quad (3.52)$$

Here, λ – wavelength, g – semi-empiric gain factor, in this case $g = 600$.

For the detection of stress-related changes of the vegetation, the distributions of NDVI (Jackson et al., 1983) and EVI (Enhanced Vegetation Index) should be analyzed (Gamon et al., 1997). These indices may be used as the indicators of impact of long-term landscape changes. To detect an impact of water stresses, a Structure Intensive Pigment Index SIPI should be analyzed (Kostyuchenko et al., 2012). According to Kostyuchenko et al. (2012), these indices can be calculated as follows:

$$\mathrm{NDVI^{MSS}} = \left(\left[\int_{0.70}^{0.80} Id\lambda - \int_{0.60}^{0.70} Id\lambda \right] \Big/ \left[\int_{0.70}^{0.80} Id\lambda + \int_{0.60}^{0.70} Id\lambda \right] \right) \Big/ g, \quad (3.53)$$

For MSS sensor of Landsat 1–5 satellites, and for TM and ETM sensors of Landsat 4–7 satellites,

$$\mathrm{NDVI^{TM/ETM}} = \left(\left[\int_{0.76}^{0.90} Id\lambda - \int_{0.63}^{0.69} Id\lambda \right] \Big/ \left[\int_{0.76}^{0.90} Id\lambda + \int_{0.63}^{0.69} Id\lambda \right] \right) \Big/ g. \quad (3.54)$$

where a gain factor $g = 200$.

$$\mathrm{EVI^{TM}} = \left(3.2 \left[\int_{0.76}^{0.90} Id\lambda - \int_{0.63}^{0.69} Id\lambda \right] \Big/ \left[\int_{0.76}^{0.90} Id\lambda \right.\right.$$
$$\left.\left. + 6 \int_{0.63}^{0.69} Id\lambda - 7.5 \int_{0.45}^{0.52} Id\lambda + 1 \right] \right) \Big/ g, \quad (3.55)$$

$$\text{EVI}^{\text{ETM}} = \left(2.5\left[\int_{0.760}^{0.900} Id\lambda - \int_{0.630}^{0.690} Id\lambda\right] \Big/ \left[\int_{0.760}^{0.900} Id\lambda\right.\right.$$
$$\left.\left. + 6\int_{0.630}^{0.690} Id\lambda - 7.5\int_{0.450}^{0.515} Id\lambda + 1\right]\right) \Big/ g, \qquad (3.56)$$

Here, a gain factor $g = 500$. Besides, some simplified algorithm can be proposed for the EVI index:

$$\text{EVI}^{\text{TM/ETM}} = \left(2.5\left[\int_{0.76}^{0.90} Id\lambda - \int_{0.63}^{0.69} Id\lambda\right] \Big/\right.$$
$$\left.\left[\int_{0.76}^{0.90} Id\lambda + 2.4\int_{0.63}^{0.69} Id\lambda + 1\right]\right) \Big/ g, \qquad (3.57)$$

Here, $g = 500$.

$$\text{SIPI}^{\text{TM}} = \left[\int_{0.76}^{0.90} Id\lambda - \int_{0.45}^{0.52} Id\lambda\right] \Big/ \left[\int_{0.76}^{0.90} Id\lambda - \int_{0.63}^{0.69} Id\lambda/g\right] - 1, \quad (3.58)$$

$$\text{SIPI}^{\text{ETM}} = \left[\int_{0.760}^{0.900} Id\lambda - \int_{0.450}^{0.515} Id\lambda\right] \Big/ \left[\int_{0.760}^{0.900} Id\lambda - \int_{0.630}^{0.690} Id\lambda/g\right] - 1, \quad (3.59)$$

Here, $g = 50$ (SIPI = 0, if SIPI < 0).

Direct water stress may be estimated by NDWI, according to algorithm proposed by Kostyuchenko et al. (2012):

$$\text{NDWI}^{\text{TM/ETM}} = \left(\left[\int_{0.76}^{0.90} Id\lambda - \int_{1.55}^{0.75} Id\lambda\right] \Big/ \left[\int_{0.76}^{0.90} Id\lambda + \int_{1.55}^{0.75} Id\lambda\right]\right) \Big/ g,$$
$$(3.60)$$

where gain factor $g = 100$. Proposed forms Equation (3.51)–(3.60) may be applied also for field measurements in corresponding spectral bands.

The detected changes of the spectral indices can be used as the parameters in method of assessment of risks of the processes that cause stress associated with changes of spectral characteristics of the vegetation.

As it was mentioned in methodological section, field spectrometric measurements should be carried out to calibrate described spectral indices and to verify calculated models. The main purpose of calibration is to obtain stable inter-correlations between satellite and in-field parameters.

It should be noted that there is a strong methodological constraint in the task of inter-calibration and verification. Because a direct comparison of in-field and satellite data is incorrect through a different acquiring methodology, and so different radiometric, spectral, spatial, and temporal characteristics of acquired data and the regularization procedures should be applied. Regularization by energy flux (spectral and radiometric regularization) Equations (3.40)–(3.42) and spatial-temporal regularization Equations (3.43)–(3.45) is necessary to obtain data distributions with stable inter-correlation (see example in Figure 3.3).

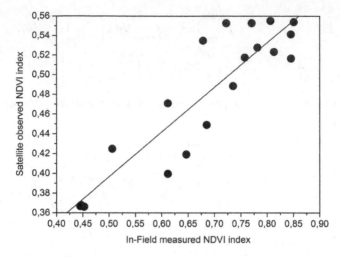

Figure 3.3 Example of inter-comparison of NDVI indices from satellite and field data ($R = 0.90$; $\sigma = 0.031$.)

To calculate correlation from Figure 3.3 on the separated field sites using spectrometer FieldSpec®3 FR in period 2007–2015 were acquired 117 spectral signatures in different ranges: in range 350–2,500 nm with bandwidth 1.4 nm; in the range 350–1,000 nm with bandwidth 2 nm; in range 1,000–2,500 with bandwidth 3 nm; and in range 1,400–2,100 nm with bandwidth 10 nm (Kostyuchenko et al., 2019).

As a result, to calibrate a data of satellite observations, the calibrations correlations in the form of linear regressions may be proposed.

$$\mathrm{NDVI}_{sat} = (0.22 + 0.38)\,\mathrm{NDVI}_{ground}^{lab}. \tag{3.61}$$

The same way may be proposed for the water indices NDWI.

It should be noted that NDWI indices have better correlations than vegetation indices, although its spatial resolution is worse (see example in Figure 3.4).

Linear approximation equations for average calculated satellite NDWI index and field measured data also may be presented:

$$\mathrm{NDWI}_{sat} = (0.65 + 2.75)\,\mathrm{NDWI}_{ground}^{lab}, \tag{3.62}$$

This form of procedures can be applied as the calibration dependencies for vegetation spectral indices to recalculate satellite and in-field data. So, the procedure of calibration of satellite-derived spectral indices on the base of field spectrometry may be proposed.

Important limitation of proposed way of inter-calibration is connected with non-linearity of indices distribution. Determined linear inter-calibration dependencies for NDVI index are correct enough only in the interval of NDVI values 0.4–0.55. If NDVI > 0.55, calibration dependency will not be linear, as it shows the analyzed data. Determination of inter-calibration dependencies for whole interval of possible values of the index is the task of future studies.

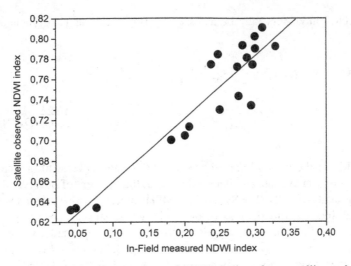

Figure 3.4 Example of inter-comparison of NDWI indices from satellite and field data ($R = 0.94$; $\sigma = 0.021$.)

Additionally, it needs to be noted that there is a substantial difference between satellite and field spectrometry, connected with scheme of survey, which might also be considered as a methodical constraint.

While a field spectrometry usually uses a limbic scheme of observation, which can lead to losses of information under anisotropic landscapes, the satellite surveys use nadir scheme, which collect full signal from the site. Therefore, signals acquired and data collected can be different. To reduce this difference, it is necessary to calculate a general function of view of the field spectrometer. Generally speaking, it is necessary to determine in the explicit form the distribution of angles field spectrometric survey $p_m(\vartheta_i)$, where ϑ – spectrometer view angle, m – measurements sites, and i – measurements points, as a function $p_m(\vartheta_i) \rightarrow \mathbf{F}(\theta(t), x_i, y_i, z_i, z_i^*)$; where θ – solar angle, t – measurement time, x, y – coordinates of measurement point, z – terrain, and z^* – effective height of plants (position of the spectrometer relative to the vegetation surface). It is necessary to reduce the uncertainties and errors, connected with differences of limbic and nadir survey schemes.

3.6 RISK MODEL: METHOD OF TO THE RISK ASSESSMENTS USING BAYES APPROACH

In the framework of proposed "information-response" formalization of risk assessment and decision support, it is necessary to estimate a probability distribution function $H_I(i|\theta)$ with a state of the studied natural system θ using acquired satellite data to make an information (I) formally completed (I^*). Practically, this task can be realized as a separate task of risks assessment using satellite data. Herein because the input satellite data has a stochastic nature, a probability of stress detection using the set of spectral indicators may be assessed by Bayes rule as (Kostyuchenko et al., 2019)

$$P(\Delta\mathrm{SRI}^*(x,y)|Q_{\text{stress}}) = \frac{P_S(x,y)\prod_N P_N(\Delta\mathrm{SRI}^*|Q_{\text{stress}})}{\int_{x,y} P_N(\Delta\mathrm{SRI}^*|Q)\, dP_s(x,y)}$$

$$= \frac{P_S(x,y)P_N(\Delta\mathrm{SRI}^*|Q_{\text{stress}})}{P_N(\Delta\mathrm{SRI}^*|Q_{\text{stress}})P_S(x,y) + P_N(\Delta\mathrm{SRI}^*|Q_0)P_0(x,y)}.$$

$$(3.63)$$

Here, $\Delta\mathrm{SRI}^*$ is the period-reduced spectral index (normalized difference between observation period average and detected value at the moment of observation), which is used for analysis (NDVI, EVI, NDWI, SIPI); index Q_{stress} marks sites impacted by stresses; index Q_0 marks classes of pixels, where stresses are not present with high reliability. Probability $P_S(x,y)$ is determined self-empirically. Ratio of probabilities $P_S(x,y)$ and $P_0(x,y)$ is determined as $\lim_{x,y,\tau}(P_S(x,y)_\tau + P_0(x,y)_\tau) = 1$ (i.e., on the enough long periods, it is possible to assume that: $P_S(x,y) = 1 - P_0(x,y)$). Probability $P_S(x,y)$ can be determined using Gauss weight function, if pixel on the satellite scene does not refer to the site where stress factors are observed and cannot be unambiguously attributed to site, in which there is no stress. So, the probability of uncertain presence of stress at location (that corresponds to position of this pixel on the surface) will be a function of the geometric distance from the closest place under the registered stress:

$$P_S(x,y) = P_{\min} + (P_{\max} - P_{\min})\, e^{d_s^2/2\sigma_p^2}.\qquad(3.64)$$

where $P_S(x,y)$ – probability of existence (or appearance in time scale of observation period) of the studied stresses; $P_{\max}$ – the maximum probability of the current presence of stress in the study site (unregistered during the interpretation and classification of the image), which depends on the type of sensor, the physical and geographical features of the region, and the type of surface (based on the general methodological principles (Kostyuchenko et al., 2019), $P_{\max}$ for TM and ETM sensors of Landsat satellite may be assessed as 0.25–0.28); $P_{\min}$ – minimum probability, depending on sensor type, physical and geographical features of the region, and type of surface ($P_{\min}$ may be assessed about 0.01); $d_s(x,y)$ – distance from the nearest place under the registered stress; σ_p – an empirical parameter to be determined, using field research data, based on the characteristics of vegetation cover of the study area and the type of sensor (e.g., for TM and ETM sensors of Landsat satellites, the parameter σ_p can be assessed as 1.1–1.5 km). Therefore, for the hydrological hazards, for TM and ETM sensors of Landsat satellite, the parameter $P_S(x,y)$ can be calculated in simplified form: $P_S(x,y) = 0.01 + 0.26\, e^{d_s^2/1.69}$. Calculated probabilities are a basis for the assessment of risks, connected with hydrological and hydro-geological threats. Assessment of complex risk is a complicated task and requires taking into account multi-scale processes and links in studied multi-component systems. In the general case, the following form of Equation (3.3) is presented (after Ermoliev and Gaivoronski, 1992):

$$R(t) = f_A\left(R^L(t), R_0(t)\right) \iint_{xy} \int_{t_0}^{t} p(v)f(x,y,v)dtdxdy. \qquad (3.65)$$

Here, $f_A\left(R^L(t), R_0(t)\right)$ is the approximation function of impact, which describes the interaction of short- and long-term impact factors to complex hydrological and hydro-geological security (depending on the used model, it may be presented in different form, even as linear composition of corresponding probabilities); $p(v)$ – probability of negative impact with determined conditions, where v – effective velocity of development of the disastrous process (modeled by the model of runoff); $R_0(t)$ – general (mean) probability of disaster (in general case, this is the probability distribution function of impacts $f^{\alpha}(\psi, I)$ over the site $\psi(x, y)$; function of losses – parametric description of negative impact $f_v(I)$; and risk function H, which is determined from regional disasters statistics), and might be calculated using remote sensing data; $R^L(t)$ – long-term risk (here should be used the impact factors Q_j from the set $j \in \mathfrak{I}$, connected with climate and environmental changes, and scenarios of change in disasters frequency $F(Q_j)$ are calculated with forecast models; $f(x,y,v)$ – function of expansion, describing by distribution of hydrological (properties of water-table) and geo-morphological (location and size of water discharge zones) parameters. It should be noted that expansion function $f(x,y,v)$ may also be determined through satellite data classification and its verification with field data (Kostyuchenko et al., 2019).

Using the approach proposed, the hydrological risks can be assessed. The set of spectral indices has been calculated to the estimation of stresses using the Landsat data for the region of upper sites of Western Bug and Pripyat rivers, Ukraine, with coordinates of the center: 490 57'15, 36" N, 240 46' 05, 31" E). The result of assessment is presented in Figure 3.5.

As the results of application (in the Figure 3.5) demonstrate, using this approach, the current state of hydrological security and the current level of corresponding risks can be assessed. But in the analysis of the long-term behavior of the investigated variables, the study of a wider set of parameters, both climatic and environmental, which should be used in predictive models, is required.

3.7 CONCLUSIONS

Modeling of geo-systems should be an integral part both of remote sensing interpretation methods, as well as of the risk assessment systems based on remote sensing data utilization. It requires abundant knowledge in the field of Earth sciences, as well as by increased requirements in the area of decision-making. New challenges define new methodological requirements.

First, the methodology proposed allows to expand the problem definition of using the satellite observations in tasks of socio-ecological security. In addition to traditional statistical analysis directed to surface change detection, it is possible to analyze and predict state of the studied systems, based on the models of geo-systems.

This certainly expands the scope and sphere of application of approach, and could positively affect the reliability of the results obtained through the using of different sources of data.

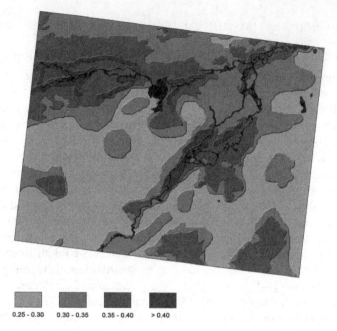

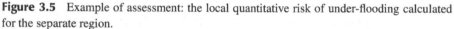

0.25 - 0.30   0.30 - 0.35   0.35 - 0.40   > 0.40

Figure 3.5 Example of assessment: the local quantitative risk of under-flooding calculated for the separate region.

Second, the proposed methodology includes feedbacks between management decisions and the systems state. Thus, it is postulated that the state of the system depends on the observer: risks depend on the decision made and management impacts (past, current, and planned) to the system.

This could positively affect the effectiveness of management decisions and the quality of risk assessment (Ermoliev et al., 2012; Schlaifer and Raiffa, 1961).

The presented results obtained in the framework of multi-model approach allow to conclude that based on satellite observation data, Bayes risk assessment techniques are an adequate basis for assessing the hydrological and hydro-geological risks both at the regional and local levels. Based on the model of hazardous processes, it is possible to construct a method for processing the satellite images and analyzing the spectral indicators.

In practice, the multi-model approach proposed and the results obtained may be used for the assessing and forecasting the risks of flooding and swamping of the ecologically sensitive areas, for the development of decision-making systems for losses minimization, as well as for the development of ecological monitoring systems.

ACKNOWLEDGMENTS

Authors are first grateful to colleagues from the International Institute for Applied Systems Analysis (IIASA), for their critical and constructive comments and suggestions, to National Academy of Sciences of Ukraine for the support of this study in the framework of multilateral program "Integrated robust management of food-energy-water-land use nexus for sustainable development," and to anonymous referees for constructive suggestions that resulted in important improvements to the chapter.

REFERENCES

Acarreta, J. R., & Stammes, P. (2005). Calibration comparison between SCIAMACHY and MERIS onboard ENVISAT. *IEEE Geoscience and Remote Sensing Letters*, 2(1), 31–35.

Blackburn, G. A. (1998). Spectral indices for estimating photosynthetic pigment concentrations: A test using senescent tree leaves. *International Journal of Remote Sensing*, 19(4), 657–675.

Campbell, J. B., & Wynne, R. H. (2011). *Introduction to Remote Sensing*. Guilford Press, New York.

Castelli, F., Entekhabi, D., & Caporali, E. (1999). Estimation of surface heat flux and an index of soil moisture using adjoint-state surface energy balance. *Water Resources Research*, 35(10), 3115–3125.

Choudhury, B. J. (2001). Estimating gross photosynthesis using satellite and ancillary data: Approach and preliminary results. *Remote Sensing of Environment*, 75(1), 1–21.

Dobrowski, S. Z., Pushnik, J. C., Zarco-Tejada, P. J., & Ustin, S. L. (2005). Simple reflectance indices track heat and water stress-induced changes in steady-state chlorophyll fluorescence at the canopy scale. *Remote Sensing of Environment*, 97(3), 403–414.

Elachi, C., & Van Zyl, J. J. (2006). *Introduction to the Physics and Techniques of Remote Sensing* (Vol. 28). John Wiley & Sons, New York.

Engle, R. F., & Russell, J. R. (1998). Autoregressive conditional duration: A new model for irregularly spaced transaction data. *Econometrica*, 66, 1127–1162.

Entekhabi, D., Rodriguez-Iturbe, I., & Bras, R. L. (1992). Variability in large-scale water balance with land surface-atmosphere interaction. *Journal of Climate*, 5(8), 798–813.

Ermoliev, Y., Makowski, M., & Marti, K. (2012). Robust management of heterogeneous systems under uncertainties. In Ermoliev, Y., Makowski, M. & Marti, K. (Eds.) *Managing Safety of Heterogeneous Systems* (pp. 1–16). Springer, Berlin Heidelberg.

Ermoliev, Y. M., & Gaivoronski, A. A. (1992). Stochastic quasigradient methods for optimization of discrete event systems. *Annals of Operations Research*, 39(1), 1–39.

Gamon, J., Serrano, L., & Surfus, J. S. (1997). The photochemical reflectance index: An optical indicator of photosynthetic radiation use efficiency across species, functional types, and nutrient levels. *Oecologia*, 112(4), 492–501.

Gao, B. C. (1995). Normalized difference water index for remote sensing of vegetation liquid water from space. In De Long, R. K., Romesser, T. E., Marmo, J., & Folkman, M. A. (Eds.), *Imaging Spectrometry* (Vol. 2480, pp. 225–237). International Society for Optics and Photonics, Orlando, FL, United States.

Gupta, H. V., Bastidas, L. A., Sorooshian, S., Shuttleworth, W. J., & Yang, Z. L. (1999). Parameter estimation of a land surface scheme using multicriteria methods. *Journal of Geophysical Research: Atmospheres*, 104(D16), 19491–19503.

Jackson, R. D., Slater, P. N., & Pinter Jr, P. J. (1983). Discrimination of growth and water stress in wheat by various vegetation indices through clear and turbid atmospheres. *Remote Sensing of Environment*, 13(3), 187–208.

Kopachevsky, I., Kostyuchenko, Y. V., & Stoyka, O. (2016). Land use drivers of population dynamics in tasks of security management and risk assessment. *International Journal of Mathematical, Engineering and Management Sciences*, 1, 18–24.

Kostyuchenko, Y. V. (2015). Geostatistics and remote sensing for extremes forecasting and disaster risk multiscale analysis. In Numerical Methods for Reliability and Safety Assessment (pp. 439-458). Springer International Publishing. DOI 10.1007/978-3-319-07167-1_16

Kostyuchenko Y. V. (2018). On the methodology of satellite data utilization in multi-modeling approach for socio-ecological risks assessment tasks: A problem formulation. *IJMEMS*, 3(1), 1–8.

Kostyuchenko, Y. V., Kopachevsky, I., Solovyov, D., Bilous, Y., & Gunchenko, V. (2010). Way to reduce the uncertainties on ecological consequences assessment of technological disasters using satellite observations. *In 4th International Workshop on Reliable Engineering Computing Robust Design–Coping with Hazards*, Risk and Uncertainty, National University of Singapore, Singapore, pp. 3–5.

Kostyuchenko, Y. V., Kopachevsky, I., Yuschenko, M., Solovyov, D., Marton, L., & Levynsky, S. (2012). Spectral reflectance indices as indirect indicators of ecological threats. In Phoon, K.-K., Beer, M., Quek, S. T., & Pang, S. D. (Eds.), *Sustainable Civil Infrastructures–Hazards, Risk, Uncertainty*. Research Publishing, Singapore, 557–562.

Kostyuchenko, Y. V., Bilous, Y., Kopachevsky, I., & Solovyov, D. (2013). Coherent risk measures assessment based on the coupled analysis of multivariate distributions of multi-source observation data. *In Proceedings of 11-th International Probabilistic Workshop*, Brno, pp. 183–192.

Kostyuchenko, Y. V., Movchan, D., Kopachevsky, I., & Bilous, Y. (2015). Robust algorithm of multi-source data analysis for evaluation of social vulnerability in risk assessment tasks. In *SAI Intelligent Systems Conference (IntelliSys), 2015* (pp. 944–949). IEEE, doi: 978-1-4673-7606-8/15.

Kostyuchenko, Y. V., Sztoyka, Y., Kopachevsky, I., Artemenko, I., & Yuschenko, M. (2017). Multisensor satellite data for water quality analysis and water pollution risk assessment: Decision making under deep uncertainty with fuzzy algorithm in framework of multimodel approach. *In Remote Sensing for Agriculture, Ecosystems, and Hydrology XIX* (Vol. 10421, p. 1042105), Warsaw, Poland, International Society for Optics and Photonics.

Kostyuchenko Y. V., Yuschenko, M., Kopachevsky, I., & Artemenko, I. (2019), Bayes decision making systems for quantitative assessment of hydrological climate-related risk using satellite data. In Das, K., & Ram, M. (Eds.) *Mathematical Modelling of System Resilience*, "Mathematical and Engineering Sciences" Series, River Publishers, Denmark, pp. 113–141.

Laurenson, E. M. (1964). A catchment storage model for runoff routing. *Journal of Hydrology*, 2(2), 141–163.

Lillesand, T., Kiefer, R. W., & Chipman, J. (2014). *Remote Sensing and Image Interpretation*. John Wiley & Sons, Hoboken, NJ.

Parker, D. J. (2014). Floods. Routledge, London.

Penuelas, J., Baret, F., & Filella, I. (1995). Semi-empirical indices to assess carotenoids/chlorophyll a ratio from leaf spectral reflectance. *Photosynthetica*, 31(2), 221–230.

Qiu, H. L., Sanchez-Azofeifa, A., & Gamon, J. A. (2007). Ecological applications of remote sensing at multiple scales. In Pugnaire, F. & Valladares, F. (Eds.), *Functional Plant Ecology*, Second Edition. CRC Press, Boca Raton, FL, 655–675.

Rodriguez-Iturbe, I., Entekhabi, D., Lee, J. S., & Bras, R. L. (1991). Nonlinear dynamics of soil moisture at climate scales: 2. Chaotic analysis. *Water Resources Research*, 27(8), 1907–1915.

Salomonson, V. V., & Appel, I. (2004). Estimating fractional snow cover from MODIS using the normalized difference snow index. *Remote Sensing of Environment*, 89(3), 351–360.

Schlaifer, R., & Raiffa, H. (1961). *Applied Statistical Decision Theory*. Clinton Press, Inc., Boston, MA.

Suter II, G. W. (2016). *Ecological Risk Assessment*. CRC Press, Boca Raton, FL.

Verma, S. B., Sellers, P. J., Walthall, C. L., Hall, F. G., Kim, J., & Goetz, S. J. (1993). Photosynthesis and stomatal conductance related to reflectance on the canopy scale. *Remote Sensing of Environment*, 44(1), 103–116.

4 Integral Transforms and Parseval–Goldstein-Type Relationships

Osman Yürekli
Ithaca College

CONTENTS

4.1 INTRODUCTION

By an integral transform, we mean a relation of the form:

$$\int_0^\infty K(x,y)f(x)\,dx = \mathscr{L}\{f(x);y\} = (\mathscr{L}f)(y) = F(y) \qquad (4.1)$$

such that a given function $f(x)$ is transformed to another function $F(y) = (\mathscr{L}f)(y)$. The new function $F(y) = (\mathscr{L}f)(y)$ is called the transform of $f(x)$, and $K(x,y)$ is called the kernel of the transform. The kernel $K(x,y)$ and $f(x)$ must satisfy certain conditions to ensure the existence and the uniqueness of the transform $(\mathscr{L}f)(y)$. The uniqueness and the existence of the integral transform are well established, and throughout the remainder of this chapter, it is assumed that all integrals involved converge absolutely.

Suppose that the kernel of the integral transform Equation (4.1) is symmetric, that is,

$$K(x,y) = K(y,x) \tag{4.2}$$

for all x and y. Using the definition (4.1),

$$\int_0^\infty g(y)\,(\mathscr{L}f)(y)\,dy = \int_0^\infty g(y)\left[\int_0^\infty K(x,y)f(x)\,dx\right]dy. \tag{4.3}$$

Formally changing the order of the integration,

$$\int_0^\infty g(y)\,(\mathscr{L}f)(y)\,dy = \int_0^\infty f(x)\left[\int_0^\infty K(x,y)g(y)\,dy\right]dx. \tag{4.4}$$

Using the definition (4.1) and the equality Equation (4.2), we obtain

$$\int_0^\infty g(y)\,(\mathscr{L}f)(y)\,dy = \int_0^\infty f(x)\,(\mathscr{L}g)(x)\,dx. \tag{4.5}$$

If, in particular, we set $K(x,y) = e^{-xy}$ in Equation (4.1), then we obtain the Laplace transform:

$$(\mathscr{L}f)(y) = \mathscr{L}\{f(x);y\} = \int_0^\infty e^{-xy}f(x)\,dx \tag{4.6}$$

and the relation Equation (4.4) takes the form

$$\int_0^\infty g(y)\,(\mathscr{L}f)(y)\,dy = \int_0^\infty f(x)\,(\mathscr{L}g)(x)\,dx. \tag{4.7}$$

The relation Equation (4.7) is obtained by Goldstein in [12]. He used the relation Equation (4.7) some representations for Whittaker's confluent hypergeometric function. The relation Equation (4.7) is called an exchange identity by Van der Pol and Bremmer [16]. We refer the identity Equation (4.7) as the Goldstein exchange identity. Similar relationships were obtained for other integral transforms such as the Hardy transform by Srivastava [19], the potential transform by Srivastava and Singh [20], and the generalized Hankel transform by Agarwal [1]. Glasser [11] used the integral transform:

$$(\mathscr{G}f)(y) = \mathscr{G}\{f(x);y\} = \int_0^\infty \frac{f(x)}{\sqrt{x^2+y^2}}\,dx, \tag{4.8}$$

and its Parseval–Goldstein-type relationships to derive integral identities. Apelblat [2] introduced a method for the evaluation of infinite integrals. The infinite integrals were derived by applying the same or different integral transform twice. The method has been used in obtaining the results in [3]. Ramanujan also presented some formula on the theory of definite integrals [13, Chapter 11].

Consider the following integral transform:

$$(\mathscr{L}_if)(y) = \mathscr{L}_i\{f(x);y\} = \int_0^\infty K_i(x,y)f(x)\,dx, \qquad i = 1,2,3. \tag{4.9}$$

and assume that $\left(\mathscr{L}_1 \mathscr{L}_2 f\right)(x) = \left[\mathscr{L}_1 \left(\mathscr{L}_2 f\right)(u)\right](x) = \left(\mathscr{L}_3 f\right)(x)$. Using the definition (4.9), changing the order of integration, and using the assumption, we have

$$\int_0^\infty \left(\mathscr{L}_1 f\right)(x) \left(\mathscr{L}_2 g\right)(x)\,dx = \int_0^\infty \left(\mathscr{L}_2 g\right)(x) \left[\int_0^\infty K_1(x,y) f(y)\,dy\right] dx$$

$$= \int_0^\infty f(y) \left[\int_0^\infty K_1(x,y)\left(\mathscr{L}_2 g\right)(x)\,dx\right] dy$$

$$= \int_0^\infty f(y) \left(\mathscr{L}_1 \mathscr{L}_2 g\right)(y)\,dy \qquad (4.10)$$

$$= \int_0^\infty f(y) \left(\mathscr{L}_3 g\right)(y)\,dy. \qquad (4.11)$$

Setting $h = \mathscr{L}_2 g$ in Equation (4.10), we obtain the exchange identity Equation (4.5). Therefore, the integral identity Equation (4.11) is a generalization of the exchange identity Equation (4.5). The identity of the type Equation (4.11) will be called a Parseval–Goldstein-type relationship.

4.2 A PARSEVAL–GOLDSTEIN-TYPE RELATIONSHIP

It is known that the Stieltjes transform:

$$\left(\mathscr{S} f\right)(y) = \mathscr{S}\{f(x); y\} = \int_0^\infty \frac{f(x)}{x+y}\,dx, \qquad (4.12)$$

arises naturally as an iteration of the Laplace transform. If we take the Laplace transform with respect to y of the function $F(u)$ which is the Laplace transform of the function $f(x)$, we have the relation:

$$\left(\mathscr{L} F\right)(y) = \left[\mathscr{L}\left(\mathscr{L} f\right)(u)\right](y)$$

$$= \left(\mathscr{L}^2 f\right)(y)$$

$$= \int_0^\infty e^{-uy} \left[\int_0^\infty e^{-ux} f(x)\,dx\right] du. \qquad (4.13)$$

Changing the order of integration in which we perform the integration, we see that

$$\left(\mathscr{L}^2 f\right)(y) = \left(\mathscr{S} f\right)(y). \qquad (4.14)$$

The iteration identity Equation (4.14) can be used to calculate and invert Stieltjes transforms; for example, see Widder [24] and Sneddon [17].

In this section, we generalize the iteration identity Equation (4.14) to the generalized Stieltjes transform:

$$\left(\mathscr{S}_\rho f\right)(y) = \mathscr{S}_\rho\{f(x); y\} = \int_0^\infty \frac{f(x)}{(x+y)^\rho}\,dx. \qquad (4.15)$$

This generalization gives a Parseval–Goldstein-type relationship for the Laplace transform and the generalized Stieltjes transform. We show how this Parseval-type relationship leads to Goldstein's exchange identity and an exchange identity for the generalized Stieltjes transform. As applications of the Parseval–Goldstein-type relation, we present identities relating the potential transform to the sine transform and the cosine transform, and the generalized Stieltjes transform to the Laguerre transform. We give some results on the Riemann–Liouville and the Weyl fractional integrals. We also present methods for solving some integral equations. We illustrate the use of our results in the last part of this section. Various Parseval–Goldstein-type identities are given in, for example [4,5,7,8,21,25,26,28,29], for a variety of integral transforms.

4.2.1 A PARSEVAL–GOLDSTEIN-TYPE RELATIONSHIP FOR LAPLACE TRANSFORMS

First, we give a lemma which is a generalization of identity Equation (4.14) to the generalized Stieltjes transform Equation (4.15).

Lemma 4.2.1 *If* $\Re(\rho) > 0$, *then*

$$\left[\mathscr{L}u^{\rho-1}(\mathscr{L}f)(u)\right](y) = \Gamma(\rho)(\mathscr{S}_\rho f)(y). \tag{4.16}$$

Proof. Using the definition of the Laplace transform Equation (4.6), we have

$$\left[\mathscr{L}u^{\rho-1}(\mathscr{L}f)(u)\right](y) = \int_0^\infty u^{\rho-1}e^{-uy}\left[\int_0^\infty e^{-ux}f(x)\,dx\right]du. \tag{4.17}$$

Changing the order of the integration, which is permissible by the absolute convergence of the integrals involved, we find from Equation (4.17) that

$$\left[\mathscr{L}u^{\rho-1}(\mathscr{L}f)(u)\right](y) = \int_0^\infty f(x)\left[\int_0^\infty u^{\rho-1}e^{-(x+y)u}\,du\right]dx. \tag{4.18}$$

Computing the inner integral on the right-hand side and using the definition of the generalized Stieltjes transform Equation (4.15), we deduce the iteration identity Equation (4.16).

The following Parseval–Goldstein-type relationship relating the Laplace transform to the generalized Stieltjes transform is the main theorem of this section.

Theorem 4.2.2 *If* $\Re(\rho) > 0$, *then*

$$\int_0^\infty x^{\rho-1}(\mathscr{L}f)(x)(\mathscr{L}g)(x)\,dx = \Gamma(\rho)\int_0^\infty g(y)(\mathscr{S}_\rho f)(y)\,dy. \tag{4.19}$$

Proof. The definition of the Laplace transform implies

$$\int_0^\infty x^{\rho-1}(\mathscr{L}f)(x)(\mathscr{L}g)(x)\,dx = \int_0^\infty x^{\rho-1}(\mathscr{L}f)(x)\left[\int_0^\infty e^{-xy}g(y)\,dy\right]dx \tag{4.20}$$

Changing the order of the integration, which is permissible by the absolute convergence of the integrals involved, we find from Equation (4.20) that

$$\int_0^\infty x^{\rho-1}(\mathscr{L}f)(x)(\mathscr{L}g)(x)\,dx = \int_0^\infty g(y)\left[\int_0^\infty x^{\rho-1}e^{-xy}(\mathscr{L}f)(x)\,dx\right]dy \quad (4.21)$$

Now, the assertion Equation (4.19) follows from the iteration identity Equation (4.16) of Lemma 4.2.1.

The convolution property of the Laplace transform:

$$(\mathscr{L}f * g)(x) = (\mathscr{L}f)(x)(\mathscr{L}g)(x), \quad (4.22)$$

where the convolution of two functions f and g for the Laplace transform is defined as

$$f * g(x) = \int_0^\infty f(x-y)g(y)\,dy. \quad (4.23)$$

Remark 4.2.3 *If we use the convolution property Equation (4.22) of the Laplace transform in the Parseval–Goldstein-type identity Equation (4.19) of Theorem 4.2.2, we obtain*

$$\int_0^\infty x^{\rho-1}(\mathscr{L}f * g)(x)\,dx = \Gamma(\rho)\int_0^\infty g(y)(\mathscr{S}_\rho f)(y)\,dy \quad (4.24)$$

where $\Re(\rho) > 0$.

Remark 4.2.4 *If we put* $h(x) = x^{\rho-1}(\mathscr{L}f)(x)$ *in the identity Equation (4.19) of Theorem 4.2.2 and use the definition of the Laplace transform Equation (4.6), we obtain the Goldstein exchange identity Equation (4.7).*

Corollary 4.2.5 *If* $\Re(\rho) > 0$, *then*

$$\int_0^\infty g(y)(\mathscr{S}_\rho f)(y)\,dy = \int_0^\infty f(x)(\mathscr{S}_\rho g)(x)\,dx. \quad (4.25)$$

Proof. Since the left-hand side of the identity Equation (4.19) of Theorem 4.2.2 is symmetrical with respect to the function f and g, we have the exchange identity for the generalized Stieltjes transform Equation (4.25).

Corollary 4.2.6 *If* $\Re(\rho) > 0$ *and* $\Re(v) > 0$, *then*

$$[\mathscr{S}_v x^{\rho-1}(\mathscr{L}f)(x)](u) = \frac{\Gamma(\rho)}{\Gamma(v)}[\mathscr{L}y^{v-1}(\mathscr{S}_v f)(y)](u). \quad (4.26)$$

Proof. In the Parseval–Goldstein-type identity Equation (4.19), we set

$$g(y) = y^{v-1}e^{-uy} \quad (4.27)$$

so that

$$(\mathscr{L}g)(x) = \frac{\Gamma(v)}{(x+u)^v}. \quad (4.28)$$

Now, the assertion Equation (4.26) follows immediately by substituting Equations (4.27) and (4.28) to the identity Equation (4.19) of Theorem Equation (4.2.2) and making use of the definition (4.15) of the generalized Stieltjes transform.

Corollary 4.2.7 *Let $g_1(x)$ and its derivatives of orders up to and including $n-1$ be continuous for $x > 0$, with $g_1(x)$ and its derivatives having limits at $x = 0$, and suppose $g_1(x)$ and its derivatives are of exponential order. If $\Re(\rho) > 0$, then*

$$\int_0^\infty x^{\rho-1} (\mathscr{L}f)(x) \left[x^n (\mathscr{L}g_1)(x) - x^{n-1}g_1(0) - \cdots - g_1^{(n-1)}(0) \right] dx$$

$$= \Gamma(\rho) \int_0^\infty g_1^{(n)}(y)(\mathscr{S}_\rho f)(y)\, dy. \quad (4.29)$$

Proof. In the Parseval–Goldstein-type identity Equation (4.19), we set

$$g(y) = g_1^{(n)}(y) \quad (4.30)$$

so that

$$(\mathscr{L}g)(x) = x^n (\mathscr{L}g_1)(x) - x^{n-1}g_1(0) - \cdots - g_1^{(n-1)}(0) \quad (4.31)$$

by the derivative property of the Laplace transform. Now, the assertion Equation (4.29) follows immediately by substituting Equation (4.30) and (4.31) to the identity Equation (4.19) of Theorem Equation (4.2.2) and making use of the definition (4.15) of the generalized Stieltjes transform.

A consequence of Equation (4.29) of Corollary 4.2.7 involving the generalized Stieltjes transform, the Laplace transform, and the Laguerre transform is contained in the next theorem. The Laguerre transform is defined as

$$(\mathscr{T}_\alpha f)(n) = \mathscr{T}_\alpha\{f(x); n\} = \int_0^\infty e^{-x} x^\alpha L_n^\alpha(x) f(x)\, dx, \quad (4.32)$$

where

$$L_n^\alpha(x) = \frac{e^x x^{-\alpha}}{n!} \frac{d^n}{dx^n}\left(e^{-x} x^{n+\alpha}\right) \quad (n = 0,1,2,\ldots;\ \alpha > -1) \quad (4.33)$$

are the associated Laguerre polynomials.

Theorem 4.2.8 *If $\Re(\rho) > 0$ and $\alpha > -1$, then*

$$\left[\mathscr{S}_{n+\alpha+1} x^{n+\rho-1} (\mathscr{L}f)(x)\right](1) = \frac{n!\,\Gamma(\rho)}{\Gamma(n+\alpha+1)} \left[\mathscr{T}_\alpha(\mathscr{S}_\rho f)(y)\right](n), \quad (4.34)$$

where $n = 0,1,2,\ldots$ and

$$\left[\mathscr{L} y^{n+\alpha}(\mathscr{S}_{n+\alpha} f)(y)\right](1) = \frac{n!\,\Gamma(\rho)}{\Gamma(n+\alpha)} \left[\mathscr{T}_\alpha(\mathscr{S}_\rho f)(y)\right](n), \quad (4.35)$$

where $n = 1,2,\ldots$

Proof. In the identity Equation (4.29) of Corollary Equation (4.2.7), we set

$$g_1(y) = y^{n+\alpha} e^{-y} \quad (4.36)$$

so that

$$(\mathscr{L}g_1)(x) = \frac{\Gamma(n+\alpha+1)}{(x+1)^{n+\alpha+1}} \tag{4.37}$$

and

$$g_1^k(0) = 0, \qquad \text{for } 0 \le k \le n-1. \tag{4.38}$$

Substituting Equations (4.36), (4.37), and (4.38) to the identity Equation (4.29) of Corollary Equation (4.2.7), we obtain

$$\int_0^\infty \frac{x^{n+\rho-1}}{(x+1)^{n+\alpha+1}} (\mathscr{L}f)(x)\,dx \tag{4.39}$$

$$= \frac{\Gamma(\rho)}{\Gamma(n+\alpha+1)} \int_0^\infty \frac{d^n}{dy^n} \left(e^{-y}y^{n+\alpha}\right) (\mathscr{S}_\rho f)(y)\,dy$$

$$= \frac{n!\,\Gamma(\rho)}{\Gamma(n+\alpha+1)} \int_0^\infty e^{-y}y^\alpha \mathscr{L}_n^\alpha(y)(\mathscr{S}_\rho f)(y)\,dy. \tag{4.40}$$

Now, the assertion Equation (4.34) follows from the definitions (4.15), (4.32), (4.32), and the Equation (4.40). The assertion Equation (4.35) easily follows from the identity Equation (4.26) of Corollary 4.2.6.

4.2.2 SOME ILLUSTRATIVE EXAMPLES

An illustration of Theorem 4.2.2 is provided by evaluating the Mellin transform of a function involving the complementary incomplete gamma function. The Mellin transform is defined as

$$(\mathscr{M}f)(y) = \mathscr{M}\{f(x);y\} = \int_0^\infty x^{y-1}f(x)\,dx, \tag{4.41}$$

Example 4.2.9 *If* $\Re(\mu+1) > \Re(\rho) > \Re(\mu) > 0$ *and* $\Re(a) > 0$, *then*

$$(\mathscr{M}e^{ay}\Gamma(1-\rho,ay))(\mu) = \frac{\pi\Gamma(\mu)a^{\rho-\mu-1}}{\Gamma(\rho)\sin[\pi(\rho-\mu)]}. \tag{4.42}$$

Proof. In the identity Equation (4.19) of Theorem Equation (4.2.2), we set

$$f(u) = e^{-au} \qquad \text{and} \qquad g(y) = y^{\mu-1} \tag{4.43}$$

so that

$$(\mathscr{L}f)(x) = \frac{1}{x+a} \qquad \text{and} \qquad (\mathscr{L}g)(x) = \frac{\Gamma(\mu)}{x^\mu}. \tag{4.44}$$

From Equation (4.16) of Lemma 4.2.1 and the formula [9, Vol. 1, p. 137, Entry (7)]

$$(\mathscr{S}_\rho f)(x) = \frac{1}{\Gamma(\rho)} \left(\mathscr{L}\frac{x^{\rho-1}}{x+a}\right)(y) = a^{\rho-1}e^{ay}\Gamma(1-\rho,ay). \tag{4.45}$$

Substituting Equations (4.43), (4.44), and (4.45) into (4.19), we obtain

$$\left(\mathcal{M}e^{ay}\Gamma(1-\rho,ay)\right)(\mu) = \frac{\Gamma(\mu)}{\Gamma(\rho)}a^{1-\rho}\left(\mathcal{S}x^{\rho-\mu-1}\right)(a). \tag{4.46}$$

Making use of Equation (4.16) of Lemma 4.2.1, the Stieltjes transform on the right-hand side of Equation (4.46) may be evaluated and our assertion Equation (4.42) follows immediately.

Example 4.2.10 *We show that*

$$\left(\mathcal{T}_{\alpha}y^{(\nu-\rho+1)/2}K_{\nu-\rho+1}\left(ay^{1/2}\right)\right)(n)$$

$$= \frac{a^{n+\alpha-\rho-1}}{n!\,2^{n+\alpha-\rho}}\Gamma(\nu+2)\Gamma(\rho)\Gamma(n+\alpha+1)\,e^{a^2/8}W_{-k,\ell}\left(\frac{a^2}{4}\right), \tag{4.47}$$

where $a>0$, $\Re(\rho)>0$, $\Re(\nu)>-1$, $\Re(\alpha)+n>0$, $2k=n+\alpha+\nu+2$, $2\ell=\nu-n-\alpha+1$, and $n=1,2,\ldots$, $K_{\nu}(x)$ denotes the modified Bessel function of the third kind, and $W_{k,\ell}(x)$ denotes Whittaker's confluent hypergeometric function.

Proof. In the identity Equation (4.35) of Theorem Equation (4.2.8), we set $f(u)=u^{\nu/2}J_{\nu}\left(au^{1/2}\right)$ so that (cf. [10, Vol. 2, p. 235, Entry (20)])

$$\left(\mathcal{S}_{\rho}f\right)(y) = \frac{a^{\rho-1}}{2^{\rho}\Gamma(\rho)}y^{(\nu-\rho+1)/2}K_{\nu-\rho+1}\left(ay^{1/2}\right), \tag{4.48}$$

and

$$\left[\mathcal{L}y^{n+\alpha}\left(\mathcal{S}_{n+\alpha}f\right)(y)\right](1)$$

$$= \frac{a^{n+\alpha-2}}{2^{n+\alpha}}\frac{\Gamma(\nu+2)\Gamma(n+\alpha+1)}{\Gamma(n+\alpha)}e^{a^2/8}W_{-k,\ell}\left(\frac{a^2}{4}\right). \tag{4.49}$$

where k and ℓ are as defined in the statement of the example. Substituting Equations (4.48) and (4.49) into (4.35) of Theorem 4.2.8, we obtain the assertion Equation (4.47).

Example 4.2.11 *We show that*

$$\left(\mathcal{L}y^{(2\mu+2\ell-3)/2}e^{ay/2}W_{k,\ell}(ay)\right)(b)$$

$$= \frac{a^{1/2-\ell}}{b^{\mu}}\frac{\Gamma(\mu-\rho+\lambda)\Gamma(\mu)}{\Gamma(\mu+\lambda)}\,{}_2F_1\left(\mu,\rho;\mu+\lambda,1-\frac{a}{b}\right), \tag{4.50}$$

where $a>0$, $b>0$, $\Re(\rho)>0$, $\Re(\mu)>0$, $\Re(\lambda)>0$, $\Re(\lambda)>\Re(\rho-\mu)$, $2k=1-\lambda-\rho$, $2\ell=\lambda-\rho$, and $n=1,2,\ldots$, $K_{\nu}(x)$ denotes the modified Bessel function of the third kind, and $W_{k,\ell}(x)$, ${}_2F_1$ denotes Gauss's hypergeometric function.

Proof. In the identity Equation (4.35) of Theorem Equation (4.2.8), we set

$$f(u) = u^{\lambda-1}e^{-au} \qquad \text{and} \qquad g(y) = y^{\mu-1}e^{-by} \qquad (4.51)$$

so that

$$(\mathscr{L}f)(x) = \frac{\Gamma(\lambda)}{(x+a)^{\lambda}} \qquad \text{and} \qquad (\mathscr{L}g)(x) = \frac{\Gamma(\mu)}{(x+b)^{\mu}} \qquad (4.52)$$

and (cf. [10, Vol. 2, p. 234, Entry (12)]

$$(\mathscr{S}_{\rho}f)(x) = \Gamma(\lambda)a^{-(2\ell+1)/2}y^{((2\ell-1)/2)/2}e^{ay/2}W_{k,l}(ay), \qquad (4.53)$$

where ℓ and k are as in the statement of the example. Substituting Equation (4.51), (4.52), and (4.53) into (4.19) of Theorem 4.2.2, we obtain the assertion Equation (4.50).

4.3 THE $\mathscr{L}_2$-TRANSFORM AND ITS APPLICATIONS

In this section, we consider the $\mathscr{L}_2$-transform

$$(\mathscr{L}_2 f)(y) = \int_0^{\infty} xe^{-x^2 y^2} f(x)\,dx. \qquad (4.54)$$

and the Widder transform

$$(\mathscr{W} f)(y) = \int_0^{\infty} \frac{xf(x)}{x^2+y^2}\,dx. \qquad (4.55)$$

The $\mathscr{L}_2$-transform is introduced by the author [27]. The Widder transform is introduced as the potential transform by Widder, and he presented a systematic account of the transform in [22,23]. Widder pointed out that the potential transform is related to the Poisson integral representation of a function which is harmonic in a half-plane and gave several inversion formulae for the transform and applied his results to harmonic functions. Srivastava and Singh [20] gave the following exchange identity

$$\int_0^{\infty} yg(y)(\mathscr{W} f)(y)\,dy = \int_0^{\infty} xf(x)(\mathscr{W} g)(x)\,dx \qquad (4.56)$$

for the Widder transform. Some generalizations of the $\mathscr{L}_2$-transform are considered in, for example, [6].

If we make a simple change of variable in the integral on the right-hand side of Equation (4.54), we obtain

$$(\mathscr{L}_2 f)(y) = \frac{1}{2} \int_0^{\infty} xe^{-x^2 y^2} f(\sqrt{x})\,dx. \qquad (4.57)$$

Comparing Equation (4.57), the definitions (4.54) of the $\mathscr{L}_2$-transform, and Equation (4.6) of the Laplace transform gives the relationship

$$(\mathscr{L}_2 f)(y) = \frac{1}{2}\left(\mathscr{L}f(\sqrt{x})\right)(y^2) \qquad (4.58)$$

There is a similar relationship between the Widder transform and the Stieltjes transform:

$$(\mathscr{W}f)(y) = \frac{1}{2}\left(\mathscr{S}f(\sqrt{x})\right)(y^2).$$ (4.59)

We introduce the idea of the convolution of two functions for the $\mathscr{L}_2$-transform:

$$f \star g(x) = \int_0^x yf\left(\sqrt{x^2 - y^2}\right)g(y)\,dy.$$ (4.60)

It can be easily shown that the convolution for the $\mathscr{L}_2$-transform satisfies all the properties of the convolution for the Laplace transform.

Theorem 4.3.1 *We have*

$$(\mathscr{L}_2 f \star g)(y) = (\mathscr{L}_2 f)(y)(\mathscr{L}_2 g)(y)$$ (4.61)

Proof. By the definitions (4.54) of the $\mathscr{L}_2$-transform and Equation (4.60) of the convolution for the $\mathscr{L}_2$-transform, we have

$$(\mathscr{L}_2 f \star g)(y) = \int_0^\infty xe^{-x^2y^2}\left[uf\left(\sqrt{x^2 - u^2}\right)g(u)\,du\right]dx$$ (4.62)

Changing the order of integration and then changing the variable of integration in the inner integral from x to $t = \sqrt{x^2 - u^2}$, we obtain

$$(\mathscr{L}_2 f \star g)(y) = \int_0^\infty ug(u)\left[\int_0^\infty xe^{-x^2y^2}f\left(\sqrt{x^2 - u^2}\right)dx\right]du$$
$$= \int_0^\infty ue^{-u^2y^2}g(u)\,du \int_0^\infty te^{-t^2y^2}f(t)\,dt.$$ (4.63)

Hence, the assertion Equation (4.61) follows from the definition (4.54) of the $\mathscr{L}_2$-transform.

In this section, we establish a Parseval–Goldstein-type relationship between the Widder transform and the $\mathscr{L}_2$-transform. We also obtain identities relating the $\mathscr{K}$-transform

$$(\mathscr{K}_v f)(y) = \int_0^\infty \sqrt{xy}\,K_v(xy)f(x)\,dx.$$ (4.64)

to the $\mathscr{L}_2$-transform, where $K_v(x)$ is the Bessel function of the third kind (it is also known as the Macdonald function), and the Laplace transform to the $\mathscr{L}_2$-transform. Using these results, we show how one can extend tables of Laplace and Hankel transforms.

4.3.1 A PARSEVAL–GOLDSTEIN-TYPE RELATIONSHIP AND ITS COROLLARIES

First, we give a lemma that shows the iteration of the $\mathscr{L}_2$-transform by itself is the Widder transform.

Lemma 4.3.2 *We have*

$$\left(\mathscr{L}_2^2 f\right)(y) = \frac{1}{2}\left(\mathscr{W} f\right)(y). \tag{4.65}$$

Proof. Using the definition of the $\mathscr{L}_2$-transform Equation (4.54), we have

$$\left(\mathscr{L}_2^2 f\right)(y) = \left[\mathscr{L}_2\left(\mathscr{L}_2 f\right)(u)\right](y)$$

$$= \int_0^\infty u e^{-y^2 u^2} \left[\int_0^\infty x e^{-x^2 u^2} f(x)\,dx\right] du. \tag{4.66}$$

Changing the order of the integration on the right-hand side of Equation (4.66), which is permissible by the absolute convergence of the integrals involved, we find from Equation (4.66) that

$$\left(\mathscr{L}_2^2 f\right)(y) = \int_0^\infty x f(x) \left[\int_0^\infty u e^{-(x^2+y^2)u^2}\,du\right] dx. \tag{4.67}$$

Computing the inner integral on the right-hand side and using the definition of the Widder transform Equation (4.55), we deduce the iteration identity Equation (4.65).

The following Parseval–Goldstein-type relationship relating the $\mathscr{L}_2$-transform to the Widder transform is the main theorem of this section.

Theorem 4.3.3 *We have*

$$\int_0^\infty x\left(\mathscr{L}_2 f\right)(x)\left(\mathscr{L}_2 g\right)(x)\,dx = \frac{1}{2}\int_0^\infty y g(y)\left(\mathscr{W} f\right)(y)\,dy. \tag{4.68}$$

Proof. The definition Equation (4.54) of the $\mathscr{L}_2$-transform transform implies

$$\int_0^\infty x\left(\mathscr{L}_2 f\right)(x)\left(\mathscr{L}_2 g\right)(x)\,dx$$

$$= \int_0^\infty x\left(\mathscr{L}_2 f\right)(x) \left[\int_0^\infty y e^{-x^2 y^2} g(y)\,dy\right] dx \tag{4.69}$$

Changing the order of the integration, which is permissible by the absolute convergence of the integrals involved, we find from Equation (4.69) that

$$\int_0^\infty x\left(\mathscr{L}_2 f\right)(x)\left(\mathscr{L}_2 g\right)(x)\,dx$$

$$= \int_0^\infty y g(y) \left[\int_0^\infty x e^{-x^2 y^2}\left(\mathscr{L} f\right)(x)\,dx\right] dy. \tag{4.70}$$

Now the assertion Equation (4.68) follows from the iteration identity Equation (4.65) of Lemma 4.3.2.

Remark 4.3.4 *We have*

$$\int_0^\infty x\left(\mathscr{L}_2 f\right)(x)\left(\mathscr{L}_2 g\right)(x)\,dx = \frac{1}{2}\int_0^\infty y f(y)\left(\mathscr{W} g\right)(y)\,dy. \tag{4.71}$$

Since the relation Equation (4.68) is symmetrical with respect to f and g. Using the Parseval–Goldstein-type relationships Equations (4.68) and (4.71), we obtain the exchange identity Equation (4.56) for the Widder transform. Thus, the identity Equation (4.68) of Theorem 4.3.3 generalizes relation Equation (4.71).

Corollary 4.3.5 *We have*

$$\int_0^\infty xh(x)\left(\mathscr{L}_2 f\right)(x)\,dx = \int_0^\infty yf(y)\left(\mathscr{L}_2 h\right)(y)\,dy. \tag{4.72}$$

Proof. The identity Equation (4.72) follows immediately after letting $h(x) = \left(\mathscr{L}_2 g\right)(x)$ in the identity Equation (4.68) of Theorem 4.3.3.

Corollary 4.3.6 *We have*

$$\left[\mathscr{W}\left(\mathscr{L}_2 f\right)\right](x) = \left[\mathscr{L}_2\left(\mathscr{W} f\right)\right](x) \tag{4.73}$$

Proof. We set $f(y) = e^{-u^2 y^2}$ in Equation (4.68) of Theorem 4.3.3. Then,

$$\left(\mathscr{L}_2 f\right)(x) = \int_0^\infty ye^{-(u^2+x^2)y^2}\,dy = \frac{1}{2(u^2+x^2)} \tag{4.74}$$

Now, the assertion Equation (4.73) follows from the identity Equation (4.68) of Theorem 4.3.3.

The following corollary to Lemma 4.3.2 gives a method to invert the Widder transform Equation (4.55):

Corollary 4.3.7 *Consider the integral equation*

$$\int_0^\infty \frac{xf(x)}{x^2+y^2}\,dx = g(y), \tag{4.75}$$

where g is known, and f is unknown. Then, the integral equation Equation (4.75) has a unique solution given by

$$f(x) = \left[\mathscr{L}_2^{-1}\left(\mathscr{L}_2^{-1} g\right)\right](x) \tag{4.76}$$

Proof. Comparing the integral equation Equation (4.75) with the identity Equation (4.65) of Lemma 4.3.2 implies

$$g(y) = \left(\mathscr{L}_2^2\right)(y) \tag{4.77}$$

The assertion Equation (4.75) of Lemma 4.3.2 follows from the result Equation (4.77).

Theorem 4.3.8 *If $\Re(v) \geq -1$*

$$\left(\mathscr{H}_{vy}^{(2v+1)/2} f(y)\right)(z) = 2^v u^{(2v+1)/2}\left[\mathscr{L}_2 x^{2v-2}\left(\mathscr{L}_2 f\right)\left(\frac{1}{2x}\right)\right](u). \tag{4.78}$$

Proof. We set $g(u) = u^\nu J_\nu(zu)$ Equation (4.68) of Theorem 4.3.3, where J_ν is the Bessel function of the first kind of order ν. Using the identity Equation (4.58) and making the use of the Laplace transform table (cf. [9, Vol. 1, p. 185, Entry (30)])

$$(\mathscr{L}_2 g)(y) = \frac{1}{2}(\mathscr{L}g(\sqrt{y}))(x^2) = \frac{1}{2}(\mathscr{L}u^{\nu/2}J_\nu(zu^{1/2}))(x^2)$$
$$= \frac{1}{2}\left(\frac{z}{2}\right)^\nu x^{-2\nu-2}\exp\left(-\frac{z^2}{4x^2}\right) \qquad (4.79)$$

Now in order to evaluate the Widder transform of the function $g(u)$, we use the identity Equation (4.65) of Lemma 4.3.2 and obtain

$$(\mathscr{W}g)(y) = \left(\frac{z}{2}\right)^\nu \left(\mathscr{L}_2 x^{-2\nu-2}\exp\left(-\frac{z^2}{4x^2}\right)\right)(y). \qquad (4.80)$$

The $\mathscr{L}_2$-transform on the right-hand side of Equation (4.80) may be evaluated by using the iteration identity Equation (4.58) and then the Laplace transform table (cf. [9, Vol. 1, p. 146, Entry (20)]). Thus,

$$(\mathscr{W}g)(y) = y^\nu K_\nu(zy). \qquad (4.81)$$

Substituting the results Equation (4.79) and (4.81) into (4.68) of Theorem 4.3.3 gives

$$\int_0^\infty y^{\nu+1}K_\nu(zy)f(y)\,dy = \left(\frac{z}{2}\right)^\nu \int_0^\infty x^{-2\nu-1}\exp\left(-\frac{z^2}{4x^2}\right)(\mathscr{L}_2 f)(x)\,dx. \qquad (4.82)$$

Now, the assertion follows by making a simple change of variable in the integral on the right-hand side of Equation (4.82) and then using the definitions (4.64) of the $\mathscr{K}$-transform and Equation (4.54) of the $\mathscr{L}_2$-transform.

It is well known that

$$K_{1/2}(x) = K_{-1/2}(x) = \left(\frac{\pi}{2x}\right)^{1/2}e^{-x}, \qquad (4.83)$$

(see [18, p. 306]). Using the identity Equation (4.78) of Theorem 4.3.8 and the special cases Equation (4.83), we obtain the identities in the following corollary:

Corollary 4.3.9 *We have*

$$(\mathscr{L}yf)(x) = \frac{2}{\sqrt{\pi}}\left[\mathscr{L}_2\frac{1}{x}(\mathscr{L}_2 f)\left(\frac{1}{2x}\right)\right](z) \qquad (4.84)$$

and

$$(\mathscr{L}yf)(x) = \frac{1}{\sqrt{\pi}}\left[\mathscr{L}_2\frac{1}{x^3}(\mathscr{L}_2 f)\left(\frac{1}{2x}\right)\right](z). \qquad (4.85)$$

In the following corollary, we present a new integral representation for the Macdonald function.

Corollary 4.3.10 *If* $\Re(v) \geq -1$, *then*

$$K_v(y) = \frac{1}{(2y)^v} \int_0^\infty x^{-2v-1} exp\left(-x^2 y^2 - \frac{1}{4x^2}\right) dx. \tag{4.86}$$

Proof. We set $f(x) = x^v J_v(x)$ in the identity Equation (4.78) of Theorem 4.3.8.

4.3.2 SOME ILLUSTRATIVE EXAMPLES

An illustration of Theorem 4.3.8 is provided by the following example:

Example 4.3.11 *If* $2\Re(\rho) > 2|\Re(v)| - 1$, *then*

$$\left(\mathcal{K}_v y^{\rho-1}\right)(z) = 2^{(2\rho-3)/2} z^{-\rho} \Gamma\left(\frac{\rho}{2} - \frac{v}{2} + \frac{1}{4}\right) \Gamma\left(\frac{\rho}{2} + \frac{v}{2} + \frac{1}{4}\right), \tag{4.87}$$

(cf. [10, p. 127, Entry (1)]).

Proof. In the identity Equation (4.78) of Theorem Equation (4.3.8) we set $f(y) = y^{(2\rho-2v-3)/2}$. Making use of the identity Equation (4.58) we obtain

$$\left(\mathcal{L}_2 y^{(2\rho-2v-3)/2}\right)\left(\frac{1}{2x}\right) = \frac{1}{2}\left(\mathcal{L} y^{(2\rho-2v-3)/4}\right)\left(\frac{1}{4x^2}\right)$$

$$= \Gamma\left(\frac{\rho}{2} - \frac{v}{2} + \frac{1}{4}\right)(2x)^{(2\rho-2v-1)/2}, \tag{4.88}$$

provided that $\Re(\rho - v) > 1/2$. Substituting Equation (4.88) into the identity Equation (4.78) of Theorem Equation (4.3.8) we find

$$\left(\mathcal{K}_v y^{\rho-1}\right)(z) = 2^\rho - \frac{1}{2}\Gamma\left(\frac{\rho}{2} - \frac{v}{2} + \frac{1}{4}\right) z^{(2v+1)/2}\left(\mathcal{L}_2 y^{(2\rho+2v-3)/2}\right)(z). \tag{4.89}$$

Now, the assertion Equation (4.87) follows after evaluating the $\mathcal{L}_2$-transform on the right side of Equation (4.89).

Example 4.3.12 *If* $\Re(v) > -2$, *then*

$$\left(\mathcal{K}_v y^{(2v+1)/2} \sin\left(ay^2\right)\right)(z) = a^{3/2}(2a)^{-v-3}\Gamma(v+2)z^{(2v+3)/2} S_{-(2v+3)/2,1/2}\left(\frac{z^2}{4a}\right), \tag{4.90}$$

where $S_{\mu,v}$ *is the Lommel function.*

Proof. In the identity Equation (4.78) of Theorem Equation (4.3.8), we set $f(y) = \sin\left(ay^2\right)$. Making use of the identity Equation (4.58) and then the formula [15, p. 54, Entry (7.1)], we obtain

$$\left(\mathcal{L}_2 \sin\left(ay^2\right)\right)\left(\frac{1}{2x}\right) = \frac{1}{2}\left(\mathcal{L} \sin\left(ay\right)\right)\left(\frac{1}{4x^2}\right) = \frac{2\alpha x^4}{x^4 + \alpha^2}. \tag{4.91}$$

where $\alpha = 1/4a$. Substituting Equation (4.91) into the identity Equation (4.78) of Theorem Equation (4.3.8), we find

$$\left(\mathscr{K}_\nu y^{(2\nu+1)/2} \sin\left(ay^2\right) \right)(z) = 2^\nu a z^{(2\nu+1)/2} \left(\mathscr{L} \frac{x^{\nu+1}}{x^2+\alpha^2} \right)(z). \tag{4.92}$$

Now, the assertion Equation (4.90) follows from the tables of Laplace transforms (see [15, p. 22, Entry 3.11]).

Example 4.3.13 *We show that*

$$\left(\mathscr{L} \sin\left(ay^2\right) \right)(z) = \sqrt{\frac{\pi}{2a}} \left[\left(\frac{1}{2} - C(t) \right) \cos t + \left(\frac{1}{2} - S(t) \right) \sin t \right], \tag{4.93}$$

where $t = z^2/(4a)$ and $C(t)$ and $S(t)$ are the Fresnel integrals.

Proof. We set $\nu = -1/2$ in Equation (4.88). Using the special cases Equation (4.83) and the definition Equation (4.64) of the $\mathscr{K}$-transform, we obtain

$$\left(\mathscr{L} \sin\left(ay^2\right) \right)(z) = \frac{z}{4a} S_{-1,1/2} \left(\frac{z^2}{4a} \right). \tag{4.94}$$

It follows from a formula for the Lommel function (see [15, p.416]) that

$$S_{-1,1/2}(t) = \pi \left[J_{1/2}(t) + J_{-1/2}(t) + \mathbf{J}_{1/2}(t) - \mathbf{J}_{-1/2}(t) \right], \tag{4.95}$$

where $\mathbf{J}_{1/2}(t)$ is the Anger–Weber function of order ν. However, we have

$$J_{1/2}(t) = \sqrt{\frac{2}{\pi t}} \sin t \quad \text{and} \quad J_{-1/2}(t) = \sqrt{\frac{2}{\pi t}} \cos t, \tag{4.96}$$

(cf. [18, p. 306]) and

$$\mathbf{J}_{1/2}(t) = \sqrt{\frac{2}{\pi t}} \left\{ [C(t) - S(t)] \cos t + [C(t) + S(t)] \sin t \right\} \tag{4.97}$$

and

$$\mathbf{J}_{-1/2}(t) = \sqrt{\frac{2}{\pi t}} \left\{ [C(t) - S(t)] \cos t - [C(t) - S(t)] \sin t \right\}, \tag{4.98}$$

(cf. [15, p. 415]). Now, substituting Equation (4.96), (4.97), and (4.98) into (4.95) and then using Equation (4.94), we obtain the assertion Equation (4.93).

Example 4.3.14 *If $|\Re(\nu)| < 1$, and $\Re(a) > 0$ then*

$$\left(\mathscr{L}(x + 4ax^2)^{(\nu-1)/2} \right)(z) = 2^\nu \Gamma \left(\frac{\nu}{2} + \frac{1}{2} \right) \sqrt{\frac{z}{\pi a}} \exp\left(\frac{z}{8a} \right) K_{\nu/2} \left(\frac{z}{8a} \right). \tag{4.99}$$

Proof. We set $f(y) = y^{\nu-1}e^{-ay^2}$ in Equation (4.78) of Theorem Equation (4.3.8). Making use of Equation (4.58) and then using tables of Laplace transforms (cf. [15, p. 37, Entry 5.3]), we obtain

$$\left(\mathcal{L}_2 y^{\nu-1}e^{-ay^2}\right)\left(\frac{1}{2x}\right) = \frac{1}{2}\left(\mathcal{L}y^{(\nu-1)/2}e^{-ay}\right)\left(\frac{1}{4x^2}\right)$$

$$= \Gamma\left(\frac{1}{2} - \frac{\nu}{2}\right)\frac{(4ax^2+1)^{(\nu-1)/2}}{(2x)^{\nu}}, \qquad (4.100)$$

for $\mathfrak{R}(\nu) < 1$. Using tables of Hankel transform (cf. [9, p. 132]), we obtain

$$\left(\mathcal{H}_\nu y^{-1/2}e^{-ay^2}\right)(z) = \frac{1}{2}\sqrt{\frac{\pi z}{a}}\sec\left(\frac{\nu\pi}{2}\right)K_{\nu/2}\left(\frac{1}{4x^2}\right), \qquad (4.101)$$

where $|\mathfrak{R}(\nu)| < 1$, and $\mathfrak{R}(a) > 0$. Now the assertion Equation (4.99) follows from substituting Equation (4.100) and (4.101) into (4.78) of Theorem Equation (4.3.8) and then using Equation (4.58).

Example 4.3.15 *We show that*

$$\left(\mathcal{L}x^{1/2}\mathrm{Erf}\left(ax^{1/2}\right)\right)(z) = \frac{2}{\sqrt{\pi z}}\arctan\left(az^{-1/2}\right) \qquad (4.102)$$

and

$$\left(\mathcal{L}x^{3/2}\mathrm{Erf}\left(ax^{1/2}\right)\right)(z) = \frac{4}{\sqrt{\pi}}\arctan\left(az^{-1/2}\right) \qquad (4.103)$$

Proof. We set $f(y) = y^{-2}\sin(ay)$ in Theorem Equation (4.3.7). Making use of Equation (4.58) and then using tables of Laplace transforms (cf. [15, p. 66, Entry 7.76]), we obtain

$$\left(\mathcal{L}_2\frac{1}{y^2}\sin(ay)\right)\left(\frac{1}{2x}\right) = \frac{1}{2}\left(\mathcal{L}\frac{1}{y}\sin(ay^{1/2})\right)\left(\frac{1}{4x^2}\right) = \frac{\pi}{2}\mathrm{Erf}(ax). \qquad (4.104)$$

Using tables of Laplace transforms (cf. [15, p. 54, Entry 7.5]), we obtain

$$\left(\mathcal{L}\frac{1}{y}\sin(ay)\right)(z) = \arctan\left(\frac{a}{z}\right). \qquad (4.105)$$

Now, the assertion Equation (4.102) follows from substituting Equation (4.104) and (4.105) into (4.75) of Corollary 4.3.7. Similarly, the assertion Equation (4.103) follows from substituting Equations (4.104) and (4.105) into (4.76) of Corollary 4.3.7.

4.4 SOLVING CLASSICAL DIFFERENTIAL EQUATIONS WITH THE $\mathscr{L}_2$-TRANSFORM

In this section, we present some properties of the $\mathscr{L}_2$-transform, and using these properties, we show how to solve the well-known Bessel's differential equation and Hermite's differential equation.

We introduce a differentiation operator δ that we call the δ-derivative and define as

$$\delta_t = \frac{1}{t}\frac{d}{dt} \tag{4.106}$$

We note that

$$\delta_t^2 = \delta_t \cdot \delta_t = \frac{1}{t}\frac{d}{dt}\left(\frac{1}{t}\frac{d}{dt}\right) = \frac{1}{t^2}\frac{d^2}{dt^2} - \frac{1}{t^3}\frac{d}{dt}. \tag{4.107}$$

The δ-derivative operator can successively applied in a similar fashion for any positive integer power. Here, we derive a relation between the $\mathscr{L}_2$-transform of the δ-derivative of a function and the $\mathscr{L}_2$-transform of the function itself.

Suppose that $f(t)$ is a continuous function with a piecewise continuous derivative $f'(t)$ on the interval $t \geq 0$. Also suppose that f and f' are of exponential order $\exp(c^2t^2)$ as $t \to \infty$, where c is a constant. Using the definitions Equation (4.54) of the $\mathscr{L}_2$-transform and Equation (4.106) of the δ-derivative, and then using integration by parts, we obtain

$$(\mathscr{L}_2\delta_t f)(s) = \int_0^\infty e^{-t^2 s^2} f'(t)\,dt$$

$$= e^{-t^2 s^2} f(t)\Big|_{t=0}^\infty + 2s^2 \int_0^\infty t e^{-t^2 s^2} f(t)\,dt. \tag{4.108}$$

Since f is of exponential order $\exp(c^2t^2)$ as $t \to \infty$, it follows that

$$\lim_{t\to\infty} e^{-t^2 s^2} f(t) = 0, \tag{4.109}$$

and consequently,

$$(\mathscr{L}_2\delta_t f)(s) = 2s^2(\mathscr{L}_2 f)(s) - f(0^+). \tag{4.110}$$

Similarly, if f and f' are continuous functions with a piecewise continuous derivative f'' on the interval $t \geq 0$, and if all three function are of exponential order $\exp(c^2t^2)$ as $t \to \infty$ for some constant c, we can use Equation (4.109)

$$(\mathscr{L}_2\delta_t^2 f)(s) = 2s^2(\mathscr{L}_2\delta_t f)(s) - (\delta_t f)(0^+). \tag{4.111}$$

Using Equations (4.110) and (4.111) we obtain

$$(\mathscr{L}_2\delta_t^2 f)(s) = 4s^4(\mathscr{L}_2 f)(s) - 2s^2 f(0^+) - (\delta_t f)(0^+). \tag{4.112}$$

By repeated application of Equations (4.110) and (4.112), we obtain the following theorem:

Theorem 4.4.1 *If $f, f', \ldots, f^{(n-1)}$ are all continuous functions with a piecewise continuous derivative $f^{(n)}$ on the interval $t \geq 0$, and if all three function are of exponential order $\exp(c^2 t^2)$ as $t \to \infty$ for some constant c, then*

$$\left(\mathscr{L}_2 \delta_t^n f\right)(s) = 2^n s^{2n} \left(\mathscr{L}_2 f\right)(s) - 2^{n-1} s^{2(n-1)} f(0^+)$$
$$- 2^{n-2} s^{2(n-2)} (\delta_t f)(0^+) - \cdots - \left(\delta_t^{n-1} f\right)(0^+). \tag{4.113}$$

for $n = 1, 2, \ldots$

When we are solving Bessel's differential equation and Hermite's differential equations, we will be required to evaluate the transform of a function that is expressed as $t^2 f(t)$. Here, we develop a general property to evaluate the transform of $t^{2n} f(t)$. If $f(t)$ is a piecewise continuous function on $x \geq 0$ and is of exponential order $\exp(c^2 x^2)$ as $x \to \infty$ for some constant c, then the $\mathscr{L}_2$-transform defined in Equation (4.54) is an analytic function in the half-plane $\Re(s) > c$. Therefore, $\left(\mathscr{L}_2 f\right)(y)$ has derivatives of all orders and the derivatives can be formally obtained by differentiating (4.54). Applying the δ-derivative with respect to the variable y, we find that

$$\delta_t \left(\mathscr{L}_2 f\right)(s) = \int_0^\infty t e^{-t^2 s^2} \left[-2t^2 f(t) \right] dt. \tag{4.114}$$

Using the definition (4.54) of the $\mathscr{L}_2$-transform, we obtain

$$\left(\mathscr{L}_2 t^2 f\right)(s) = -\frac{1}{2} \delta_t \left(\mathscr{L}_2 f\right)(s). \tag{4.115}$$

If we keep taking the δ-derivative of Equation (4.54) with respect y, then we deduce that

$$\delta_t^n \left(\mathscr{L}_2 f\right)(s) = \int_0^\infty t e^{-t^2 s^2} \left[(-2t^2)^n f(t) \right] dt. \tag{4.116}$$

for $n = 1, 2, 3, \ldots$ As a result, we obtain the following theorem:

Theorem 4.4.2 *If f is a piecewise continuous on $x \geq 0$ and is of exponential order $\exp(c^2 t^2)$ as $t \to \infty$ for some constant c, then*

$$\left(\mathscr{L}_2 t^{2n} f\right)(s) = \frac{(-1)^n}{2^n} \delta_s^n \left(\mathscr{L}_2 f\right)(s) \tag{4.117}$$

for $n = 1, 2, \ldots$

4.4.1 A TECHNIQUE FOR SOLVING BESSEL'S DIFFERENTIAL EQUATION USING THE $\mathscr{L}_2$-TRANSFORM

In this section, we introduce a new technique for solving Bessel's differential equation:

$$t^2 y''(t) + t y'(t) + (t^2 - v^2) y(t) = 0. \tag{4.118}$$

The solution of Bessel's equation is

$$J_\nu(t) = \left(\frac{t}{2}\right)^\nu \sum_{n=0}^{\infty} \frac{(-1)^n}{n!\Gamma(n+\nu+1)} \left(\frac{t}{2}\right)^{2n} \tag{4.119}$$

where $\Re(\nu) \geq 0$, and J_ν is called the Bessel function of the first kind of order ν. Changing the dependent variable of Bessel's Equation (4.118) from y to z where $y(t) = t^{-\nu}z(t)$, we find that Bessel's equation becomes

$$tz''(t) - (2\nu - 1)z'(t) + tz(t) = 0. \tag{4.120}$$

Dividing both sides of Equation (4.120) by t, and adding and subtracting the term z'/t, we obtain

$$\left[z''(t) - \frac{1}{t}z'(t)\right] - \frac{2(\nu-1)}{t}z'(t) + z(t) = 0. \tag{4.121}$$

Using the definition of the δ-derivative Equations (4.106) and (4.107), we can express Equation (4.121) as

$$t^2\delta_t^2 z(t) - 2(\nu-1)\delta_t z(t) + z(t) = 0. \tag{4.122}$$

Applying the $\mathscr{L}_2$-transform to Equation (4.122), we find

$$\left(\mathscr{L}_2 t^2 \delta_t^2 z(t)\right)(s) - 2(\nu-1)\left(\mathscr{L}_2 \delta_t z(t)\right)(s) + \left(\mathscr{L}_2 z(t)\right)(s) = 0. \tag{4.123}$$

Using Equation (4.117) of Theorem 4.4.2 for $n=1$ in Equation (4.123), we obtain

$$-\frac{1}{2}\delta_s\left(\mathscr{L}_2 \delta_t^2 z(t)\right)(s) - 2(\nu-1)\left(\mathscr{L}_2 \delta_t z(t)\right)(s) + \left(\mathscr{L}_2 z(t)\right)(s) = 0. \tag{4.124}$$

Using Equation (4.113) of Theorem 4.4.1 for $n=1$ and $n=2$ in Equation (4.124) and performing the calculations, we obtain the first-order differential equation

$$2s^3 Z'(s) + \left[4s^2(\nu+1) - 1\right]Z(s) = 2\nu z(0^+), \tag{4.125}$$

where $Z(s) = \left(\mathscr{L}_2 z(t)\right)(s)$. We assume that $z(0^+) = 0$; that is,

$$z(0^+) = \begin{cases} \text{arbitrary} & \text{for } \nu = 0, \\ 0 & \text{for } \nu \neq 0. \end{cases} \tag{4.126}$$

Solving the first-order differential equation after substituting Equation (4.126) into, we have

$$Z(s) = Cs^{-2(\nu+1)}\exp\left(-\frac{1}{4s^2}\right). \tag{4.127}$$

Calculating the Taylor expansion of the exponential function in Equation (4.127), we have

$$Z(s) = \left(\mathscr{L}_2 z(t)\right)(s) = C\sum_{n=0}^{\infty} \frac{(-1)^n}{n!2^{2n}s^{2(n+\nu+1)}} \tag{4.128}$$

Using the value

$$\left(\mathscr{L}_2 t^{2(n+\nu)}\right)(s) = \frac{\Gamma(n+\nu+1)}{2s^{2(n+\nu+1)}} \tag{4.129}$$

in Equation (4.128), we have

$$z(t) = C \sum_{n=0}^{\infty} \frac{(-1)^n t^{2(n+\nu)}}{n! 2^{2n-1} \Gamma(n+\nu+1)} \tag{4.130}$$

Setting $C = 2^{-\nu-1}$ in Equation (4.131) we get

$$z(t) = t^\nu \sum_{n=0}^{\infty} \frac{(-1)^n}{n! \Gamma(n+\nu+1)} \left(\frac{t}{2}\right)^{2n+\nu}. \tag{4.131}$$

Substituting $y(t) = t^{-\nu} z(t)$ into the Equation (4.131), we obtain the Bessel function (4.119) of the first kind of order ν as solution of Bessel's Equation (4.118).

4.4.2 A TECHNIQUE FOR SOLVING HERMITE'S DIFFERENTIAL EQUATION USING THE $\mathscr{L}_2$-TRANSFORM

In this section, we introduce a new technique for solving Hermite's differential equation:

$$y''(t) - 2ty'(t) + 2ny(t) = 0 \tag{4.132}$$

for $n = 0, 1, 2, \ldots$ The solutions of Hermite's equation are

$$H_{2k}(t) = \sum_{i=0}^{\infty} \frac{(-1)^i (2k)!}{i!(2k-2i)!} (2t)^{2k-2i}, \tag{4.133}$$

and

$$H_{2k+1}(t) = \sum_{i=0}^{\infty} \frac{(-1)^i (2k+1)!}{i!(2k-2i+1)!} (2t)^{2k-2i+1}, \tag{4.134}$$

for $k = 0, 1, 2, \ldots$ and H_n is called Hermite's polynomial of order n (cf. [14, pp. 60–66]). We have the following initial cases:

$$H_{2k}(0) = (-1)^k \frac{(2k)!}{k!} \qquad H_{2k+1}(0) = 0, \tag{4.135}$$

for $k = 0, 1, 2, \ldots$

Using the definition of the δ-derivative Equations (4.106) and (4.107), we can express Hermite's differential equation (4.132) as

$$t^2 \delta_t^2 y(t) - 2t^2 \delta_t y(t) + \delta_t y(t) + 2ny(t) = 0. \tag{4.136}$$

Applying the $\mathscr{L}_2$-transform to Equation (4.136), we find that

$$\left(\mathscr{L}_2 t^2 \delta_t^2 y(t)\right)(s) - 2\left(\mathscr{L}_2 \delta_t t^2 y(t)\right)(s)$$
$$+ \left(\mathscr{L}_2 \delta_t y(t)\right)(s) + 2n\left(\mathscr{L}_2 y(t)\right)(s) = 0. \tag{4.137}$$

Using Equation (4.117) of Theorem 4.4.2 for $n = 1$ in Equation (4.137), we obtain

$$-\frac{1}{2}\delta_s\left(\mathscr{L}_2\delta_t^2 y(t)\right)(s) + \delta_s\left(\mathscr{L}_2\delta_t t^2 y(t)\right)(s)$$
$$+ \left(\mathscr{L}_2\delta_t y(t)\right)(s) + 2n\left(\mathscr{L}_2 y(t)\right)(s) = 0. \tag{4.138}$$

Using Equation (4.113) of Theorem 4.4.1 for $n = 1$ and $n = 2$ in Equation (4.138) and performing the calculations, we obtain the first-order differential equation:

$$Y'(s) + \frac{3s^2 - n - 2}{s^3 - s} Y(s) = \frac{Y(0)}{2(s^3 - s)} \tag{4.139}$$

where $Y(s) = \left(\mathscr{L}_2 y(t)\right)(s)$. Solving the first-order differential Equation (4.139), we have

$$Y_n(s) = Y(0)\frac{(s^2 - 1)^{(n-1)/2}}{2s^{n+2}} \int \frac{s^{n+1}}{(s^2 - 1)^{(n+1)/2}} ds + \frac{C(s^2 - 1)^{(n-1)/2}}{s^{n+2}} \tag{4.140}$$

where $n = 0, 1, 2, \ldots$, and C is an arbitrary constant.

We consider two cases where n is an odd positive integer or an even positive integer. Let $n = 2k + 1$, $k = 0, 1, 2, \ldots$ Using the initial cases Equation (4.135), we may assume $Y_{2k+1}(0) = 0$. In this case, Equation (4.140) becomes

$$Y_{2k+1}(s) = \left(\mathscr{L}_2 y(t)\right)(s) = C_k\frac{(-1)^k(1 - s^2)^k}{s^{2k+3}}. \tag{4.141}$$

Using the binomial formula, we obtain

$$Y_{2k+1}(s) = \left(\mathscr{L}_2 y(t)\right)(s) = C_k(-1)^k \sum_{i=0}^{k} \binom{k}{i}(-1)^i s^{2i-2k-3}. \tag{4.142}$$

Applying the inverse $\mathscr{L}_2$-transform, we find

$$y_{2k+1}(t) = C_k(-1)^k \sum_{i=0}^{k} \binom{k}{i}(-1)^i \left(\mathscr{L}_2^{-1} s^{2i-2k-3}\right)(t). \tag{4.143}$$

The inverse $\mathscr{L}_2$-transform on the right-hand side of Equation (4.143) can be evaluated by using the definition Equation (4.54) of the $\mathscr{L}_2$-transform

$$y_{2k+1}(t) = C_k(-1)^k \sum_{i=0}^{k} \binom{k}{i}(-1)^i \frac{2t^{2k-2i+1}}{\Gamma\left(k - i + \frac{3}{2}\right)}. \tag{4.144}$$

Using the duplication formula for the gamma function (cf. [14, p.3])

$$2^{2z-1}\Gamma(z)\Gamma\left(z + \frac{1}{2}\right) = \sqrt{\pi}\Gamma(2z) \tag{4.145}$$

with $z = k - i + 1$, the result (4.144) becomes

$$y_{2k+1}(t) = C_k \frac{(-1)^k k!}{\sqrt{\pi}} \sum_{i=0}^{k} \frac{(-1)^i (2t)^{2k-2i+1}}{i!(2k-2i+1)}.$$ (4.146)

If we set the arbitrary constant C_k as

$$C_k = \frac{(-1)^k (2k+1)!}{2k!}$$ (4.147)

in the formula (4.146), we obtain Hermite's polynomial Equation (4.134).

Let $n = 2k$, $k = 0, 1, 2, \ldots$ Using the initial cases Equation (4.135), we may assume

$$Y_{2k}(0) = \frac{(-1)^k (2k)!}{k!}.$$ (4.148)

Now, if we set $C = 0$, then the solution (4.140) becomes

$$Y_{2k}(s) = (-1)^k \frac{(2k)!}{2k!} \frac{(s^2-1)^{(2k-1)/2}}{s^{2k+2}} \int \frac{s^{2k+1}}{(s^2-1)^{(2k+1)/2}} \, ds$$ (4.149)

The integral on the right-hand site of Equation (4.149) can be evaluated by making the substitution $u = s^2 - 1$ and then using the binomial formula:

$$\int \frac{s^{2k+1}}{(s^2-1)^{(2k+1)/2}} \, ds = \frac{1}{2} \int u^{-1/2} \left(\frac{1}{u} + 1 \right)^k \, du$$

$$= \frac{1}{2} \int \sum_{i=0}^{k} \binom{k}{i} u^{(2i-2k+1)/2} \, du$$ (4.150)

$$= \sum_{i=0}^{k} \binom{k}{i} \frac{(s^2-1)^{(2i-2k+1)/2}}{2i-2k+1}$$ (4.151)

Substituting the result (4.151) into (4.149) and using the binomial theorem, we find

$$Y_{2k}(s) = (-1)^k \frac{(2k)!}{k!} \left(\frac{1}{s^2} + \sum_{i=1}^{k} (-1)^i \frac{2^i k(k-1)\cdots(k-i+1)}{1 \times 3 \times \cdots \times (2i-1)} s^{-2i-2} \right)$$ (4.152)

Applying the inverse $\mathcal{L}_2$-transform to Equation (4.151) and using the definition (4.54) of the $\mathcal{L}_2$-transform, we find

$$y_{2k}(t) = (-1)^k \frac{(2k)!}{k!} \left(1 + \sum_{i=1}^{k} (-1)^i \frac{2^i k(k-1)\cdots(k-i+1)}{i![1 \times 3 \times \cdots \times (2i-1)]} t^{2i} \right)$$ (4.153)

It is clear that the formula (4.153) is the same as Equation (4.133).

REFERENCES

1. R. P. Agarwal. Some properties of generalised Hankel transform. *Bull. Calcutta Math. Soc.*, 43:153–167, 1951.
2. A. Apelblat. Repeating use of integral transforms: A new method for evaluation of some infinite integrals. *IMA J. Appl. Math.*, 27(4):481–496, 1981.
3. A. Apelblat. Table of definite and infinite integrals, volume 13. *Physical Sciences Data*. Elsevier Scientific Publishing Co., Amsterdam, Netherlands, 1983.
4. A. Dernek, N. Dernek, and O. Yürekli. Identities for the Glasser transform and their applications. *Contemp. Anal. Appl. Math.*, 2(1):146–160, 2014.
5. A. Dernek, N. Dernek, and O. Yürekli. Identities for the Hankel transform and their applications. *J. Math. Anal. Appl.*, 354(1):165–176, 2009.
6. N. Dernek and F. Aylı kçı. Some results on the $\mathscr{P}_{v,2n}$, $\mathscr{K}_{v,n}$, and $\mathscr{H}_{v,n}$-integral transforms. *Turkish J. Math.*, 41(2):337–349, 2017.
7. N. Dernek, A. Dernek, and O. Yürekli. A generalization of the Krätzel function and its applications. *J. Math.*, 1–7, 2017.
8. N. Dernek, E. Ö. Ölçüçü, and F. Aylı kçı. New identities and Parseval type relations for the generalized integral transforms $\mathscr{L}_{4n}$, $\mathscr{P}_{4n}$, $\mathscr{F}_{s,2n}$ and $\mathscr{F}_{c,2n}$. *Appl. Math. Comput.*, 269:536–547, 2015.
9. A. Erdélyi, W. Magnus, F. Oberhettinger, and F. G. Tricomi. *Tables of Integral Transforms*, Vol. 1. McGraw-Hill Book Company, Inc., New York-Toronto-London, 1954.
10. A. Erdélyi, W. Magnus, F. Oberhettinger, and F. G. Tricomi. *Tables of Integral Transforms*, Vol. 2. McGraw-Hill Book Company, Inc., New York-Toronto-London, 1954.
11. M. L. Glasser. Some Bessel function integrals. *Kyungpook Math. J.*, 13:171–174, 1973.
12. S. Goldstein. Operational representations of Whittaker's Confluent Hypergeometric Function and Weber's Parabolic Cylinder Function. *Proc. London Math. Soc.*, 34(2):103–125, 1932.
13. G. H. Hardy. *Ramanujan: Twelve Lectures on Subjects Suggested by his Life and Work*. Cambridge University Press and Macmillan Company, Cambridge, England and New York, 1940.
14. N. N. Lebedev. *Special Functions and their Applications*. Dover Publication, Inc., New York, 1972.
15. F. Oberhettinger and K. Badii. *Tables of Integral Transforms*. Springer-Verlag, New York, 1973.
16. B. van der Pol and H. Bremmer. *Operational Calculus, Based on the Two-Sided Laplace Integral*. Cambridge University Press, Cambridge, 1950.
17. I. N. Sneddon. *The Use of Integral Transforms*. McGraw-Hill Book Company, Inc., New York-Toronto-London, 1972.
18. J. Spanier and K. B. Oldham. *An Atlas of Functions*. Hemisphere Publishing Corporation, Washington, DC, 1987.
19. H. M. Srivastava. Some theorems on Hardy transform. *Nederl. Akad. Wetensch. Proc. Ser. A 71 = Indag. Math.*, 30:316–320, 1968.
20. H. M. Srivastava and S. P. Singh. A note on the Widder transform related to the Poisson integral for a half-plane. *Int. J. Math. Ed. Sci. Tech.*, 16(6):675–677, 1985.
21. H. M. Srivastava and O. Yürekli. A theorem on a Stieltjes-type integral transform and its applications. *Complex Var. Theory Appl.*, 28(2):159–168, 1995.
22. D. V. Widder. A transform related to the Poisson integral for a half-plane. *Duke Math. J.*, 33:355–362, 1966.
23. D. V. Widder. *An Introduction to Transform Theory*. Academic Press, New York, 1971.

24. D. V. Widder. What is the Laplace transform? *Am. Math. Monthly*, 52:419–425, 1945.
25. O. Yürekli. A parseval-type theorem applied to certain integral transforms. *IMA J. Appl. Math.*, 42(3):241–249, 1989.
26. O. Yürekli. A theorem on the generalized Stieltjes transform and its applications. *J. Math. Anal. Appl.*, 168(1):63–71, 1992.
27. O. Yürekli. Identities, inequalities, Parseval type relations for integral transforms and fractional integrals. Ph.D. Thesis, University of California, Santa Barbara, CA. ProQuest LLC, Ann Arbor, MI, 1988.
28. O. Yürekli. New identities involving the laplace and the $\mathscr{L}_2$-transforms and their applications. *Appl. Math. Comput.*, 99(2–3):141–151, 1999.
29. O. Yürekli. Theorems on $\mathscr{L}_2$-transform and its applications. *Complex Var. Theory Appl.*, 38(2):95–107, 1999.

5 Numerical Solution of Cauchy and Hypersingular Integral Equations

Vaishali Sharma
BITS Pilani - K K Birla Goa Campus

CONTENTS

5.1 INTRODUCTION SINGULAR INTEGRAL EQUATIONS WITH CAUCHY KERNEL

Singular integral equations (SIEs) with Cauchy kernel occur frequently in mixed boundary value problems for partial differential equations. Many fluid dynamics problems such as the stationary linear problem of ideal fluid flow is reducible to Cauchy singular integral equation (CSIE) (see [5] and the references therein). An interesting and comprehensive survey of applications of SIEs can be found in [14,42]. It is worth noting that the methods of exact and approximate solutions of SIEs have been still a challenging problem for the research community. In this chapter, we find the approximate solution of Cauchy singular integral equations (CSIEs)

$$\int_{-1}^{1} \frac{\chi^{[c]}(t)}{t-x} dt - \int_{-1}^{1} k^{[c]}(x,t)\chi^{[c]}(t)dt = g^{[c]}(x), |x| < 1, \tag{5.1}$$

where $\chi^{[c]}(t)$ is an unknown function which vanishes at $t = 1$ and becomes unbounded at $t = -1$. The functions $g^{[c]}(x)$ and $k^{[c]}(x,t)$ are known real-valued Hölder continuous over the interval $[-1,1]$ and $[-1,1] \times [-1,1]$, respectively. The first integral in Equation (5.1) is understood to be exist in the sense of CPV. Also, the function $\chi^{[c]}(t)$ is assumed to be a Hölder continuous in order to ensure the existence [9] of Cauchy principal value. CSIE equations have various applications in the field of aerodynamics [28], fracture mechanics [16], neutron transport [35], etc. They are also used in various areas of mathematical physics such as potential theory [10], elasticity problems as well as electromagnetic scattering [43]. The numerical methods which are developed for one-dimensional Cauchy-type singular integral equations include: Galerkin's method [4,38], collocation method [32], quadrature method [23,41], inverse method [24], Sinc approximations [1] etc. In this chapter, we propose a residual-based Galerkin's method with Legendre polynomial as basis function in order to find the numerical solution of Equation (5.1).

5.2 METHOD OF SOLUTION FOR CSIES OVER [−1,1]

To find the approximate solution of CSIE (5.1), we write the unknown function [18] as

$$\chi^{[c]}(t) = \sqrt{\frac{1-t}{1+t}} \xi^{[c]}(t), \tag{5.2}$$

where $\xi^{[c]}(t)$ is an unknown function of $t \in [-1,1]$. Now using Equation (5.2) in Equation (5.1), we obtain

$$\int_{-1}^{1} \sqrt{\frac{1-t}{1+t}} \frac{\xi^{[c]}(t)}{t-x} dt - \int_{-1}^{1} \sqrt{\frac{1-t}{1+t}} k^{[c]}(x,t)\xi^{[c]}(t)dt = g^{[c]}(x), |x| < 1. \tag{5.3}$$

We approximate the function $\xi^{[c]}(t)$ by orthonormalized Legendre polynomials as follows:

$$\xi^{[c]}(t) \approx \xi_n^{*[c]}(t) = \sum_{j=0}^{n} a_j^{[c]} e_j(t), \tag{5.4}$$

where $a_j^{[c]}; j = 0,1,2,\ldots,n$ are unknown constant coefficients and $\{e_j(t)\}_{j=0}^{n}$ is the set of $(n+1)$ orthonormalized Legendre polynomials on $[-1,1]$. To get the values of unknown coefficients $a_j^{[c]}$, we use residual-based Galerkin's method. On using the above approximation for $\xi^{[c]}(t)$ in Equation (5.3), the residual error $\mathcal{R}^{[c]}(x, a_0^{[c]}, a_1^{[c]}, a_2^{[c]}, \ldots, a_n^{[c]})$ will be

$$\mathcal{R}^{[c]}(x, a_0^{[c]}, a_1^{[c]}, a_2^{[c]}, \ldots, a_n^{[c]}) = \int_{-1}^{1} \sqrt{\frac{1-t}{1+t}} \frac{\xi_n^{*[c]}(t)}{t-x} dt$$

$$- \int_{-1}^{1} \sqrt{\frac{1-t}{1+t}} k^{[c]}(x,t) \xi_n^{*[c]}(t) dt - g^{[c]}(x), |x| < 1. \qquad (5.5)$$

In Galerkin's method, this residual error $\mathcal{R}^{[c]}(x, a_0^{[c]}, a_1^{[c]}, a_2^{[c]}, \ldots, a_n^{[c]})$ is assumed to be orthogonal to the space spanned by orthonormal polynomials $\{e_j(x)\}_{j=0}^{n}$, that is, we have

$$\langle \mathcal{R}^{[c]}(x, a_0^{[c]}, a_1^{[c]}, a_2^{[c]}, \ldots, a_n^{[c]}), e_j \rangle_{L^2} = 0, \ \forall j = 0, 1, 2, \cdots, n. \qquad (5.6)$$

The explicit form of Equation (5.6) is as follows:

$$\sum_{r=0}^{n} a_r^{[c]} \left(\int_{-1}^{1} \int_{-1}^{1} \sqrt{\frac{1-t}{1+t}} \frac{1}{t-x} e_r(t) e_q(x) dt dx \right.$$

$$\left. - \int_{-1}^{1} \int_{-1}^{1} \sqrt{\frac{1-t}{1+t}} k^{[c]}(x,t) e_r(t) e_q(x) dt dx \right)$$

$$= \int_{-1}^{1} g^{[c]}(x) e_q(x) dx, \ q = 0, 1, 2, \ldots, n, \qquad (5.7)$$

where

$$b_{rq}^{[c]} = \int_{-1}^{1} \int_{-1}^{1} \sqrt{\frac{1-t}{1+t}} \frac{1}{x-t} e_r(x) e_q(t) dt dx - \int_{-1}^{1} \int_{-1}^{1} \sqrt{\frac{1-t}{1+t}} k^{[c]}(x,t) e_r(t) e_q(x) dt dx,$$

$$r, q = 0, 1, 2, \ldots, n,$$

$$g_q^{[c]} = \int_{-1}^{1} g^{[c]}(x) e_q(x) dx, \quad q = 0, 1, 2, \ldots, n.$$

Finally, the system (5.7) can be written in matrix form as

$$B^{[c]^T} A^{[c]} = B_1^{[c]} A^{[c]} = G^{[c]}, \qquad (5.8)$$

where

$$B_1^{[c]} = B^{[c]^T}, \quad B^{[c]} = \begin{pmatrix} b_{00}^{[c]} & b_{01}^{[c]} & \cdots & b_{0n}^{[c]} \\ b_{10}^{[c]} & b_{11}^{[c]} & \cdots & b_{1n}^{[c]} \\ \vdots & \vdots & \ddots & \vdots \\ b_{n0}^{[c]} & b_{n1}^{[c]} & \cdots & b_{nn}^{[c]} \end{pmatrix}, \quad A^{[c]} = \begin{pmatrix} a_0^{[c]} \\ a_1^{[c]} \\ \vdots \\ a_n^{[c]} \end{pmatrix}, \quad G^{[c]} = \begin{pmatrix} g_0^{[c]} \\ g_1^{[c]} \\ \vdots \\ g_n^{[c]} \end{pmatrix}.$$

$$(5.9)$$

After solving the system (5.8) which is obtained as a result of approximation of Equation (5.3), we get the values of a_j; $j = 0, 1, 2, \ldots, n$.

5.3   ERROR ANALYSIS

We show the convergence of sequence of approximate solutions, and we also derive the error bound, in this section. We write the Equation (5.3) in operator form

$$(S^{[c]} - K^{[c]})\xi^{[c]}(x) = g^{[c]}(x), \quad |x| < 1. \tag{5.10}$$

In the above Equation (5.10), the operators $S^{[c]}$ and $K^{[c]}$ are defined as

$$S^{[c]}\xi^{[c]}(x) = \int_{-1}^{1} \sqrt{\frac{1-t}{1+t}} \frac{\xi^{[c]}(t)}{t-x} dt, \tag{5.11}$$

$$K^{[c]}\xi^{[c]}(x) = \int_{-1}^{1} \sqrt{\frac{1-t}{1+t}} k^{[c]}(x,t)\xi^{[c]}(t) dt. \tag{5.12}$$

We assume that

$$\int_{-1}^{1}\int_{-1}^{1}\left(\sqrt{\frac{1-t}{1+t}} k^{[c]}(x,t)\right)^2 dt dx < \infty. \tag{5.13}$$

Now, we define the Hilbert space $L^2[-1,1]$ as

$$L^2[-1,1] = \left\{u : [-1,1] \to \mathbb{R} : \int_{-1}^{1}(u(t))^2 dt < \infty\right\}. \tag{5.14}$$

with the following norm $\|.\|_{L^2}^2$ and inner product $\langle .,.\rangle_{L^2}$

$$\|u\|_{L^2} = \left(\int_{-1}^{1}\left(u(t)\right)^2 dt\right)^{1/2}, \text{ for } u(t) \in L^2, \tag{5.15}$$

$$\langle u,v\rangle_{L^2} = \int_{-1}^{1} u(t)v(t) dt, \text{ for } u(t),v(t) \in L^2. \tag{5.16}$$

We define another function space, say $M^{[c]}$ such that

$$M^{[c]} = \{u(t) \in L^2 : \sum_{i=0}^{\infty}(d_i^{[c]})^2\langle u,e_i\rangle_{L^2}^2 < \infty\}. \tag{5.17}$$

where $d_i^{[c]} = \sqrt{\sum_{j=0}^{\infty}\langle S^{[c]}e_i,e_j\rangle_{L^2}^2}$. Following results [6], the function $S^{[c]}e_i(x)$ is a polynomial of degree at the most of i; hence, $d_i^{[c]}$ $\forall i$ will be a finite number. $M^{[c]}$ is a subspace of L^2 space which is actually a Hilbert space with respect to the following norm $\|\cdot\|_{M^{[c]}}$ and inner product $\langle \cdot,\cdot\rangle_{M^{[c]}}$

$$\| u \|_{M^{[c]}}^2 = \sum_{i=0}^{\infty}\left(d_i^{[c]}\right)^2\langle u,e_i\rangle_{L^2}^2, \text{ for } u(t) \in M^{[c]}, \tag{5.18}$$

$$\langle u,v\rangle_{M^{[c]}} = \sum_{i=0}^{\infty}\left(d_i^{[c]}\right)^2\langle u,e_i\rangle_{L^2}\langle v,e_i\rangle_{L^2}, \text{ for } u(t),\ v(t) \in M^{[c]}. \tag{5.19}$$

Now with the aid of results in [6], we obtain

$$S^{[c]} e_n(x) = \sum_{i=0}^{n} \alpha_i^{[c]} e_i(x), \qquad (5.20)$$

where the coefficients $\alpha_i^{[c]} = \langle S^{[c]} e_n, e_i \rangle_{L^2}$, $i = 0, 1, 2, \ldots, n$. Using the above result in Equation (5.20), the operator $S^{[c]} : M^{[c]} \to L^2$, can be extended as a bounded linear operator and defined as

$$S^{[c]} \xi^{[c]}(x) = \sum_{i=0}^{\infty} \langle \xi^{[c]}, e_i \rangle_{L^2} \sum_{j=0}^{i} \langle Se_i^{[c]}, e_j \rangle_{L^2} e_j(x) \in L^2[-1, 1]. \qquad (5.21)$$

Using the orthogonal property of Legendre polynomial, we find the norm of operator $S^{[c]}$

$$\| S^{[c]} \xi^{[c]} \|_{L^2}^2 = \sum_{i=0}^{\infty} (d_i^{[c]})^2 \langle \xi^{[c]}, e_i \rangle_{L^2}^2 = \| \xi \|_{M^{[c]}}^2 . \qquad (5.22)$$

Therefore, using Equation (5.22), we obtain

$$\| S^{[c]} \| = 1. \qquad (5.23)$$

Also, the operator $S^{[c]}$ from $M^{[c]} \to L^2$ is one-one and onto [18]. Hence, the operator $(S^{[c]})^{-1} : L^2 \to M^{[c]}$ exists as a bounded linear operator by using *Bounded Inverse Theorem* [27]. This operator $(S^{[c]})^{-1}$ is defined as

$$(S^{[c]})^{-1} \xi^{[c]}(x) = \sum_{i=0}^{\infty} \frac{\langle \xi^{[c]}(x), e_i(x) \rangle_{L^2}}{d_i^{[c]}} e_i(x). \qquad (5.24)$$

The Equation (5.10) will have a unique solution if and only if the operator $(S^{[c]} - K^{[c]})^{-1}$ is bounded. We assume that this condition exists. We define the mapping $Q_n^{[c]} : L^2 \to L^2$ as

$$Q_n^{[c]} \xi^{[c]}(x) = \sum_{i=0}^{n} \langle \xi^{[c]}, e_i \rangle_{L^2} e_i(x), \qquad (5.25)$$

where $Q_n^{[c]}$ is the operator of orthogonal projection. With the aid of Equation (5.6), we obtain

$$Q_n^{[c]} \left((S^{[c]} - K^{[c]}) \xi_n^{*[c]}(x) - g^{[c]}(x) \right) = 0. \qquad (5.26)$$

Since the function $\xi^{*[c]}(x)$ defined in Equation (5.4), is a polynomial. Therefore, with the help of the formulas given in [6], the function $S^{[c]} \xi_n^{*[c]}(x)$ will be a polynomial; therefore, we obtain

$$Q_n^{[c]} S^{[c]} \xi_n^{*[c]}(x) = S^{[c]} \xi_n^{*[c]}(x), \qquad (5.27)$$

and hence, the Equation (5.26) becomes

$$S^{[c]} \xi_n^{*[c]}(x) - Q_n^{[c]} K^{[c]} \xi_n^{*[c]}(x) = Q_n^{[c]} g^{[c]}(x). \qquad (5.28)$$

Due to the boundedness of $(S^{[c]})^{-1}$ and the compactness Equation (5.13) of $K^{[c]}$ for all $n \geq n_0$, the operator $(S^{[c]} - Q_n^{[c]} K^{[c]})^{-1}$ exists as a bounded linear operator [20]. Hence, the Equation (5.28) has a unique solution which is defined as

$$\xi_n^{*[c]}(x) = (S^{[c]} - Q_n^{[c]} K^{[c]})^{-1} Q_n^{[c]} g^{[c]}(x). \tag{5.29}$$

From Equations (5.10) and (5.29), for all $n \geq n_0$, we have

$$\xi^{[c]}(x) - \xi_n^{*[c]}(x) = (S^{[c]} - Q_n^{[c]} K^{[c]})^{-1}$$
$$\left(g^{[c]}(x) - Q_n^{[c]} g^{[c]}(x) + K^{[c]} \xi^{[c]}(x) - Q_n^{[c]} K^{[c]} \xi^{[c]}(x) \right). \tag{5.30}$$

Now taking $M^{[c]}$ norm on both the sides of Equation (5.30), we obtain

$$\| \xi^{[c]} - \xi_n^{*[c]} \|_{M^{[c]}} \leq \| (S^{[c]} - Q_n^{[c]} K^{[c]})^{-1} \| \| g^{[c]} - Q_n^{[c]} g^{[c]} \|_{L^2}$$
$$+ \| (S^{[c]} - Q_n^{[c]} K^{[c]})^{-1} \| \| K^{[c]} \xi^{[c]}(x) - Q_n^{[c]} K^{[c]} \xi^{[c]}(x) \|_{L^2}. \tag{5.31}$$

Due the compactness of operator $K^{[c]}$, we have $\| K^{[c]} - Q_n^{[c]} K^{[c]} \|_{L^2} \to 0$ as $n \to \infty$ [20]. Also, $\| g^{[c]} - Q_n^{[c]} g^{[c]} \|_{L^2} \to 0$ as $n \to \infty$. Therefore, $\| \xi^{[c]} - \xi_n^{*[c]} \|_M^{[c]} \to 0$ as $n \to \infty$. Further, it is noticed that if $\xi^{[c]} \in M^{[c]}$, then we have

$$\| \xi^{[c]} \|_{L^2} \leq \| \xi^{[c]} \|_{M^{[c]}}. \tag{5.32}$$

Using Equation (5.32) in (5.31), we finally obtain

$$\| \xi^{[c]} - \xi_n^{*[c]} \|_{L^2} \leq \| (S^{[c]} - Q_n^{[c]} K^{[c]})^{-1} \| \| g^{[c]} - Q_n^{[c]} g^{[c]} \|_{L^2}$$
$$+ \| (S^{[c]} - Q_n^{[c]} K^{[c]})^{-1} \| \| K^{[c]} \xi^{[c]}(x) - Q_n^{[c]} K^{[c]} \xi^{[c]}(x) \|_{L^2}. \tag{5.33}$$

5.3.1 WELL POSEDNESS

In this subsection, we verify the Hadamard well posedness of problem (5.28). The existence the operator $(S^{[c]} - Q_n^{[c]} K^{[c]})^{-1}$, which is already shown above, implies that the problem (5.28) has a solution. We now show the uniqueness of the solution to the problem (5.28) with the aid of principle of contradiction.

Let us assume that the system (5.28) has two distinct solutions, say y_1 and y_2. Then, we have

$$S^{[c]} y_1(x) - Q_n^{[c]} K^{[c]} y_1(x) = Q_n^{[c]} g^{[c]}(x), \tag{5.34}$$

and

$$S^{[c]} y_2(x) - Q_n^{[c]} K^{[c]} y_2(x) = Q_n^{[c]} g^{[c]}(x). \tag{5.35}$$

Taking the difference of Equations (5.34) and (5.35), we obtain

$$(S^{[c]} - Q_n^{[c]} K^{[c]})(y_1(x) - y_2(x)) = 0. \tag{5.36}$$

In Section 5.3, the existence of bounded linear inverse operator $(S^{[c]} - Q_n^{[c]} K^{[c]})^{-1}$ is already shown. Therefore on applying the inverse operator $(S^{[c]} - Q_n^{[c]} K^{[c]})^{-1}$ on both the sides of Equation (5.36), we get

$$y_1(x) = y_2(x), \ |x| < 1. \tag{5.37}$$

Equation (5.37) contradicts our assumption. Hence, the problem (5.28) has a unique solution. Also, the boundedness of the inverse operator $(S^{[c]} - Q_n^{[c]} K^{[c]})^{-1}$ implies the continuity of $(S^{[c]} - Q_n^{[c]} K^{[c]})^{-1}$. This means that solution is stable. Since the problem (5.28) satisfies all the conditions of well-posedness therefore it is a well-posed problem.

5.3.2 EXISTENCE AND UNIQUENESS

In this subsection, we show the existence and uniqueness of solution for linear system (5.8). We start the proof by defining the prolongation operator [20] $P_n^{[c]} : \mathbb{R}^{n+1} \to E$ as follows:

$$P_n^{[c]} G^{[c]} = \sum_{j=0}^{n} \langle g^{[c]}, e_j \rangle_{L^2} e_j(x) \in E, \tag{5.38}$$

where $\mathbb{R}^{n+1}$ is a real vector space [26] having $(n+1)$-tuples of real numbers as its vectors, $E = span\{e_j(x)\}_{j=0}^{n}$ and $G^{[c]}$ is same as defined in Equation (5.9). Now using the definition of orthogonal projection Q_n, we obtain

$$Q_n^{[c]} g^{[c]}(x) = \sum_{j=0}^{n} \langle g^{[c]}, e_j \rangle_{L^2} e_j(x). \tag{5.39}$$

From Equations (5.38) and (5.39), we have

$$P_n^{[c]} G^{[c]} = Q_n^{[c]} g^{[c]}(x), \ g^{[c]}(x) \in L^2, \ G^{[c]} \in \mathbb{R}^{n+1}, \ |x| < 1. \tag{5.40}$$

We further define a restriction operator [20] $R_n^{[c]} : E \to \mathbb{R}^{n+1}$ as follows:

$$R_n^{[c]} \xi_n^{*[c]}(x) = (\langle \xi_n^{*[c]}, e_0 \rangle_{L^2}, \langle \xi_n^{*[c]}, e_1 \rangle_{L^2}, \dots \langle \xi_n^{*[c]}, e_n \rangle_{L^2})^T \in \mathbb{R}^{n+1}. \tag{5.41}$$

By orthogonal property of Legendre polynomials in Equation (5.4), we get

$$a_j^{[c]} = \langle \xi_n^{*[c]}, e_j \rangle_{L^2}, \ j = 0, 1, \dots, n. \tag{5.42}$$

Therefore, from Equations (5.41) and (5.42), we obtain

$$R_n^{[c]} \xi_n^{*[c]}(x) = A^{[c]}, \tag{5.43}$$

where the matrix $A^{[c]}$ is same as defined in Equation (5.9). Since system (5.29) has a unique solution $\xi_n^{*[c]}(x)$ due to the existence of bounded linear operator $(S^{[c]} - Q_n^{[c]} K^{[c]})^{-1}$. Therefore, from Equation (5.43), the solution $A^{[c]}$ of system (5.8) exists uniquely.

5.4 ILLUSTRATIVE EXAMPLES

In this section, we find the approximate solution of numerical examples by using the proposed method discussed in Section 5.2.

Example 5.1 Consider the Cauchy singular integral equation

$$\int_{-1}^{1} \sqrt{\frac{1-t}{1+t}} \frac{\xi^{[c]}(t)}{t-x} dt - \frac{1}{2} \int_{-1}^{1} \sqrt{\frac{1-t}{1+t}} (x-xt^2)\xi^{[c]}(t)dt = \frac{1}{\pi}\left(xe^x - J_1(1)x^2 + 7x^5\right),$$

$$(5.44)$$

where $J_1(1)$ is the Bessel function of first kind of order one. The exact solution is not known. Figure 5.1 shows that as n is increasing, the approximate solutions are coming closer to each other which means that the error between the approximate solution and the exact solution is keep on decreasing.

Further, it can be seen from Table 5.1 that the error bound is decreasing as n is increasing which shows the convergence of sequence of approximate solutions to the exact solution.

Example 5.2 Consider the singular integral equation with Cauchy kernel

$$\int_{-1}^{1} \sqrt{\frac{1-t}{1+t}} \frac{\xi^{[c]}(t)}{t-x} dt - \frac{1}{5} \int_{-1}^{1} \sqrt{\frac{1-t}{1+t}} (1+x^4)(1+t)^2 \xi^{[c]}(t)dt$$

$$= \frac{1}{\pi}\left(\frac{5179}{40} + 4x - 5x^2 + 7x^3 + \frac{97x^4}{20}\right).$$

$$(5.45)$$

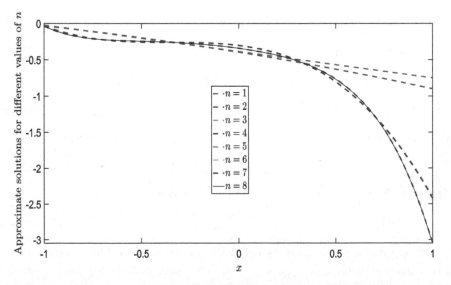

Figure 5.1 Comparison of approximate solution for different values of n in case of Example 5.1.

Table 5.1
The Theoretical Error Bound in Case of Example 5.1

n	Error Bound for $\|\xi^{[c]} - \xi_n^{*[c]}\|_{L^2}$
$n = 1$	1.21436
$n = 2$	1.19832
$n = 3$	2.51361×10^{-1}
$n = 4$	2.51042×10^{-1}
$n = 5$	1.5630×10^{-4}
$n = 6$	1.2961×10^{-5}
$n = 7$	9.22537×10^{-7}
$n = 8$	5.75069×10^{-8}

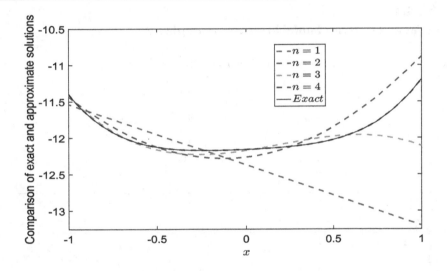

Figure 5.2 Comparison of exact solution and approximate solutions in case of Example 5.2.

$\chi^{[c]}(x) = \dfrac{1}{\pi^2} \sqrt{\dfrac{1-x}{1+x}} \left(-120 + x + \dfrac{3x^2}{2} + 7x^4 \right)$ is exact solution. It can be seen in Figure 5.2 that the approximate solution is the exact solution for $n = 4$. It is also shown in Table 5.2 that actual error satisfies the error bound which is calculated by using Equation (5.33).

Example 5.3 Consider a Cauchy singular integral equation

$$\int_{-1}^{1} \sqrt{\frac{1-t}{1+t}} \frac{\xi^{[c]}(t)}{t-x} dt - \frac{1}{2} \int_{-1}^{1} \sqrt{\frac{1-t}{1+t}} x(1+t) \xi^{[c]}(t) dt = \frac{1}{\pi} \left(x - 5x^2 + 7x^3 \right) \sin x.$$

$$(5.46)$$

Example 5.3 has not known exact solution. We find its approximate solution by proposed method and the results are detailed in Table 5.3. In Figure 5.3, the closeness of the approximate solutions to each other with the increase in the value of can be

Table 5.2
The Actual Error and Theoretical Error Bound in Case of Example 5.2

n	Actual Error $\|\xi^{[c]} - \xi_n^{*[c]}\|_{L^2}$	Error Bound for $\|\xi - \xi^{*[c]}\|_{L^2}$
$n = 1$	1.00668	3.58307
$n = 2$	2.40175×10^{-1}	1.62093
$n = 3$	2.10905×10^{-1}	1.02621
$n = 4$	0	0

Table 5.3
The Theoretical Error Bound in Case of Example 5.3

n	Error Bound for $\|\xi^{[c]} - \xi_n^{*[c]}\|_{L^2}$
$n = 1$	1.66512
$n = 2$	6.33927×10^{-1}
$n = 3$	3.47631×10^{-1}
$n = 4$	3.05371×10^{-2}
$n = 5$	1.75190×10^{-2}
$n = 6$	2.09064×10^{-4}
$n = 7$	1.19283×10^{-5}
$n = 8$	1.00000×10^{-8}

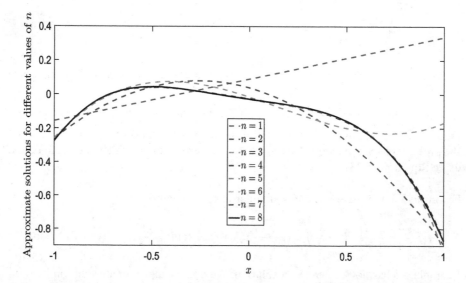

Figure 5.3 Comparison of approximate solution for different values of n in case of Example 5.3.

seen which verifies the results (5.33). Table 5.3 implies the convergence of sequence of approximate solutions to the exact solution.

5.5 INTRODUCTION OF HYPERSINGULAR INTEGRAL EQUATIONS

Hypersingular integral equations (HSIEs) have great importance in the field of aeronautics [3,28,29,33]. Many problems occur in the field of aeronautics such as wing and tail surfaces problem; pairs or collections of wings problems [3] are reducible to HSIEs. These HSIEs also appeared during the mathematical modeling of vortex wakes behind aircraft at altitude, near the ground at the time of takeoff and landing operations [17]. Many two-dimensional problems of aerodynamics can be modeled as singular integral equation such as for an inviscid incompressible fluid flow past a rectangular airfoil problem reduces into a hypersingular integral equation [5]. Apart from problems of aeronautics, the problems of electromagnetic scattering [17], acoustics [21], fluid dynamics [37], electromagnetic diffraction [44], elasticity [15], and fracture mechanics [8] are modeled as hypersingular integral equations. In early nineties, Parsons and Martin [36] used HSIE to study the problem of water wave scattering. Further, these equations for crack problems in the field of fracture mechanics [2,11,12] have been explored by many researchers. Various methods such as complex variable function method [7], boundary element method [25], polynomial approximation method [8,30,31], reproducing kernel method [13], and piecewise linear approximations on a nonuniform grid [39] for solving SIEs have already been explored. However, to find the approximate solution of HSIEs, search for a better method in some sense is always welcomed. In this chapter, we propose a residual-based Galerkin's method with Legendre polynomial as basis function to find the approximate solution of HSIEs. The HSIEs of practical interest which occur during the formulation of many boundary value problems are of the form:

$$\fint_{-1}^{1} \frac{\chi^{[h]}(t)}{(t-x)^2} dt - \int_{-1}^{1} k^{[h]}(x,t) \chi^{[h]}(t) dt = g^{[h]}(x), \ |x| < 1, \tag{5.47}$$

with $\chi^{[h]}(\pm 1) = 0$. The functions $g^{[h]}(x)$ and $k^{[h]}(x,t)$ are known real-valued Hölder continuous over the interval $[-1,1]$ and $[-1,1] \times [-1,1]$, respectively. $\chi^{[h]}(x)$ is an unknown function defined over the interval $[-1,1]$. The first integral in Equation (5.47) is understood to be exist in the sense of Hadamard finite part integral. Also, the derivative of unknown function $\chi^{[h]}(x)$ is assumed to be Hölder continuous in order to ensure the existence of finite-part integral [34].

5.6 METHOD OF SOLUTION TO THE PROBLEM

A function $\chi^{[h]}(t)$ defined over the interval $[-1,1]$ in Equation (5.47) with $\chi^{[h]}(\pm 1) = 0$ can be represented [19] as follows:

$$\chi^{[h]}(t) = \sqrt{1-t^2} \xi^{[h]}(t), \tag{5.48}$$

where $\xi^{[h]}(t)$ is an unknown function. Using Equation (5.48) in Equation (5.47), we obtain

$$\fint_{-1}^{1} \frac{\xi^{[h]}(t)\sqrt{1-t^2}}{(t-x)^2}dt - \int_{-1}^{1} \sqrt{1-t^2}k^{[h]}(x,t)\xi^{[h]}(t)dt = g^{[h]}(x), \ |x| < 1. \quad (5.49)$$

Now, we approximate the function $\xi^{[h]}(t)$ by orthonormalized Legendre polynomials as follows:

$$\xi^{[h]}(t) \approx \xi_n^{*[h]}(t) = \sum_{j=0}^{n} a_j^{[h]} e_j(t), \quad (5.50)$$

where $a_j^{[h]}$; $j = 1, 2, \ldots, n$ are unknown constant coefficients.

Using Equation (5.50) for $\xi^{[h]}(t)$ in Equation (5.49), we get the residual error $\mathcal{R}^{[h]}(x, a_0^{[h]}, a_1^{[h]}, a_2^{[h]}, \ldots, a_n^{[h]})$, where

$$\mathcal{R}^{[h]}(x, a_0^{[h]}, a_1^{[h]}, a_2^{[h]}, \ldots, a_n^{[h]}) = \fint_{-1}^{1} \frac{\xi_n^{*[h]}(t)\sqrt{1-t^2}}{(t-x)^2}dt$$

$$- \int_{-1}^{1} \sqrt{1-t^2}k^{[h]}(x,t)\xi_n^{*[h]}(t)dt - g^{[h]}(x), \ |x| < 1. \quad (5.51)$$

In Galerkin's method, the space $E = span\{e_j(x)\}_{j=0}^{n}$ is assumed to be orthogonal to $\mathcal{R}^{[h]}(x, a_0^{[h]}, a_1^{[h]}, a_2^{[h]}, \ldots, a_n^{[h]})$, so we have

$$\langle \mathcal{R}^{[h]}(x, a_0^{[h]}, a_1^{[h]}, a_2^{[h]}, \ldots, a_n^{[h]}), e_j \rangle_{L^2} = 0, \ \forall j = 0, 1, 2, \ldots, n. \quad (5.52)$$

Using Equation (5.51) for $j = 0, 1, 2, \ldots, n$, Equation (5.52) becomes

$$\left\langle \fint_{-1}^{1} \frac{\xi_n^{*[h]}(t)\sqrt{1-t^2}}{(t-x)^2}dt - \int_{-1}^{1} k^{[h]}(x,t)\xi_n^{*[h]}(t)\sqrt{1-t^2}dt - g^{[h]}(x), e_0 \right\rangle_{L^2} = 0,$$

$$\left\langle \fint_{-1}^{1} \frac{\xi_n^{*[h]}(t)\sqrt{1-t^2}}{(t-x)^2}dt - \int_{-1}^{1} k^{[h]}(x,t)\xi_n^{*[h]}(t)\sqrt{1-t^2}dt - g^{[h]}(x), e_1 \right\rangle_{L^2} = 0,$$

$$\left\langle \fint_{-1}^{1} \frac{\xi_n^{*[h]}(t)\sqrt{1-t^2}}{(t-x)^2}dt - \int_{-1}^{1} k^{[h]}(x,t)\xi_{sn}^{*[h]}(t)\sqrt{1-t^2}dt - g^{[h]}(x), e_n \right\rangle_{L^2} = 0. \quad (5.53)$$

In order to evaluate singular integral in each integral equation of system (5.53), we use the results of [22] and we obtain linear system of order $(n+1) \times (n+1)$.

The above system (5.53) can be written in matrix form as

$$B^{[h]T} A^{[h]} = B_1^{[h]} A^{[h]} = G^{[h]}, \quad (5.54)$$

where

$$B_1^{[h]} = B^{[h]\,T}, \quad B^{[h]} = \begin{pmatrix} b_{00}^{[h]} & b_{01}^{[h]} & \cdots & b_{0n}^{[h]} \\ b_{10}^{[h]} & b_{11}^{[h]} & \cdots & b_{1n}^{[h]} \\ \vdots & \vdots & \ddots & \vdots \\ b_{n0}^{[h]} & b_{n1}^{[h]} & \cdots & b_{nn}^{[h]} \end{pmatrix}, \quad A^{[h]} = \begin{pmatrix} a_0^{[h]} \\ a_1^{[h]} \\ \vdots \\ a_n^{[h]} \end{pmatrix}, \quad G^{[h]} = \begin{pmatrix} g_0^{[h]} \\ g_1^{[h]} \\ \vdots \\ g_n^{[h]} \end{pmatrix},$$

$$\tag{5.55}$$

$$b_{rq}^{[h]} = \int_{-1}^{1} \left(\fint_{-1}^{1} \frac{\sqrt{1-t^2}\,e_r(t)}{(t-x)^2}\,dt - \int_{-1}^{1} k^{[h]}(x,t)e_r(t)\sqrt{1-t^2}dt \right) e_q(x)dx,$$

$$r,q = 0,1,2,\ldots,n,$$

$$g_q^{[h]} = \int_{-1}^{1} g^{[h]}(x)e_q(x)dx, \quad q = 0,1,2,\ldots,n.$$

Now, we solve the linear system (5.54) which gives the value of unknown coefficients $a_j^{[h]}$; $j = 0,1,2,\ldots,n$.

5.7 CONVERGENCE

This section shows the convergence of sequence $\{\xi_n^{*[h]}\}_{n=0}^{\infty}$ to the exact solution $\xi_n^{*[h]}(x)$ in L^2 space.

5.7.1 FUNCTION SPACES

We first define a Hilbert space $L^2[-1,1] = \{u(t) : [-1,1] \to \mathbb{R} : \int_{-1}^{1}(u(t))^2 dt < \infty\}$ with the norm $\|\cdot\|_{L^2}^2$ and inner product $\langle.,.\rangle_{L^2}$

$$\|u(t)\|_{L^2} = \left(\int_{-1}^{1}(u(t))^2 dt \right)^{1/2} \text{ for } u(t) \in L^2[-1,1], \tag{5.56}$$

$$\langle u,v\rangle_{L^2} = \int_{-1}^{1} u(t)v(t)dt \text{ for } u(t),v(t) \in L^2[-1,1]. \tag{5.57}$$

Now, we define the set of functions

$$M^{[h]} = \{u(t) \in L^2 : \sum_{j=0}^{\infty} (d_j^{[h]})^2 \langle u,e_j\rangle_{L^2}^2 < \infty\}, \tag{5.58}$$

where

$$(d_j^{[h]})^2 = \| S^{[h]} e_j \|_{L^2}^2, \tag{5.59}$$

$$S^{[h]} e_j(x) = \fint_{-1}^{1} \frac{\sqrt{1-t^2}}{(t-x)^2} e_j(t)dt. \tag{5.60}$$

The set $M^{[h]}$ which is a subspace of L^2 is actually a Hilbert space with respect to the norm $\| \cdot \|_{M^{[h]}}$ and inner product $\langle .,. \rangle_{M^{[h]}}$

$$\| u \|^2_{M^{[h]}} = \sum_{j=0}^{\infty} \left(d_j^{[h]} \right)^2 \langle u, e_j \rangle^2_{L^2} \text{ for } u(t) \in M^{[h]}, \qquad (5.61)$$

$$\langle u, v \rangle_{M^{[h]}} = \sum_{j=0}^{\infty} \left(d_j^{[h]} \right)^2 \langle u, e_j \rangle_{L^2} \langle v, e_j \rangle_{L^2} \text{ for } u(t), v(t) \in M^{[h]}. \qquad (5.62)$$

Now operating the operator $S^{[h]}$, on $e_j(x); j = 0, 1, 2, \ldots, n$ and using the results of [22], we obtain

$$S^{[h]} e_0(x) = -\pi e_0(x),$$

$$S^{[h]} e_1(x) = -2\pi e_1(x), \ S^{[h]} e_2(x) = -\pi \left[\frac{\sqrt{5}}{4\sqrt{3}} e_0(x) + 3e_2(x) \right],$$

$$S^{[h]} e_n(x) = \sum_{i=0}^{n} c_i^{[h]} e_i(x); \text{ where } c_i^{[h]} = \langle S^{[h]} e_n, e_i \rangle_{L^2}, i = 0, 1, 2, \ldots, n. \qquad (5.63)$$

5.7.2   ERROR ANALYSIS

Using Equation (5.63), we extend the operator $S^{[h]} : M^{[h]} \to L^2$ as a bounded linear operator

$$S^{[h]} \xi^{[h]}(x) = \sum_{j=0}^{\infty} \langle \xi^{[h]}, e_j \rangle_{L^2} \sum_{i=0}^{j} \langle S^{[h]} e_j, e_i \rangle_{L^2} e_i(x) \in L^2[-1, 1]. \qquad (5.64)$$

Using Equation (5.64), the norm of the operator $S^{[h]}$

$$\| S^{[h]} \xi^{[h]} \|^2_{L^2} = \sum_{j=0}^{\infty} \left(d_j^{[h]} \right)^2 \langle \xi^{[h]}, e_j \rangle^2_{L^2} = \| \xi^{[h]} \|^2_{M^{[h]}}. \qquad (5.65)$$

Hence, using Equation (5.65), we obtain

$$\| S^{[h]} \| = 1. \qquad (5.66)$$

Moreover, the mapping $S^{[h]} : M^{[h]} \to L^2$ is bijection mapping. Therefore, following *Bounded Inverse Theorem*, the operator $(S^{[h]})^{-1} : L^2 \to M^{[h]}$ exists as a bounded linear operator which is defined as

$$(S^{[h]})^{-1} \xi^{[h]}(x) = \sum_{j=0}^{\infty} \frac{\langle \xi^{[h]}(x), e_j(x) \rangle_{L^2}}{d_j} e_j(x). \qquad (5.67)$$

Now, with Equation (5.67), the norm for linear operator $(S^{[h]})^{-1}$

$$\| (S^{[h]})^{-1} \xi^{[h]}(x) \|_{M^{[h]}} = \| \xi^{[h]}(x) \|_{L^2}. \tag{5.68}$$

Now, we define the mapping $Q_n^{[h]} : L^2 \to L^2$ as

$$Q_n^{[h]} \xi^{[h]}(x) = \sum_{j=0}^{n} \langle \xi^{[h]}, e_j \rangle_{L^2} e_j(x). \tag{5.69}$$

After defining all the operators and function spaces, we can finally estimate the error bound for the error which occurs in approximating the exact solution of Equation (5.47) by taking its projection from Hilbert space onto a vector space $E = \text{span}\{e_j(x)\}_{j=0}^n$. Writing Equation (5.49) in an operator equation from the spaces $M^{[h]}$ to L^2

$$(S^{[h]} - K^{[h]}) \xi^{[h]}(x) = g^{[h]}(x), \ g^{[h]}(x) \in L^2, \ \xi^{[h]}(x) \in M^{[h]}, \tag{5.70}$$

where the operator $K^{[h]} : M^{[h]} \to L^2$ is as follows:

$$K^{[h]} \xi^{[h]}(x) = \int_{-1}^{1} \sqrt{1-t^2} k^{[h]}(x,t) \xi^{[h]}(t) dt. \tag{5.71}$$

In order to prove that the operator $K^{[h]} : M^{[h]} \to L^2$ as a compact operator, we need to assume:

$$\int_{-1}^{1} \int_{-1}^{1} \left(\sqrt{1-t^2} k^{[h]}(x,t) \right)^2 dt dx < \infty. \tag{5.72}$$

The Equation (5.70) will have a unique solution if and only if the operator $(S^{[h]} - K^{[h]})^{-1}$ exists as a bounded linear operator. We assume that $(S^{[h]} - K^{[h]})^{-1}$ exists as bounded operator. From Equation (5.52), we have

$$Q_n^{[h]} \left((S^{[h]} - K^{[h]}) \xi_n^{*[h]}(x) - g^{[h]}(x) \right) = 0. \tag{5.73}$$

Since the function $S^{[h]} \xi_n^{*[h]}(x)$ is a polynomial therefore, we get

$$Q_n^{[h]} S^{[h]} \xi_n^{*[h]}(x) = S^{[h]} \xi_n^{*[h]}(x). \tag{5.74}$$

Using the above fact, Equation (5.73) becomes

$$S^{[h]} \xi_n^{*[h]}(x) - Q_n^{[h]} K^{[h]} \xi_n^{*[h]}(x) = Q_n^{[h]} g^{[h]}(x). \tag{5.75}$$

Due to the boundedness of $S^{[h]^{-1}}$ and compactness of $K^{[h]}$, for all n arbitrarily large, say $n > n_0$, $(S^{[h]} - Q_n^{[h]} K^{[h]})^{-1}$ exists as a bounded linear operator [20]. Hence, there exists a unique solution for Equation (5.75) which is as follows:

$$\xi_n^{*[h]}(x) = (S^{[h]} - Q_n^{[h]} K^{[h]})^{-1} Q_n^{[h]} g^{[h]}(x). \tag{5.76}$$

Now from Equations (5.70) and (5.76), we have

$$\xi^{[h]}(x) - \xi_n^{*[h]}(x) = (S^{[h]} - Q_n^{[h]} K^{[h]})^{-1}$$
$$\times \left(g^{[h]}(x) - Q_n^{[h]} g^{[h]}(x) + K^{[h]} \xi^{[h]}(x) - Q_n^{[h]} K^{[h]} \xi^{[h]}(x) \right). \qquad (5.77)$$

The norm of Equation (5.77)

$$\| \xi^{[h]} - \xi_n^{*[h]} \|_{M^{[h]}} \leq \| (S^{[h]} - Q_n^{[h]} K^{[h]})^{-1} \| \| g^{[h]} - Q_n^{[h]} g^{[h]} \|_{L^2}$$
$$+ \| (S^{[h]} - Q_n^{[h]} K^{[h]})^{-1} \| \| K^{[h]} \xi^{[h]}(x) - Q_n^{[h]} K^{[h]} \xi^{[h]}(x) \|_{L^2}. \qquad (5.78)$$

The assumption made in Equation (5.72); the operator $S^{[h]}$ is a Hilbert–Schmidt operator [20] and so, $\| K^{[h]} - Q_n^{[h]} K^{[h]} \|_{L^2} \to 0$ as $n \to \infty$. Also, we have $\| \xi^{[h]} - \xi_n^{*[h]} \|_{M^{[h]}} \to 0$ as $n \to \infty$ as $\| g^{[h]} - Q_n^{[h]} g^{[h]} \|_{L^2} \to 0$ as $n \to \infty$.

Further, due to the fact that if $\xi^{[h]} \in M^{[h]}$, then we have

$$\| \xi^{[h]} \|_{L^2} \leq \| \xi^{[h]} \|_{M^{[h]}}. \qquad (5.79)$$

Using Equation (5.79), Equation (5.78) can be written as follows:

$$\| \xi^{[h]} - \xi_n^{*[h]} \|_{L^2} \leq \| (S^{[h]} - Q_n^{[h]} K^{[h]})^{-1} \| \| g^{[h]} - Q_n g^{[h]} \|_{L^2}$$
$$+ \| (S^{[h]} - Q_n^{[h]} K^{[h]})^{-1} \| \| K \xi^{[h]}(x) - Q_n^{[h]} K^{[h]} \xi^{[h]}(x) \|_{L^2}. \qquad (5.80)$$

Also, we have

$$\| \xi^{[h]} - \xi_n^{*[h]} \|_{L^2} \to 0 \text{ as } n \to \infty. \qquad (5.81)$$

Hence, the convergence of the sequence $\{\xi_n^{*[h]}\}_{n=0}^{\infty}$ is shown.

5.7.3 WELL POSEDNESS OF LINEAR SYSTEM

Here, we show that well-posedness of the problem (5.76) in the sense of Hadamard. The problem (5.76) has a solution this is due to the existence of inverse operator $(S^{[h]} - Q_n^{[h]} K^{[h]})^{-1}$ which is already proved in the previous section. Now to prove the uniqueness of solution to the problem (5.76), we use contradiction principle. Let us assume that problem (5.76) has two solutions, say y_1 and y_2, which are distinct from each other. Now, we have

$$S^{[h]} y_1(x) - Q_n^{[h]} K^{[h]} y_1(x) = Q_n^{[h]} g^{[h]}(x), \qquad (5.82)$$

and

$$S^{[h]} y_2(x) - Q_n^{[h]} K^{[h]} y_2(x) = Q_n^{[h]} g^{[h]}(x). \qquad (5.83)$$

From Equations (5.82) and (5.83), we get

$$(S^{[h]} - Q_n^{[h]} K^{[h]})(y_1(x) - y_2(x)) = 0. \tag{5.84}$$

Since the bounded operator $(S^{[h]} - Q_n^{[h]} K^{[h]})^{-1}$ exists, from Equation (5.84), we get

$$y_1(x) = y_2(x), \ |x| < 1. \tag{5.85}$$

Equation (5.85) contradicts our assumption. Hence, we have proved that solution to the problem (5.76) exists uniquely. Moreover, the continuity of the inverse operator $(S^{[h]} - Q_n^{[h]} K^{[h]})^{-1}$ proves that a minor change in the given data results a minor change in the solution. As it is shown that the problem (5.76) satisfies all the well-posedness conditions therefore it is a well-posed problem.

5.7.4 EXISTENCE AND UNIQUENESS

This subsection shows that the solution of the linear system (5.54) has a unique solution. We start the proof by defining the prolongation operator [20] $\mathcal{P}_n^{[h]} : \mathbb{R}^{n+1} \to E$ as follows:

$$\mathcal{P}_n^{[h]} G^{[h]} = \sum_{j=0}^{n} \langle g^{[h]}, e_j \rangle_{L^2} e_j(x) \in E, \tag{5.86}$$

where $\mathbb{R}^{n+1}$ is a real vector space [26] whose elements are $(n+1)$-tuples of real numbers, $E = span\{e_j(t)\}_{j=0}^{n}$ and $G^{[h]}$ is already defined in Equation (5.55). Now from the definition of orthogonal projection $Q_n^{[h]}$, we get

$$Q_n^{[h]} g^{[h]}(x) = \sum_{j=0}^{n} \langle g^{[h]}, e_j \rangle_{L^2} e_j(x). \tag{5.87}$$

Following Equations (5.86) and (5.87), we have

$$\mathcal{P}_n^{[h]} G^{[h]} = Q_n^{[h]} g^{[h]}(x), \ g^{[h]}(x) \in L^2, \ G^{[h]} \in \mathbb{R}^{n+1}, \ |x| < 1. \tag{5.88}$$

Also, we define restriction operator [20] $R_n^{[h]} : E \to \mathbb{R}^{n+1}$ as follows:

$$R_n^{[h]} \xi_n^{*[h]}(x) = (\langle \xi_n^{*[h]}(x), e_0 \rangle_{L^2}, \langle \xi_n^{*[h]}(x), e_1 \rangle_{L^2}, \dots \langle \xi_n^{*[h]}(x), e_n \rangle_{L^2})^T \in \mathbb{R}^{n+1}, \tag{5.89}$$

where the function $\xi_n^{*[h]}(x)$ is already defined in Equation (5.50). From Equation (5.50), we get

$$a_j^{[h]} = \langle \xi_n^{*[h]}(x), e_j \rangle_{L^2}, \ j = 0, 1, \dots, n. \tag{5.90}$$

Therefore, from Equations (5.89) and (5.90), we obtain

$$R_n^{[h]} \xi_n^{*[h]}(x) = A^{[h]}, \tag{5.91}$$

where the matrix $A^{[h]}$ is already defined in Equation (5.55). Since operator $(S^{[h]} - Q_n^{[h]} K^{[h]})^{-1}$ exists as bounded linear operator therefore $\xi_n^{*[h]}(x)$ also exists uniquely. Hence, from Equation (5.91), the solution $A^{[h]}$ of system (5.54) exists uniquely for every given $G^{[h]}$. Also, the inverse of matrix $B_1^{[h]}$ exists [40].

5.8   ILLUSTRATIVE EXAMPLES

The efficiency of our proposed numerical method and verification of the theoretical results is shown in this section, with the aid of numerical illustrations.

Example 5.4   We consider an integral equation:

$$\fint_{-1}^{1} \frac{\xi^{[h]}(t)\sqrt{1-t^2}}{(t-x)^2}dt + \int_{-1}^{1} \frac{\xi^{[h]}(t)\sqrt{1-t^2}\exp(t+x)}{12}dt = \pi g^{[h]}(x), \ |x| < 1, \quad (5.92)$$

where $g^{[h]}(x) = -\frac{9}{8}x + 6x^3 - 6x^5 + \frac{81}{64}\exp(x)I_2(1) - \frac{31}{4}\exp(x)I_3(1)$, and I_2, I_3 are modified Bessel functions of first kind of order two and three, respectively. $\xi^{[h]}(x) = \frac{3x}{16} - x^3 + x^5$ is the exact solution.

The numerical results for actual error and error bound are detailed in Table 5.4. Figure 5.4 shows the comparison between approximate solutions and exact solution for $n = 1, 2, \ldots, 5$.

Table 5.4

The Theoretical Error Bound in Case of Example 5.4

n	Actual Error (In L^2 norm)	Error Bound
1	0.06481	1.43228
2	0.06480	1.43183
3	0.05688	1.31111
4	0.05687	1.31110
5	2.19155×10^{-14}	1.19141×10^{-10}

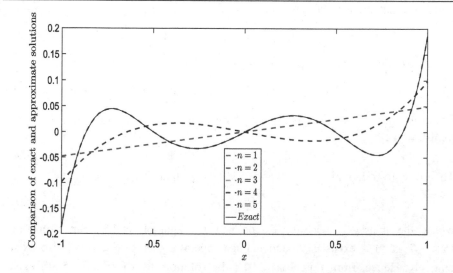

Figure 5.4   Comparison of exact solution with approximate solutions of Example 5.4.

Example 5.5 Consider one more hypersingular integral equation:

$$\fint_{-1}^{1} \frac{\xi^{[h]}(t)\sqrt{1-t^2}}{(t-x)^2}dt + \int_{-1}^{1} \frac{(x+x^2)\xi^{[h]}(t)\sqrt{1-t^2}}{36+12s}dt = \pi g^{[h]}(x), \quad |x| < 1, \quad (5.93)$$

where

$$g^{[h]}(x) = \frac{1326099}{655360} - \frac{1469711672063x}{7864320} + \frac{84573531x}{320\sqrt{2}} - \frac{1470155415887x^2}{7864320}$$
$$+ \frac{84573531x^2}{320\sqrt{2}} + \frac{115527x^3}{10240} + \frac{4953727x^4}{16384} - \frac{88851x^5}{2560} - \frac{5394557x^6}{10240}$$
$$+ \frac{7571x^7}{320} + \frac{1453239x^8}{5120} + \frac{327x^9}{64} - \frac{1793x^{10}}{256} - \frac{45x^{11}}{16} + \frac{1885x^{12}}{128}.$$

This example has an exact solution

$$\xi^{[h]}(x) = \frac{1}{640}\Big(-252 + 45x + 4510x^2 - 725x^3 - 22258x^4 + 2680x^5 + 38000x^6$$
$$- 2000x^7 - 20252x^8 - 252x^9 + 45x^{10} + 150x^{11} - 725x^{12}\Big).$$

Table 5.5 shows that the error is decreasing with the increase in the value of n which verifies the result Equation (5.81). The comparison of approximate solutions for $n = 1, 2, \ldots, 12$ with the exact solution is shown in Figure 5.5. The actual error is also calculated for Example 5.5 with respect to norm in L^2.

Table 5.5
The Theoretical Error Bound in Case of Example 5.5

n	Actual Error (In L^2 norm)	Error Bound
1	0.38064	8.84521
2	0.37161	8.74921
3	0.37160	8.74689
4	0.28343	7.76302
5	0.27503	7.64867
6	0.25688	7.24213
7	0.25274	7.15094
8	0.00570	0.19899
9	0.00564	0.19739
10	0.00054	0.02196
11	0.00049	0.02053
12	5.95198×10^{-16}	1.07677×10^{-10}

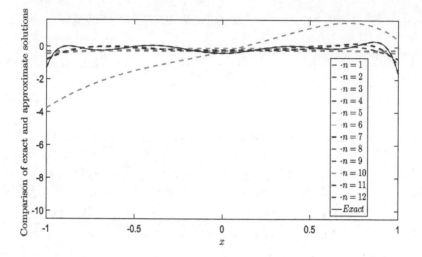

Figure 5.5 Comparison of exact solution with approximate solutions of Example 5.5.

5.9 CONCLUSION

In this chapter, we have considered the numerical solution of CSIEs and HSIEs over the interval $[-1, 1]$. A residual-based Galerkin's method is proposed to find the numerical solution of CSIEs and HSIEs over the finite interval. This method that converts the SIE into a system of linear equations is shown. The existence and uniqueness of solution for the system of linear algebraic equations is proved. The convergence of sequence of approximate solutions is proved, and the error bound is also obtained. With the aid of numerical examples, all the derived theoretical results are verified.

REFERENCES

1. Akel, M. S. and Hussein, H. S. (2011). Numerical treatment of solving singular integral equations by using sinc approximations. *Applied Mathematics and Computation*, 218(7):3565–3573.
2. Ang, W., Clements, D., and Cooke, T. (1999). A hypersingular boundary integral equation for a class of antiplane multiple crack problems for inhomogeneous elastic materials. *International Journal for Numerical Methods in Biomedical Engineering*, 15(3): 183–191.
3. Ashley, H. and Landahl, M. (1965). *Aerodynamics of Wings and Bodies*. Courier Corporation, Chelmsford, MA.
4. Bhattacharya, S. and Mandal, B. (2008). Numerical solution of a singular integro-differential equation. *Applied Mathematics and Computation*, 195(1):346–350.
5. Bisplinghoff, R. L., Ashley, H., and Halfman, R. L. (1996). *Aeroelasticity*. Dover Publications, Mineola, NY.
6. Chakrabarti, A. and Berghe, G. V. (2004). Approximate solution of singular integral equations. *Applied Mathematics Letters*, 17(5):553–559.

7. Chakrabarti, A., Mandal, B., Basu, U., and Banerjea, S. (1997). Solution of a hypersingular integral equation of the second kind. *ZAMM-Journal of Applied Mathematics and Mechanics/Zeitschrift für Angewandte Mathematik und Mechanik*, 77(4):319–320.

8. Chan, Y. S., Fannjiang, A. C., and Paulino, G. H. (2003). Integral equations with hypersingular kernels theory and applications to fracture mechanics. *International Journal of Engineering Science*, 41(7):683–720.

9. Chan, Y. S., Fannjiang, A. C., Paulino, G. H., and Feng, B. F. (2007). Finite part integrals and hypersingular kernels. *Advances in Dynamical Systems and Applications*, 14(S2):264–269.

10. Chandler, G. (1992). Midpoint collocation for Cauchy singular integral equations. *Numerische Mathematik*, 62(1):483–509.

11. Chen, Y. (1998). Hypersingular integral equation approach for plane elasticity crack problem with circular boundary. *International Journal for Numerical Methods in Biomedical Engineering*, 14(5):451–461.

12. Chen, Y. (2004). Hypersingular integral equation method for three-dimensional crack problem in shear mode. *International Journal for Numerical Methods in Biomedical Engineering*, 20(6):441–454.

13. Chen, Z. and Zhou, Y. (2011). A new method for solving hypersingular integral equations of the first kind. *Applied Mathematics Letters*, 24(5):636–641.

14. Cuminato et al. (2007). A review of linear and nonlinear Cauchy singular integral and integro-differential equations arising in mechanics. *Journal of Integral Equations and Applications*, 19(2):163–207.

15. De Lacerda, L. and Wrobel, L. (2001). Hypersingular boundary integral equation for axisymmetric elasticity. *International Journal for Numerical Methods in Engineering*, 52(11):1337–1354.

16. Erdogan, F., Gupta, G. D., and Cook, T. (1973). Numerical solution of singular integral equations. In: Sih, G. (ed.) *Methods of Analysis and Solutions of Crack Problems*, pp. 368–425. Springer, Berlin, Heidelberg.

17. Ginevsky, A., Vlasov, Y. V., and Karavosov, R. (2009). *Foundations of Engineering Mechanics*. Springer, Berlin, Heidelberg.

18. Golberg, M. (1990). Introduction to the numerical solution of Cauchy singular integral equations. In: Miele, A. (ed.) *Numerical Solution of Integral Equations*, pp. 183–308. Springer, Berlin, Heidelberg.

19. Golberg, M. A. (1987). The convergence of several algorithms for solving integral equations with finite part integrals. II. *Applied Mathematics and Computation*, 21(4): 283–293.

20. Golberg, M. A., Chen, C.-S., and Fromme, J. (1997). Functional analysis. In: *Discrete Projection Methods for Integral Equations*, pp. 51–115. American Society of Mechanical Engineers, New York.

21. Iovane, G., Lifanov, I., and Sumbatyan, M. (2003). On direct numerical treatment of hypersingular integral equations arising in mechanics and acoustics. *Acta Mechanica*, 162(1):99–110.

22. Kaya, A. C. and Erdogan, F. (1987). On the solution of integral equations with strongly singular kernels. *Quarterly of Applied Mathematics*, 45(1):105–122.

23. Kim, S. (1998). Solving singular integral equations using gaussian quadrature and overdetermined system. *Computers & Mathematics with Applications*, 35(10):63–71.

24. Kim, S. (1999). Numerical solutions of Cauchy singular integral equations using generalized inverses. *Computers & Mathematics with Applications*, 38(5):183–195.

25. Kim-Chuan, T. and Mukherjee, S. (1994). Hypersingular and finite part integrals in the boundary element method. *International Journal of Solids and Structures*, 31(17): 2299–2312.
26. Kreyszig, E. (1989). *Introductory Functional Analysis with Applications*. Wiley, New York.
27. Kubrusly, C. S. (2012). *Spectral Theory of Operators on Hilbert Spaces*. Springer Science & Business Media, Berlin, Heidelberg.
28. Ladopoulos, E. (2013). *Singular Integral Equations: Linear and Non-Linear Theory and its Applications in Science and Engineering*. Springer Science & Business Media, Berlin, Heidelberg.
29. Lifanov, I., Setukha, A., Tsvetinsky, Y. G., and Zhelannikov, A. (1997). Mathematical modelling and the numerical analysis of a nonstationary flow around the deck of a ship. *Russian Journal of Numerical Analysis and Mathematical Modelling*, 12(3):255–270.
30. Mahiub, M. A., Long, N. N., and Eshkuvatov, Z. (2011). Numerical solution of hypersingular integral equations. *International Journal of Pure and Applied Mathematics*, 69(3):265–274.
31. Mandal, B. and Bera, G. (2006). Approximate solution for a class of hypersingular integral equations. *Applied Mathematics Letters*, 19(11):1286–1290.
32. Mandal, B. and Bera, G. (2007). Approximate solution of a class of singular integral equations of second kind. *Journal of Computational and Applied Mathematics*, 206(1):189–195.
33. Mangler, K. (1951). *Improper Integrals in Theoretical Aerodynamics*. Royal Aircraft Establishment, Farnborough.
34. Martin, P. (1991). End-point behaviour of solutions to hypersingular integral equations. *Proceedings of the Royal Society of London A: Mathematical, Physical and Engineering Sciences*, 1885:301–320.
35. Mohankumar, N. and Natarajan, A. (2008). On the numerical solution of Cauchy singular integral equations in neutron transport. *Annals of Nuclear Energy*, 35(10):1800–1804.
36. Parsons, N. and Martin, P. (1992). Scattering of water waves by submerged plates using hypersingular integral equations. *Applied Ocean Research*, 14(5):313–321.
37. Ryzhakov, G. (2013). On the numerical method for solving a hypersingular integral equation with the computation of the solution gradient. *Differential Equations*, 49(9): 1168–1175.
38. Setia, A. (2014). Numerical solution of various cases of Cauchy type singular integral equation. *Applied Mathematics and Computation*, 230:200–207.
39. Setukha, A. (2017). Convergence of a numerical method for solving a hypersingular integral equation on a segment with the use of piecewise linear approximations on a nonuniform grid. *Differential Equations*, 53(2):234–247.
40. Shirali, S. and Vasudeva, H. L. (2010). *Multivariable Analysis*. Springer Science & Business Media, Berlin, Heidelberg.
41. Srivastav, R. and Zhang, F. (1991). Solving Cauchy singular integral equations by using general quadrature-collocation nodes. *Computers & Mathematics with Applications*, 21(9):59–71.
42. Venturino, E. (1986). Recent developments in the numerical solution of singular integral equations. *Journal of Mathematical Analysis and Applications*, 115(1):239–277.
43. Yu, D.-H. (2002). *Natural Boundary Integral Method and its Applications*, vol. 539. Springer Science & Business Media, Berlin, Heidelberg.
44. Zakharov, E., Setukha, A., and Bezobrazova, E. (2015). Method of hypersingular integral equations in a three-dimensional problem of diffraction of electromagnetic waves on a piecewise homogeneous dielectric body. *Differential Equations*, 51(9):1197–1210.

6 Krylov Subspace Methods for Numerically Solving Partial Differential Equations

Santosh Dubey and Sanjeev K. Singh
University of Petroleum & Energy Studies

Anter El-Azab
Purdue University

CONTENTS

6.1   INTRODUCTION

Since their conception, partial differential equations (PDEs) have been considered as one of the greatest intellectual achievements of researchers in their attempt to understand the physical world. PDEs arise naturally whenever we deal with systems having a continuous extent; for instance, solids (mass and heat transport) and fluid mechanics (Navier–Stokes equations), quantum mechanics (Schrodinger equation), general relativity, electricity and magnetism (Maxwell equations), and optics. Several more applications of PDEs in Engineering may be found in [1]. PDEs contain partial derivatives in which an unknown function depends upon several variables unlike in ordinary differential equations (ODEs) in which the unknown function depends only on one variable.

6.1.1   TYPES OF PDES

Partial differential equations are classified into linear and non-linear. In linear PDEs, the dependent variable and all the derivatives occur in linear fashion, and vice versa in case of non-linear PDEs. A general second-order linear PDE for $f(x,y)$ may be written as

$$A\frac{\partial^2 f}{\partial x^2} + B\frac{\partial^2 f}{\partial x \partial y} + C\frac{\partial^2 f}{\partial y^2} + D\frac{\partial f}{\partial x} + E\frac{\partial f}{\partial y} + Ff = Q \tag{6.1}$$

where $A, B, C, D, E, F,$ and Q may be either constants or functions of independent variables. Based on various values of $A, B, C, D, E, F,$ and Q, there are three types of linear PDEs:

a. Parabolic PDE: In this case, $B^2 - 4AC = 0$. Parabolic PDEs describe flow of information through diffusion process. Examples of parabolic equations are

  I.   The diffusion equation: $\frac{\partial C}{\partial t} = D\frac{\partial^2 C}{\partial x^2}$, where C and D are the concentration field and diffusion coefficient, respectively.

  II. Heat conduction equation: $\frac{\partial T}{\partial t} = \alpha\frac{\partial^2 T}{\partial x^2}$, where T and α are the temperature field and thermal diffusivity, respectively.

b. Hyperbolic PDE: These PDEs describe transport and wave propagation phenomena.

  I.   First-order Hyperbolic PDE: The first-order hyperbolic PDE is: $\frac{\partial f}{\partial t} + a\frac{\partial f}{\partial x} = 0, a > 0$. The examples are
  - Wave equation: $\frac{\partial f}{\partial t} + a\frac{\partial f}{\partial x} = 0, a > 0$.
  - Euler equations governing adiabatic and inviscid flow:

$$\frac{\partial \rho}{\partial t} + u \cdot \nabla \rho + \rho \nabla \cdot u = 0$$

$$\frac{\partial u}{\partial t} + u \cdot \nabla u + \frac{\nabla p}{\rho} = g$$

$$\frac{\partial e}{\partial t} + u \cdot \nabla e + \frac{p}{\rho}\nabla \cdot u = 0$$

where $\rho, u, p,$ and e are the fluid mass density, flow velocity, pressure, and specific internal energy, respectively.

II. Second-order Hyperbolic PDE: In this case, $B^2 - 4AC > 0$. The examples are

- Wave equation: Wave equation representing propagation of, say, sound waves in fluids, water waves, oscillations in solid structures, and electromagnetic waves: $\frac{\partial^2 f}{\partial t^2} = u^2 \frac{\partial^2 f}{\partial x^2}$, where u is the wave velocity, and $f(x,t)$ is the displacement.

c. Elliptic PDE: In this case, $B^2 - 4AC < 0$. Elliptic PDEs describe steady-state phenomena. Examples of elliptic PDEs are

I. Steady-state heat equation: $\nabla^2 T = 0$, where $\nabla^2 = \frac{\partial^2}{\partial x^2} + \frac{\partial^2}{\partial y^2} + \frac{\partial^2}{\partial z^2}$ is the Laplacian operator.

II. Poisson equation: $\nabla^2 u = f$.

III. Laplace equation: $\nabla^2 u = 0$.

6.2 SOLUTION OF PDES

There are several methods to solve PDEs, which may be either analytical or numerical. Depending on the type of PDE, the differential operator may depend only on space or on space and time. In case the differential operator is space dependent only, boundary conditions are required in order to obtain the unique solution; such problems are called boundary value problems (BVP). In case the PDE includes both the space- and time-dependent differential operators, initial and boundary conditions would be needed; such problems are called initial boundary value problems (IBVP). There are three types of boundary conditions:

a. *Dirichlet boundary conditions*: In this case, the value of the dependent variable is assigned at the boundary; that is, if u is the dependent variable, then $u(\Omega) = u_\Omega$, where Ω represents the boundary.

b. *Neumann boundary conditions*: In this case, the value of the normal derivative of the dependent variable at the boundary is specified, i.e., $\frac{\partial u}{\partial n}\big|_\Omega = g$.

c. *Robin boundary conditions (or mixed boundary conditions)*: In this case, a linear combination of Dirichlet and Neumann boundary conditions, $u(\Omega) + \frac{\partial u}{\partial n}\big|_\Omega$, is specified at the boundary.

Analytical solution of a PDE results in a function, which satisfies the PDE at every point in the space as well as at the boundaries. Techniques such as separation of variables, integral transforms [2,3], perturbation methods [4], calculus of variations [5], and eigenfunction expansion [6] are used to solve PDEs analytically. For IBVPs, initial conditions of the solution must be specified everywhere at time $t = 0$.

6.3 NUMERICAL SOLUTIONS

In most of the PDEs, especially which mimic physical situations, the analytical solution becomes hard to obtain due to the complexity of the PDE and boundary conditions. In such cases, numerical solutions are the only way to obtain unique solution. The availability of high-performance computers has made it possible to solve a variety of complex problems by developing reliable numerical methods, which were otherwise difficult to solve using analytical schemes. The most commonly used methods to solve PDEs numerically are the finite volume method (FVM), finite element method (FEM), and the finite difference method (FDM). Other methods such as spectral and method of lines are also used.

6.3.1 FINITE VOLUME METHOD

Finite volume method (FVM) is a conservative numerical technique used to solve PDEs on structured as well as non-structured and non-uniform meshes. It approximates the values of conserved variables averaged across a volume; the values of the conserved variables are located inside the volume element, and not at nodes like in FMD or FEM. In FVM, volume integrals are replaced by surface integrals using the Gauss divergence theorem, which are then evaluated as fluxes at the surfaces of each volume element. FVM is a very powerful technique when the geometries (where PDEs are solved) involved are irregular and also in situations where the mesh moves to track interfaces or shocks. For more details on FVM, the reader is referred to [7].

6.3.2 FINITE ELEMENT METHODS

Finite element methods (FEMs) are a class of numerical methods employed to obtain approximate solutions of PDEs. The domain over which the function is approximated is divided into smaller parts/regions called as finite elements. The solution is modeled over the finite elements via interpolation in terms of the values of the solution at the vertices (nodes) and basis functions or piecewise polynomials. Upon using this representation in a week form of the problem, a discrete form of the governing PDE is obtained in the form of an algebraic system in which the values of the solution are to be solved for at a finite number of points (the finite element nodes). The weak form of the equation has an integral form, which is converted into sum of integrals over the finite elements, leading to the discrete form of the PDE by numerical integrations involving the interpolation functions and their derivatives. Some FEM methods instead use variational methods to approximate the solution by minimizing the associated error function. One of the benefits of using the FEM is that selection of discretization as well as the basis functions is flexible. The FEM theory is well developed, which provides useful error estimates and bounds for the error while attempting numerical solution to the problem. More details on FEM can be found here [8].

6.3.3 FINITE DIFFERENCE METHODS

Finite difference methods (FDMs) are a class of numerical methods used for solving differential equations. In FDMs, finite differences are used to approximate the derivatives with the help of Taylor's expansion. In order to approximate the derivatives using FDM, the domain is divided into discrete points called *grid* points (Figure 6.1). The derivatives are approximated at these grid points using finite difference approximation. This process converts ODE/PDE into a system of algebraic equations, which may be solved with the help of variety of matrix algebra techniques. Just to illustrate the process, let us consider a 2D domain divided into several grid points (Figure 6.1); let us take the distance between two grid points along $x-$ direction as Δx and Δy along $y-$ direction. Let us also represent the nodes by i, j notation, where i index varies along x, and j index varies along y directions.

Now, let us find out the finite difference approximation of various derivatives like $\frac{\partial f}{\partial x}, \frac{\partial f}{\partial y}, \frac{\partial^2 f}{\partial x^2}$, etc. at the node (i, j) with the help of Taylor's expansion:

$$f_{i+1,j} = f_{i,j} + \left(\frac{\partial f}{\partial x}\right)_{i,j} \frac{\Delta x}{1!} + \left(\frac{\partial^2 f}{\partial x^2}\right)_{i,j} \frac{(\Delta x)^2}{2!} + \left(\frac{\partial^3 f}{\partial x^3}\right)_{i,j} \frac{(\Delta x)^3}{3!}$$
$$+ \text{higher order terms}$$

Since Δx is very small, higher powers of Δx (e.g. $(\Delta x)^2$, $(\Delta x)^3$) will be very small and may be neglected. Therefore, we may write

$$f_{i+1,j} \approx f_{i,j} + \left(\frac{\partial f}{\partial x}\right)_{i,j} \Delta x + \left(\frac{\partial^2 f}{\partial x^2}\right)_{i,j} \frac{(\Delta x)^2}{2}$$

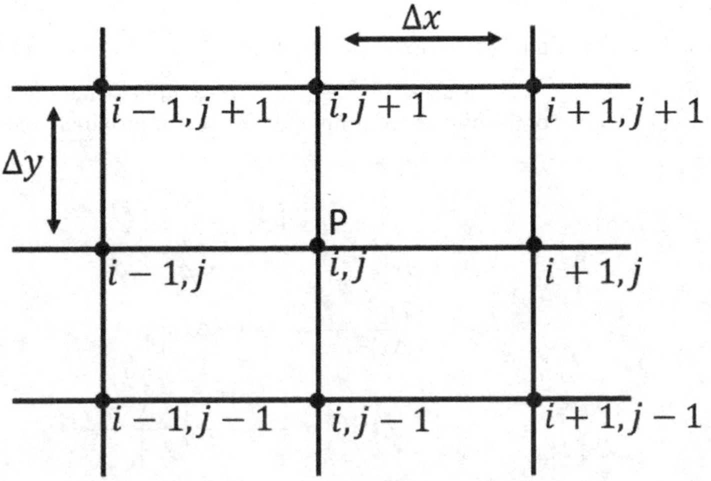

Figure 6.1 Grid points in 2D domain.

which may be written as

$$\frac{f_{i+1,j}-f_{i,j}}{\Delta x} \approx \left(\frac{\partial f}{\partial x}\right)_{i,j} + \left(\frac{\partial^2 f}{\partial x^2}\right)_{i,j}\frac{(\Delta x)}{2}$$

$$\left(\frac{\partial f}{\partial x}\right)_{i,j} = \frac{f_{i+1,j}-f_{i,j}}{\Delta x} - \left(\frac{\partial^2 f}{\partial x^2}\right)_{i,j}\frac{(\Delta x)}{2} = \frac{f_{i+1,j}-f_{i,j}}{\Delta x} + \mathfrak{O}(\Delta x)$$

Thus, $\left(\frac{\partial f}{\partial x}\right)_{i,j}$ may be approximated by the following finite difference approximation:

$$\left(\frac{\partial f}{\partial x}\right)_{i,j} = \frac{f_{i+1,j}-f_{i,j}}{\Delta x} + \mathfrak{O}(\Delta x)$$

The finite difference approximation as above is called as *first-order forward difference*.

First-order backward difference approximation for $\left(\frac{\partial f}{\partial x}\right)_{i,j}$ may also be obtained using Taylor's expansion in the similar manner:

$$\left(\frac{\partial f}{\partial x}\right)_{i,j} = \frac{f_{i,j}-f_{i-1,j}}{\Delta x} + \mathfrak{O}(\Delta x)$$

Higher order approximations for $\left(\frac{\partial f}{\partial x}\right)_{i,j}$, e.g., central difference approximation, may also be found using Taylor's expansion:

$$\left(\frac{\partial f}{\partial x}\right)_{i,j} = \frac{f_{i+1,j}-f_{i-1,j}}{2\Delta x} + \mathfrak{O}(\Delta x)^2$$

The central difference finite difference expression for $\left(\frac{\partial f}{\partial x}\right)_{i,j}$ is correct up to second order. In the same way, we can also get finite difference approximation for higher order derivatives. Some of the higher order finite difference approximations are given as follows:

$$\left(\frac{\partial^2 f}{\partial x^2}\right)_{i,j} = \left(\frac{\partial}{\partial x}\left(\frac{\partial f}{\partial x}\right)\right)_{i,j} = \frac{\left(\frac{\partial f}{\partial x}\right)_{i+1,j}-\left(\frac{\partial f}{\partial x}\right)_{i,j}}{\Delta x}$$

$$= \frac{f_{i+1,j}-2f_{i,j}+f_{i-1,j}}{(\Delta x)^2} + \mathfrak{O}(\Delta x)^2$$

$$\left(\frac{\partial^2 f}{\partial y^2}\right)_{i,j} = \left(\frac{\partial}{\partial y}\left(\frac{\partial f}{\partial y}\right)\right)_{i,j} = \frac{\left(\frac{\partial f}{\partial y}\right)_{i,j+1}-\left(\frac{\partial f}{\partial y}\right)_{i,j}}{\Delta y}$$

$$= \frac{f_{i,j+1}-2f_{i,j}+f_{i,j-1}}{(\Delta y)^2} + \mathfrak{O}(\Delta y)^2$$

Mixed derivatives may also be approximated in the similar manner:

$$\left(\frac{\partial^2 f}{\partial x \partial y}\right)_{i,j} = \left(\frac{\partial}{\partial x}\left(\frac{\partial f}{\partial y}\right)\right)_{i,j} = \frac{\left(\frac{\partial f}{\partial y}\right)_{i+1,j} - \left(\frac{\partial f}{\partial y}\right)_{i,j}}{\Delta x} = \frac{\frac{f_{i+1,j+1}-f_{i+1,j}}{\Delta y} - \frac{f_{i,j}-f_{i,j-1}}{\Delta y}}{\Delta x}$$

$$= \frac{f_{i+1,j+1} - f_{i+1,j} - f_{i,j} + f_{i,j-1}}{(\Delta x)(\Delta y)} + \mho(\Delta x \Delta y)$$

In order to illustrate the methodology used in FDM, let us solve a 2D linear parabolic PDE using FDM. Consider following time-dependent 2D heat equation:

$$\frac{\partial T}{\partial t} = \alpha\left(\frac{\partial^2 T}{\partial x^2} + \frac{\partial^2 T}{\partial y^2}\right) \quad x_0 < x < x_l \text{ and } y_0 < y < y_l \qquad (6.2)$$

The domain over which this equation is to be solved is shown in Figure 6.2.

The boundary conditions for the above equation are

$$T(x=x_0,y,t) = T_L; \quad T(x=x_l,y,t) = T_R, \quad T(x,y=y_0,t) = T_B, \quad T(x,y=y_l,t) = T_T$$

and the initial condition is given as

$$T(x,y,t=0) = T_0$$

In Equation (6.2), α is thermal diffusivity; T_L is the temperature fixed at left boundary, T_R is the temperature at the right boundary, T_B is the temperature at the bottom boundary, and T_T is the temperature at the top boundary; T_0 is the initial temperature of the 2D domain.

In order to get the numerical solution for Equation (6.2), we need to discretize the domain into a set of nodes having following coordinates (x_i, y_j):

$$x_i = x_0 + (i-1)\Delta x \quad i = 1, 2, \ldots n_x$$
$$y_j = y_0 + (j-1)\Delta y \quad j = 1, 2, \ldots n_y$$

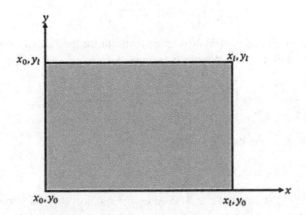

Figure 6.2 The domain over which PDE has to be solved.

where

$$n_x = \frac{x_l - x_0}{\Delta x} + 1$$

$$n_y = \frac{y_l - y_0}{\Delta y} + 1$$

The finite difference approximation of $\frac{\partial T}{\partial t}$ may be written as forward difference in time as follows:

$$\left(\frac{\partial T}{\partial t}\right)^t_{i,j} \approx \frac{T(i,j,t+\Delta t) - T(i,j,t)}{\Delta t}$$

where Δt is the time step. For the sake of convenience, let us write $t + \Delta t$ as $n+1$ and t as n. With this, we can write

$$\left(\frac{\partial T}{\partial t}\right)^n_{i,j} \approx \frac{T^{n+1}_{i,j} - T^n_{i,j}}{\Delta t} \qquad (6.3)$$

The above discretization of the time derivative is first-order accurate. Using Equation (6.3) and finite difference approximation for spatial differential operators, we can convert the PDE in Equation (6.2) into a set of linear algebraic equations corresponding to each node. Now, depending on whether we know the right-hand side (RHS) of Equation (6.2) at the current time (t or in the present notation, n) or not, we have two possibilities: explicit time discretization and implicit time discretization.

6.3.3.1 Explicit Method

In this case, the RHS of the PDE is known at the current time, and solution at the next time step ($t + \Delta t$ or $n+1$ in the current notation) may be known by the following equation:

$$\left(\frac{\partial T}{\partial t}\right)^n_{i,j} = \alpha \left[\frac{\partial^2 T}{\partial x^2} + \frac{\partial^2 T}{\partial y^2}\right]^n_{i,j}$$

Using Equation (6.3) and central difference formula for the spatial differential operators (using five-point stencil, Figure 6.3), we get

$$\frac{T^{n+1}_{i,j} - T^n_{i,j}}{\Delta t} = \alpha \left[\frac{T^n_{i+1,j} - 2T^n_{i,j} + T^n_{i-1,j}}{(\Delta x)^2} + \frac{T^n_{i,j+1} - 2T^n_{i,j} + T^n_{i,j-1}}{(\Delta y)^2}\right]$$

Considering $\Delta x = \Delta y$, we may write

$$T^{n+1}_{i,j} = \beta T^n_{i+1,j} + (1 - 4\beta)T^n_{i,j} + \beta T^n_{i-1,j} + \beta T^n_{i,j+1} + \beta T^n_{i,j-1} \qquad (6.4)$$

where $\beta = \frac{\alpha \Delta t}{(\Delta x)^2}$. We see that the RHS side in Equation (6.4) is known and hence the solution of Equation (6.2) at the next time step may be calculated easily from Equation (6.4) with the help of boundary conditions and initial condition.

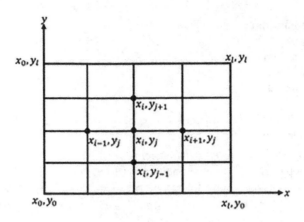

Figure 6.3   Five-point stencil to approximate the second-order spatial differential operators.

Pseudocode:

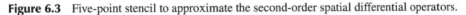

— *Enter input parameters*: $\alpha, \Delta x, \Delta t$
— *Discretize the domain into computational nodes*:
$x_i = x_0 + (i-1)\Delta x, \ i = 1, 2, \ldots, n_x$
$y_i = y_0 + (i-1)\Delta y, \ i = 1, 2, \ldots, n_y$
where $n_x = \frac{x_l - x_0}{\Delta x} + 1, n_y = \frac{y_l - y_0}{\Delta y} + 1$.
— *Initialize the domain*: Specify the initial value of temperature at all computational nodes.
$T(i, j, t = 0) = T_0$
— *Apply Dirichlet boundary conditions* at all the boundaries:
  • **Left boundary**
    $i = 1$
    Loop from $j = 1$ to n_y
      $T(i, j, t) = T_L$
    Loop End
  • **Bottom boundary**
    $j = 1$
    Loop from $i = 1$ to n_x
      $T(i, j, t) = T_B$
    Loop End
  • **Right boundary**
    $i = n_x$
    Loop from $j = 1$ to n_y
      $T(i, j, t) = T_R$
    Loop End

- **Top boundary**
  $j = n_y$
  Loop from $i = 1$ to n_x
  $\quad T(i,j,t) = T_T$
  Loop End

– *Update the dependent variable values at the inner nodes*
  Calculate: $\beta = \frac{\alpha \Delta t}{(\Delta x)^2}$
  Loop over time (*instep = 1 to nstep*)
  Loop over first space variable ($i = 1$ to $n_x - 1$)
  Loop over second space variable ($j = 2$ to $n_y - 1$)
  $\quad T_{i,j}^{t+\Delta t} = \beta T_{i+1,j}^{t} + (1 - 4\beta)T_{i,j}^{t} + \beta T_{i-1,j}^{t} + \beta T_{i,j+1}^{t} + \beta T_{i,j-1}^{t}$
  End Loop (second space variable)
  End Loop (first space variable)
  End Loop (time)

6.3.3.2 Implicit Method

In this method, the RHS of Equation (6.2) is not known at the current time. This may be written as

$$\left(\frac{\partial T}{\partial t} \right)_{i,j}^{n} = \alpha \left[\frac{\partial^2 T}{\partial x^2} + \frac{\partial^2 T}{\partial y^2} \right]_{i,j}^{n+1}$$

Using finite difference approximation (with five-point stencil for second-order spatial differential operators as in Figure 6.3 and first-order forward difference approximation for the temporal differential operator), we get

$$\frac{T_{i,j}^{n+1} - T_{i,j}^{n}}{\Delta t} = \alpha \left[\frac{T_{i+1,j}^{n+1} - 2T_{i,j}^{n+1} + T_{i-1,j}^{n+1}}{(\Delta x)^2} + \frac{T_{i,j+1}^{n+1} - 2T_{i,j}^{n+1} + T_{i,j-1}^{n+1}}{(\Delta y)^2} \right]$$

Higher order approximations for differential operators may also be derived using Taylor's expansion, which, however, will increase the number of functional evaluations to approximate the differential operator.

Moving the unknown terms to the left-hand side and known terms on the RHS, we get

$$(1 + 4\beta) T_{i,j}^{n+1} - \beta T_{i+1,j}^{n+1} - \beta T_{i-1,j}^{n+1} - \beta T_{i,j+1}^{n+1} - \beta T_{i,j-1}^{n+1} = T_{i,j}^{n} \qquad (6.5)$$

In order to devise a solution mechanism, let us first consider that the domain is discretized such that $n_x = n_y = 4$. For this case, the domain is shown in Figure 6.4.

Plugging in $i = 2,3,4$ and $j = 2,3,4$ in Equation (6.5), we get four algebraic equations corresponding to the inner nodes of the domain. The unknowns of these equations are $T_{2,2}^{n+1}, T_{2,3}^{n+1}, T_{3,2}^{n+1}, T_{3,3}^{n+1}$. The resulting algebraic equations may be written as using matrix notations as

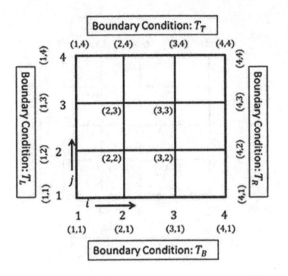

Figure 6.4 Domain discretization for $n_x = n_y = 4$. Boundary conditions at each boundary are shown.

$$\left(\begin{pmatrix} 1+4\beta & -\beta & -\beta & 0 \\ -\beta & 1+4\beta & 0 & -\beta \\ -\beta & 0 & 1+4\beta & -\beta \\ 0 & -\beta & -\beta & 1+4\beta \end{pmatrix} \begin{pmatrix} T_{2,2}^{n+1} \\ T_{2,3}^{n+1} \\ T_{3,2}^{n+1} \\ T_{3,3}^{n+1} \end{pmatrix} = \begin{pmatrix} T_{1,2}^{n+1} + T_{2,1}^{n+1} + T_{2,2}^{n} \\ T_{1,3}^{n+1} + T_{2,4}^{n+1} + T_{2,3}^{n} \\ T_{4,2}^{n+1} + T_{3,1}^{n+1} + T_{3,2}^{n} \\ T_{4,3}^{n+1} + T_{3,4}^{n+1} + T_{3,3}^{n} \end{pmatrix}\right) \tag{6.6}$$

Using the boundary conditions, we may write

$$T_{1,2}^{n+1} = T_L = T_{1,3}^{n+1}$$

$$T_{2,1}^{n+1} = T_B = T_{3,1}^{n+1}$$

$$T_{4,2}^{n+1} = T_R = T_{4,3}^{n+1}$$

$$T_{2,4}^{n+1} = T_T = T_{3,4}^{n+1}$$

Therefore, Equation (6.6) may be written as

$$\begin{pmatrix} 1+4\beta & -\beta & -\beta & 0 \\ -\beta & 1+4\beta & 0 & -\beta \\ -\beta & 0 & 1+4\beta & -\beta \\ 0 & -\beta & -\beta & 1+4\beta \end{pmatrix} \begin{pmatrix} T_{2,2}^{n+1} \\ T_{2,3}^{n+1} \\ T_{3,2}^{n+1} \\ T_{3,3}^{n+1} \end{pmatrix} = \begin{pmatrix} T_L + T_B + T_{2,2}^{n} \\ T_L + T_T + T_{2,3}^{n} \\ T_R + T_B + T_{3,2}^{n} \\ T_R + T_T + T_{3,3}^{n} \end{pmatrix} \tag{6.7}$$

We see that the coefficient matrix has five diagonals: a main diagonal with entries as $1+4\beta$, two sub-diagonals above, and two sub-diagonals below the main diagonal with entries $-\beta$. The solution of the Equation (6.7) depends upon the structure of

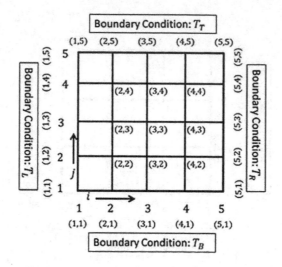

Figure 6.5 Domain discretization with $n_x = n_y = 5$.

the coefficient matrix. In order to understand the structure of the coefficient matrix
further, let us increase discretization to $n_x = n_y = 5$. With this number of nodes, the
domain discretization is shown in Figure 6.5.

The PDE will be solved at the inner nodes. The unknowns are $T_{2,2}^{n+1}$, $T_{2,3}^{n+1}$, $T_{2,4}^{n+1}$,
$T_{3,2}^{n+1}$, $T_{3,3}^{n+1}$, $T_{3,4}^{n+1}$, $T_{4,2}^{n+1}$, $T_{4,3}^{n+1}$, $T_{4,4}^{n+1}$. The set of algebraic equations in this case
(considering the boundary conditions) may be written as

$$
\begin{pmatrix}
1+4\beta & -\beta & 0 & -\beta & 0 & 0 & 0 & 0 & 0 \\
-\beta & 1+4\beta & -\beta & 0 & -\beta & 0 & 0 & 0 & 0 \\
0 & -\beta & 1+4\beta & 0 & 0 & -\beta & 0 & 0 & 0 \\
-\beta & 0 & 0 & 1+4\beta & -\beta & 0 & -\beta & 0 & 0 \\
0 & -\beta & 0 & -\beta & 1+4\beta & -\beta & 0 & -\beta & 0 \\
0 & 0 & -\beta & 0 & -\beta & 1+4\beta & 0 & 0 & -\beta \\
0 & 0 & 0 & -\beta & 0 & 0 & 1+4\beta & -\beta & 0 \\
0 & 0 & 0 & 0 & -\beta & 0 & -\beta & 1+4\beta & -\beta \\
0 & 0 & 0 & 0 & 0 & -\beta & 0 & -\beta & 1+4\beta
\end{pmatrix}
$$

$$
\times
\begin{pmatrix}
T_{2,2}^{n+1} \\
T_{2,3}^{n+1} \\
T_{2,4}^{n+1} \\
T_{3,2}^{n+1} \\
T_{3,3}^{n+1} \\
T_{3,4}^{n+1} \\
T_{4,2}^{n+1} \\
T_{4,3}^{n+1} \\
T_{4,4}^{n+1}
\end{pmatrix}
=
\begin{pmatrix}
T_L + T_B + T_{2,2}^n \\
T_L + T_{2,2}^n \\
T_L + T_T + T_{2,4}^n \\
T_B + T_{3,2}^n \\
T_{3,3}^n \\
T_T + T_{3,4}^n \\
T_B + T_R + T_{4,2}^n \\
T_R + T_{4,3}^n \\
T_T + T_R + T_{4,4}^n
\end{pmatrix}
\tag{6.8}
$$

We see that with increase in the discretization, the sparsity of the coefficient matrix changes. Although the number of non-zero diagonals (main diagonal + sub-diagonals) does not change, the placement of sub-diagonals with respect to the main diagonal changes with discretization. All these details should be kept in mind while solving Equation (6.8) computationally.

Pseudocode:

- *Enter input parameters*: $\alpha, \Delta x, \Delta t$
- *Discretize the domain into computational nodes*:
  $x_i = x_0 + (i-1)\Delta x, i = 1, 2, \ldots, n_x$
  $y_i = y_0 + (i-1)\Delta y, i = 1, 2, \ldots, n_y$
  where $n_x = \frac{x_l - x_0}{\Delta x} + 1, n_y = \frac{y_l - y_0}{\Delta y} + 1$.
- *Initialize the domain*: Specify the initial value of temperature at all computational nodes.
  $T(i, j, t = 0) = T_0$
- *Apply Dirichlet boundary conditions* at all the boundaries:
  - **Left boundary**
    $i = 1$
    Loop from $j = 1$ to n_y
    $\quad T(i, j, t) = T_L$
    Loop End
  - **Bottom boundary**
    $j = 1$
    Loop from $i = 1$ to n_x
    $\quad T(i, j, t) = T_B$
    Loop End
  - **Right boundary**
    $i = n_x$
    Loop from $j = 1$ to n_y
    $\quad T(i, j, t) = T_R$
    Loop End
  - **Top boundary**
    $j = n_y$
    Loop from $i = 1$ to n_x
    $\quad T(i, j, t) = T_T$
    Loop End
- *Update the dependent variable values at the inner nodes*
  Calculate: $\beta = \frac{\alpha \Delta t}{(\Delta x)^2}$

Assemble the matrix A *(coefficient matrix)*
Loop over time *(istep = 1 to nstep)*
  Loop over first space variable $(i = 1$ to $n_x - 1)$
    Loop over second space variable $(j = 2$ to $n_y - 1)$
      Assemble the RHS column vector (b)
      Solve: $Ax = b$ (x is unknown vector; here it is temperature)
    End Loop (second space variable)
  End Loop (first space variable)
End Loop (time)

There are several ways to solve linear system of equations represented by $Ax = b$. These methods may be categorized into following heads:

  I. Direct methods
  II. Iterative methods.

6.3.3.2.1   Direct Methods

Direct techniques usually are the methods that give exact solution of the system of equations in a finite number of steps. However, the solutions obtained by direct methods are also contaminated by round-off errors, which may be minimized by careful analysis of the system and thereafter devising novel schemes to contain round-off error. Gauss elimination with back substitution is the most important direct technique which is employed to solve $Ax = b$. Various pivoting strategies are used in Gauss elimination method to take care of the round-off errors in the solution [9].

Gauss elimination requires $O(n^3/3)$ arithmetic operations to solve an arbitrary system of linear equations, $Ax = b$, where n is the size of the system. In order to make the solution efficient, i.e., to reduce the number of arithmetic operations in finding out the unknown vector x, the coefficient matrix is factored into two matrices: lower triangular (L) and upper triangular (U), i.e., $A = LU$. This is known as LU decomposition. Thereafter, the solution is accomplished in two steps:

- First, one takes $y = Ux$, and solve the lower triangular system $Ly = b$. Solving $Ly = b$ needs only $O(n^2)$ operations as back substitution is the only step needed for getting the solutions.
- Thereafter, $Ux = y$ is solved, which again takes only $O(n^2)$ operations to solve the system. In total, the number of operations needed is $O(2n^2)$ only.

For well-conditioned systems, direct methods always arrive at the same answer. Even in case of ill-conditioned problems, direct methods may be used for the solution. All these methods require a lot of random access memory (RAM) for computation. Direct methods fail in the following cases:

- If the matrix A is very large, e.g., n is on the order of a few million, the solution will require $O(n^3)$ operations to get the solution.

— If the matrix is a sparse matrix, it is meaningless to use the computing resources to solve such a system using direct solver like Gaussian elimination. However, if the direct solver takes care of the sparsity while attempting the solution, it is preferable.

6.3.3.2.2   Iterative Methods

Iterative methods are suitable for solving large size algebraic systems (in terms of computational requirements) as the relevant algorithms may be parallelized easily using shared as well as distributed memory architectures. These methods approach the solution in a gradual fashion, contrary to direct methods in which solution is obtained in one computational step. The error in the solution decreases monotonically (or the solution converges to the actual solution) in each iteration if the problem is well conditioned; if the problem is not well conditioned, the convergence is slow and sometimes the solution does not converge at all. In ill-conditioned problems, before iterative methods are employed, the matrix if first "conditioned" using preconditioners (see Appendix A.1 for a short introduction on preconditioners), and then, the solution is attempted. Preconditioners improve the condition number of the matrix to ensure better convergence [9,10].

Iterative schemes start with an initial guess to the solution and gradually approach towards the correct solution in subsequent steps called iterations. The number of iterations depends upon a tolerance value, which serves as the stop criterion. Iteration continues until the stop criterion is satisfied. The selection of tolerance depends on the problem at hand: if faster solutions are required, the tolerance may be made smaller (i.e., the stop criterion may be made less restrictive), and in case accurate solution is needed, tolerance may be taken bigger (i.e., the stop criterion may be more restrictive). Usually, tolerance is decided based on the inputs to the model one is solving. If the input is couple of digits accurate, there is no point in taking tolerance too much tight. In any case, the tolerance should always be greater than the number that depends on the machine precision and the condition number [9,10].

There are two main categories of iterative methods used to solve system of linear equations:

  I. Stationary iterative methods
 II. Non-stationary or Krylov subspace methods.

6.4   STATIONARY ITERATIVE METHODS

In *stationary iterative method*, $Ax = b$ is converted to $x = Mx + c$, where M is the iteration matrix, and c is any vector. After the initial vector $x^{(0)}$ is selected, the sequence of approximate solution vector is estimated by

$$x^{(k+1)} = Mx^{(k)} + c \qquad (6.9)$$

where $k = 1, 2, 3, \ldots$. This method is called stationary because the formula does not change as a function of $x^{(k)}$. There are four main stationary iterative methods: Jacobi, Gauss–Seidel, successive over relaxation (SOR), and symmetric SOR.

6.5   NON-STATIONARY METHODS (KRYLOV SUBSPACE METHODS)

Non-stationary methods (Krylov subspace methods) involve information that changes at each iteration. In these methods, constants are evaluated by taking the inner products of residuals or other vectors arising from the iterative method.

Given a nonsingular matrix $A \in R^{N \times N}$ and a vector $b \in R^N$, a linear system of equations may be written as

$$Ax = b$$

Let us suppose that the exact solution of the above system is a vector x^*. The iterative solution of the above system is attempted by taking an initial guess $x^{(0)}$, and compute a sequence of iterates $x^{(k)}$ which approximates x^* moving in an affine subspace $x^{(0)} + K \subset R^N$. The goal is to construct the subspace K using simple operations such as matrix-by-vector products, which could minimize some norm of $r^{(k)} = b - Ax^{(k)}$. The subspace K constructed using these operations (e.g., matrix-vector products) is called the Krylov subspace. Most of the iterative methods that are employed nowadays to solve large-scale problems are Krylov subspace solvers.

Krylov subspace-based algorithms approximate x by transforming an N-dimensional vector space into an n-dimensional subspace ($n \le N$) by matrix-vector multiplications. This approach of forming the subspace avoids explicit estimation of A^{-1}, which is quite expensive for large sparse matrices. The algorithm starts with a guess $x^{(0)}$, which is used to obtain $x = x^{(0)} + x^{(n)}$. The estimate for $x^{(n)}$ may be extracted from the Krylov subspace:

$$K_n\left(A, r^{(0)}\right) = \text{span}\left(r^{(0)}, Ar^{(0)}, A^2 r^{(0)}, \ldots, A^{n-1} r^{(0)}\right), \quad r^{(0)} = b - Ax^{(0)}$$

In the above equation, span means that every vector in the subspace may be expressed as a linear combination of the basis $\left\{r^{(0)}, Ar^{(0)}, A^2 r^{(0)}, \ldots, A^n r^{(0)}\right\}$. Therefore, it is possible to find $x^{(n)}$ as linear combination of the basis:

$$x^{(n)} \approx \sum_{i=1}^{n-1} c_i A^i r^{(0)}$$

where c_i are scalars.

Krylov subspace is a suitable space to look for the approximate solution of the system of linear equations because it has a natural representation as a member of Krylov subspace. In fact, it is possible to find the solution in very less number of iterations if the dimension of this space is small.

There are two very famous methods based on Krylov subspace: the conjugate gradient (CG) method and the generalized minimum residual (GMRES) method. The two methods are briefly outlined in the rest of this chapter.

6.5.1 CONJUGATE GRADIENT (CG) METHOD

Hestenes and Stiefel [11] proposed the CG method. This method solves a system of linear equations:

$$Ax = b$$

for the vector x, where the matrix A is symmetric ($A^T = A$) and positive definite ($x^T A x > 0$ for all $x \in R^n$). In this method, a sequence of conjugate/orthogonal vectors (two vectors u and v are conjugate with respect to A if $u^T A v = 0$) are generated. These vectors are the residual of the iterates as well as the gradients of a quadratic functional, the minimization of which is equivalent to solve the linear system; that is, CG method solves the following optimization problem:

$$\min f(x) = \frac{1}{2} x^T A x - b x^T$$

which is equivalent to finding the solution of the original system

$$\nabla f(x) = Ax - b = 0$$

CG method proceeds by generating vector sequences of iterates, residuals, and search directions to update iterates and residuals. Two inner products are performed in every iteration to compute the scalars, which make the sequence of iterates and residuals satisfy some orthogonality conditions. In the case of a symmetric positive definite matrix, the orthogonality conditions involve some kind of norm in which the distance to the true solution is minimized.

Just to illustrate in details, we start with initial guess $x^{(0)}$ and take initial direction as $p^{(0)} = - \nabla f|_{x^{(0)}} = b - Ax^{(0)} = r^{(0)}$. The negative gradient of the function should give a direction of steepest descent, which is also the residual of the problem. However, in CG method, we know that the directions $p^{(k)}$ are conjugate to each other; therefore, $p^{(k)}$ is computed from not only the residual but also all previous directions:

$$p^{(k)} = r^{(k)} - \sum_{i<k} \frac{\left(p^{(i)}\right)^T A r^{(k)}}{\left(p^{(i)}\right)^T A p^{(i)}} p^{(i)} \tag{6.10}$$

where $r^{(k)} = b - Ax^{(k)}$. With this direction, the location of the solution may be obtained from

$$x^{(k+1)} = x^{(k)} + \alpha^{(k)} p^{(k)} \tag{6.11}$$

where

$$\alpha^{(k)} = \frac{\left(p^{(k)}\right)^T r^{(k)}}{\left(p^{(k)}\right)^T A p^{(k)}}$$

The residuals are also updated as

$$r^{(k+1)} = b - Ax^{(k+1)} = b - A\left(x^{(k)} + \alpha^{(k)} p^{(k)}\right) = b - Ax^{(k)} - \alpha^{(k)} A p^{(k)} = r^{(k)} - \alpha^{(k)} q^{(k)}$$

where $q^{(k)} = A p^{(k)}$.

Thereafter, the search directions are updated with the new residuals $r^{(k+1)}$

$$p^{(k+1)} = r^{(k+1)} + \beta^{(k)} p^{(k)}$$

where

$$\beta^{(k)} = \frac{\left(r^{(k+1)}\right)^T \left(r^{(k+1)}\right)}{\left(r^{(k)}\right)^T \left(r^{(k)}\right)}$$

The pseudocode to implement CG method is given below. This algorithm requires less computational space than expected since the search direction depends upon the residual as well as previous search directions. However, because of the fact that $p^{(k)}$ and $r^{(k)}$ span same Krylov subspace, the space requirement is significantly taken care of.

Pseudocode:

- Start with initial guess $x^{(0)}$
- Calculate the residual: $r^{(0)} = b - Ax^{(0)}$
- Take initial search direction: $p^{(0)} = r^{(0)}$
- $k = 0$
- Repeat
- Calculate $\alpha^{(k)} = \dfrac{\left(r^{(k)}\right)^T \left(r^{(k)}\right)}{\left(p^{(k)}\right)^T \left(Ap^{(k)}\right)}$
- Next iterate: $x^{(k+1)} = x^{(k)} + \alpha^{(k)} p^{(k)}$
- Update the residual: $r^{(k+1)} = r^{(k)} - \alpha^{(k)} Ap^{(k)}$
- If $r^{(k+1)}$ is sufficiently small (as per the tolerance), exit the loop
- Compute $\beta^{(k)} = \dfrac{\left(r^{(k+1)}\right)^T \left(r^{(k+1)}\right)}{\left(r^{(k)}\right)^T \left(r^{(k)}\right)}$
- Estimate the new search directions: $p^{(k+1)} = r^{(k+1)} + \beta^{(k)} p^{(k)}$
- Increment: $k = k + 1$
- End Repeat
- Return $x^{(k+1)}$ as the result

6.5.2   GENERALIZED MINIMUM RESIDUAL (GMRES) METHOD

As discussed above, one of the most accepted iterative methods for solving large sparse symmetric positive definite linear system of equations is CG method and in some cases a combination of CG method with some preconditioning techniques [12,13]. For a short introduction on preconditioning, see Appendix (A.1). However, for solving non-symmetric linear systems, GMRES method is mostly used. Saad and Schultz [14] developed this method in 1986. This method generates a sequence of orthogonal vectors with the help of all previously computed vectors in the orthogonal sequence, which, therefore, necessitates the use of only restarted versions of GMRES. In case no restarts are used, GMRES will converge in almost n steps like any other Krylov subspace method, which will be of no practical use if n is large (due to prohibitive storage computational requirements). Therefore, the success of GMRES method depends on the decision of when to restart.

Contrary to CG method where the residuals form an orthogonal basis for the space span $\left(r^{(0)}, Ar^{(0)}, A^2 r^{(0)}, \ldots \right)$, the basis in GMRES method is formed using modified Gram–Schmidt orthogonalization, which becomes Arnoldi method [15] when it is applied to Krylov sequence $(A^n r^{(0)})$. The following algorithm is used to construct the basis using Arnoldi method:

$$
\begin{aligned}
&w^{(i)} = Av^{(i)} \\
&\textbf{for } j = 1, \ldots, i \\
&\quad w^{(i)} = w^{(i)} - \left(w^{(i)}, v^{(j)} \right) v^{(j)} \\
&\textbf{end} \\
&v^{(i+1)} = w^{(i)} / \left\| w^{(i)} \right\|
\end{aligned}
$$

In the above algorithm, the inner product coefficients $(w^{(i)}, v^{(j)})$ and the norm $\left\| w^{(i)} \right\|$ are stored in an upper Hessenberg matrix. After the basis is formed, the GMRES iterates are constructed as

$$
x^{(k)} = x^{(0)} + y_1 v^{(1)} + \ldots + y_k v^{(k)}
$$

In the above equation, y_j's have been chosen to minimize the residual norm $\left\| b - Ax^{(k)} \right\|$. The complete algorithm may be found here [14]. A routine to implement restarted GMRES algorithm in FORTRAN 90 has been given here [16]. The routine is given as *subroutine mgmres_st* (N, NZ_NUM, IA, JA, A, X, RHS, ITR_MAX, MR, TOL_ABS, & TOL_REL)

In order to use the subroutine, the elements of the matrix are assumed to be stored in *sparse triplet* form. In these triplets, only non-zero entries of matrix A need to be stored. The sparse triplets are

1. $A(k)$: this vector stores the non-zero value of the elements of sparse matrix A.
2. $IA(k)$: this vector stores the row indices of the non-zero entry.
3. $JA(k)$: this vector stores the column indices of the non-zero entry.

Other arguments in the subroutine are

1. N: Order of the linear system (i.e., number of equations)
2. NZ_NUM: The number of non-zero matrix entries
3. $IA(NZ\_NUM)$, $JA(NZ\_NUM)$: Row and column indices of non-zero matrix values
4. $A(NZ\_NUM)$: non-zero matrix values

5. $X(N)$: The solution array: on input, it will contain an approximation to the solution, and on output, it will contain an improved approximation. At the end of iteration, this array will give the final solution.
6. $RHS(N)$: The right-hand side of the linear system
7. ITR_MAX: The maximum number of (outer) iterations
8. MR: The maximum number of (inner) iterations $0 < MR \leq N$.
9. TOL_ABS: Absolute tolerance applied to the current residual
10. TOL_REL: Relative tolerance comparing the current residual to the initial residual.

The above GMRES subroutine (*mgmres_st*) has been used to solve following 2D parabolic PDE on a rectangular domain:

$$\frac{\partial T}{\partial t} = \alpha \left(\frac{\partial^2 T}{\partial x^2} + \frac{\partial^2 T}{\partial y^2} \right) \text{ for } x_L < x < x_R \text{ and } y_B < y < y_T$$

where α is thermal diffusivity. The boundary conditions are (Dirichlet)

$$T(x_L, y, t) = T_L \quad \text{Left boundary}$$
$$T(x_R, y, t) = T_R \quad \text{Right boundary}$$
$$T(x, y_B, t) = T_B \quad \text{Bottom boundary}$$
$$T(x, y_T, t) = T_T \quad \text{Top boundary}$$

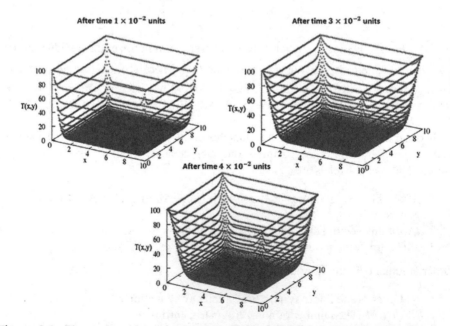

Figure 6.6 The contour plots of temperature field after different time intervals. The time step in this simulation is taken as $\Delta t = 0.001$; the spatial discretization has been taken as $\Delta x = 0.1 = \Delta y$. Other details may be seen from the code given in the Appendix A.2.

and the initial condition is

$$T(x,y,t=0) = T_0$$

The FORTRAN 90 code to solve the above IBVP using GMRES method is given in Appendix (A.2). The *mgmres_st* subroutine is not appended here, and it may be obtained from [16]. The evolution of the $T(x,y,t)$ is shown in Figure 6.6.

6.6 CONCLUSION

This chapter deals with a brief summary of various schemes employed to solve linear PDEs. In particular, the chapter has elaborated more on various important schemes based on Krylov subspace methods like CG and GMRES methods. At the end, GMRES method has been employed to solve a 2D heat transport equation.

APPENDIX

A.1 PRECONDITIONING

Sometimes, the condition of the matrix A in the linear system of equations $Ax = b$ is bad (which is understood from the condition number (κ) of the matrix: $\kappa = \|A\| \|A^{-1}\|$, where $\|A\|$ is Euclidean norm of the matrix A. The matrix is considered bad when it has large condition number). Preconditioning is done to improve the condition of the matrix. In preconditioning, the matrix A is multiplied by a preconditioner as follows:

$$M^{-1}Ax = M^{-1}b \rightarrow M^{-1}(Ax - b) = 0$$

Here, $M^{-1}A$ is the preconditioned matrix, and M^{-1} is the preconditioner.

There are no fixed rules for choosing a preconditioner. The cheapest choice of preconditioner may be $M = I$, which when applied on the linear system will result in the same system, which therefore will not serve the purpose. However, if we take $M = A \rightarrow M^{-1} = A^{-1} \rightarrow M^{-1}A = I$, which has the condition number of 1, results will be obtained in a single iteration. This is the other extreme for selection of a preconditioner. Therefore, M is chosen between these two extremes (i.e., between the extremes $M = I$ and $M = A$), to achieve a minimal number of iterations and at the same time keeping P^{-1} as simple as possible.

A.2 2D PDE CODE USING GMRES METHOD (FOR THE DETAILS OF DISCRETIZATION, PLEASE SEE 6.3.3.2 AND 6.5.2)

```
!==============(Declaration of various variables)=============

module variables
implicit none
integer, parameter:: ntsteps = 40 !number of time steps
double precision, parameter:: dt = 0.001D0, dx = 0.1D0, dy = dx
```

```fortran
double precision, parameter:: alpha = 5.0D0
double precision, parameter:: x0 = 0.0, xL = 10.0, y0 = 0.0,
                              yL = 10.0
!Temperature values at boundaries
double precision, parameter:: TB = 100.0, TT = TB, TL = TB,
                              TR = TB
!Considering dx = dy
double precision, parameter:: lambda = (alpha*dt)/(dx*dx)
integer, parameter:: nx = ((xL-x0)/dx)+1 !number of nodes
                                          along x
integer, parameter:: ny = ((yL-y0)/dy)+1 !number of nodes
                                          along y
integer, parameter:: nz_diag = nx - 4 !number of zeros diagonal
integer, parameter:: neq = (nx-2)*(ny-2)
                              !number of linear equations
integer, parameter:: nz_num_cen = neq
                  !non-zero elements in central diagonal
integer, parameter:: nz_num_d1t = (nx-3)*(mod(nx-3,neq-1)+1)
                                          !nonzero elements
                                          !in top d1 sub-diag
integer, parameter:: nz_num_d1b = (ny-3)*(mod(ny-3,neq-1)+1)
                                          !nonzero elements
                                          !in bottom d1 sub-dia
integer, parameter:: nz_num_d1 = nz_num_d1t + nz_num_d1t
                                  !total non-zero
                                      !elements in top and bottom
                                      d1 sub-diag
integer, parameter:: nz_num_d2 = 2*(neq-nz_diag-2)
                      !nonzero elements in sub-diagonals 2
integer, parameter:: nz_num = nz_num_cen + nz_num_d1 + nz_num_d2
integer, parameter:: z_num_d1t = nx-3
                          !number of zeros in top d1 sub-diag
integer, parameter:: z_num_d1b = ny-3
                          !number of zeros in bottom d1 sub-diag
end module variables


!================================================================


program pde_solve
use variables
implicit none

integer:: i, itstep, j, k
```

```fortran
integer, parameter:: itr_max = 1
integer, parameter:: mr = 10
double precision:: time, tol_abs, tol_rel
integer, dimension(1:nz_num):: ia, ja
double precision, dimension(1:nz_num):: val_nz
double precision, dimension(1:neq):: rhs, temp_est
double precision, dimension(1:nx, 1:ny):: temp_all
                                  !include boundary temp also
double precision,dimension(1:nx):: xcor
double precision,dimension(1:ny):: ycor

!Node coordinates
do i = 1, nx
   xcor(i) = x0 + (i-1)*dx
enddo
do j = 1, ny
   ycor(j) = y0 + (j-1)*dy
enddo

!Temperatures at the boundary
!Bottom boundary
j = 1
do i = 1, nx
  temp_all(i,j) = TB
enddo

!Left boundary
i = 1
do j = 1, ny
  temp_all(i,j) = TL
enddo

!Right boundary
i = nx
do j = 1, ny
  temp_all(i,j) = TR
enddo

!Top
j = ny
do i = 1, nx
  temp_all(i,j) = TT
enddo
```

```
k = 0
do i = 2, nx-1
  do j = 2, ny-1
    k = k+1
    temp_all(i,j) = 0.0D0
    temp_est(k) = 0.0D0   !temperature at inner nodes
  enddo
enddo

tol_abs = 1.0D-8
tol_rel = 1.0D-8

!Set the matrix
k = 0
do i = 1, neq
  if(i>1) then
    if(mod(i-1,nx-2) /= 0) then !for lower d1 sub-diag
      k = k+1
      ia(k) = i
      ja(k) = i - 1
      val_nz(k) = -lambda
    endif
  endif
  if(i>nz_diag+2) then !for lower d2 sub-diag
    k = k+1
    ia(k) = i
    ja(k) = i - (nz_diag+2)
    val_nz(k) = -lambda
  endif
  k = k+1 !for the main diag
  ia(k)= i
  ja(k) = i
  val_nz(k) = 4*lambda + 1
  if(i<neq) then
    if(mod(i,nx-2) /= 0) then !for upper d1 sub-diag
      k = k + 1
      ia(k) = i
      ja(k) = i+1
      val_nz(k) = -lambda
    endif
  endif
  if(i<neq-(nz_diag+1)) then !for upper d2 sub-diag
    k = k + 1
    ia(k) = i
```

```
      ja(k) = i+(nz_diag+2)
      val_nz(k) = -lambda
   endif
enddo

!Evolution of Temp by solving GMRES routine for sparse system
open(unit=10,file='data.txt')
do itstep = 1, ntsteps
  time = itstep*dt
  !set the right hand side vector
  k = 0
  do i = 2, nx-1
    do j = 2, ny-1
      if(i==2 .and. j==2) then
          k = k+1
          rhs(k) = lambda*temp_all(i-1,j)+lambda*temp_all(i,j-1)
                  +temp_est(k)
      endif
      if(i==2 .and. j==ny-1) then
          k = k+1
          rhs(k) = lambda*temp_all(i-1,j)+lambda*temp_all(i,j+1)
                  +temp_est(k)
      endif
      if(i==nx-1 .and. j==2) then
          k = k+1
          rhs(k) = lambda*temp_all(i+1,j)+lambda*temp_all(i,j-1)
                  +temp_est(k)
      endif
      if(i==nx-1 .and. j==ny-1) then
          k = k+1
          rhs(k) = lambda*temp_all(i+1,j)+lambda*temp_all(i,j+1)
                  +temp_est(k)
      endif
      if(j==2) then
        if(i>2 .and. i<nx-1) then
          k = k+1
          rhs(k) = lambda*temp_all(i,j-1)+temp_est(k)
        endif
      endif
      if(j==ny-1) then
        if(i>2 .and. i<nx-1) then
          k = k+1
          rhs(k) = lambda*temp_all(i,j+1)+temp_est(k)
        endif
```

```
      endif
      if(i==2) then
        if(j>2 .and. j<ny-1) then
          k = k+1
          rhs(k) = lambda*temp_all(i-1,j)+temp_est(k)
        endif
      endif
      if(i==nx-1) then
        if(j>2 .and. j<ny-1) then
          k = k+1
          rhs(k) = lambda*temp_all(i+1,j)+temp_est(k)
        endif
      endif
      if(i>2 .and. i<nx-1 .and. j>2 .and. j<ny-1) then
          k = k+1
          rhs(k) = temp_est(k)
      endif
    enddo
  enddo
 print*, k, neq
!Call the GMRES routine to solve the sparse system
  call mgmres_st(neq, nz_num, ia, ja, val_nz, temp_est, rhs, &
            itr_max, mr,tol_abs, tol_rel)
  k = 0
  do i = 2, nx-1
    do j = 2, ny-1
      k = k+1
      temp_all(i,j) = temp_est(k)
    enddo
  enddo
  write(10,*)"After time = ", time
  do i = 1, nx
    do j = 1, ny
      write(10,*) xcor(i), ycor(j), temp_all(i,j)
    enddo
  enddo
enddo
close(10)
end
```

REFERENCES

1. Epstein, M., *Partial Differential Equations: Mathematical Techniques for Engineers.* Springer (2017). doi: 10.1007/978-3-319-55212-5.
2. Negero, N. T., Fourier transform methods for partial differential equations. *International Journal of Partial Differential Equations and Applications* 2(3) (2014): 44–57.
3. Handibag, S. & Karande, B. D., Laplace substitution method for solving partial differential equations involving mixed partial derivatives. *International Journal of Pure and Applied Mathematics* 78(7) (2012): 973–979.
4. Shivamoggi, B., *Perturbation Methods for Differential Equations.* Birkhäuser Basel (2003). doi: 10.1007/978-1-4612-0047-5.
5. Struwe, M., *Variational Methods.* Springer-Verlag, Berlin Heidelberg (2008). doi: 10.1007/978-3-540-74013-1.
6. Young, D. L., et al., The methods of fundamental solutions with eigen function expansion method for 3D nonhomogeneous diffusion equations. *Numerical Methods for Partial Differential Equations* 25(1) (2008): 195–211. doi: 10.1002/num.20336.
7. Eymard, R. et al., The finite volume method. In: *Handbook of Numerical Analysis.* P. G. Ciarlet and J. L. Lions (Eds.) vol. 7, pp. 713–1020 (2000).
8. Tinsley Oden, J. et al., *Finite Elements: An Introduction.* Prentice Hall, Englewood Cliffs, NJ (1981).
9. Burden, R. L. & Faires, J. D., *Numerical Analysis.* Brooks/Cole, Cenage Learning, Canada (2011).
10. Saad, Y., *Iterative Methods for Sparse Linear Systems.* SIAM, Philadelphia, PA.
11. Hestenes, R. & Stiefel, E., Methods of conjugate gradients for solving linear systems. *Journal of Research of the National Bureau of Standards* 49 (1952): 409–435.
12. Chandra, R., Conjugate gradient methods for partial differential equations. Ph.D. thesis, Computer Science Department, Yale University, New Haven, CT (1978).
13. Hageman, A. L. & Young, D. M., *Applied Iterative Methods.* Academic Press, New York (1981).
14. Saad, Y. & Schultz, M., GMRES: A generalized minimal residual algorithm for solving nonsymmetric linear systems. *SIAM Journal on Scientific Computing* 7 (1986): 856–869.
15. Arnoldi, W., The principle of minimized iterations in the solution of the matrix eigenvalue problem. *Quarterly of Applied Mathematics* 9 (1951), 17–29.
16. https://people.sc.fsu.edu/~jburkardt/f_src/mgmres/mgmres.html.

7 The (2+1) Dimensional Nonlinear Sine–Gordon Soliton Waves and its Numerical Simulations

Mohammad Tamsir
Graphic Era (Deemed to be University) Dehradun

Neeraj Dhiman
Graphic Era Hill University Dehradun

CONTENTS

7.1   INTRODUCTION

The nonlinear sine–Gordon equation (SGE)has been found in numerous physical applications, viz., optical fiber signals, tsunamis, atmospheric waves, superconductivity, and gravitational fields. Due to a large number of applications, the simulation of this equation has enormous significance in research.

Consider the 2D nonlinear SGE,

$$u_{tt} + \gamma u_x = u_{xx} + u_{yy} - f(x,y)\sin(u), \ x \in R, \ t \geq 0, \tag{7.1}$$

where $R = \{(x,y) : a_1 \leq x \leq a_2, b_1 \leq y \leq b_2\}$, with initial conditions (ICs):

$$u = \varphi_1(x,y) \text{ and } u_t = \varphi_2(x,y), \text{ at } t = 0, \tag{7.2}$$

and Neumann boundary conditions (BCs):

$$u_n = g_1(x,y,t), \; x = a \text{ and } b, c \leq y \leq d,$$
$$u_n = g_2(x,y,t), \; y = c \text{ and } d, a \leq x \leq b. \tag{7.3}$$

where $u = u(x,y,t)$, $h = h(x,y)$, $\varphi_1 = \varphi_1(x,y)$, $\varphi_2 = \varphi_2(x,y)$, and $\gamma \geq 0$ is a dissipative term. The Equation (7.1) turns out into undamped SGE for $\gamma = 0$ and into a damped one for $\gamma > 0$. The functions φ_1 and φ_2 describe kinks and velocity, respectively, whereas f represents Josephson current density.

The SGE appears in the dispersion of the fluxons in Josephson's junction, disruptions in crystals, stability of fluid motions, motion of a rigid pendulum [1–5], etc. Josephson junction model [4] consists of two separated superconducting layers which can be described by 2D SGE and has numerous applications in physical science, electronics, etc. A soliton-like structure has been modeled by Djidjeli et al. [6] in higher dimensions.

The various methods have been proposed in recent years for the solution of 2D SGE. Christiansen and Lomdahl [7] used a leapfrog method in generalized form, whereas Argyris et al. [8]used finite element method. Both methods were successfully used, the latter giving slightly more accurate results. Xin [9] modeled the light bullets using this equation and viewed that sine–Gordon pulse envelopes undertake focus and defocus cycles. The "evolution of lump and ring solitons" of the 2D SGE and that of the breather-type waves while standing and on the move were examined in [10,11]. Sheng et al. [12] used a "split cosine scheme" to solve 2D SGE. Bratsos [13–17] proposed explicit and improved finite-difference methods, modified predictor–corrector method, method of lines, and a numerical scheme of order 3, to solve2D SGE.

Dehghan and Shokri [18] proposed a thin-plate radial basis splines collocation method, whereas Dehghan and Mirzaei [19] proposed a dual reciprocity boundary element method to solve 2D SGE. Mirzaei and Dehghan [20] studied the boundary element solution by applying continuous linear element approximation, while meshless LBIE and MLPG methods have been proposed in [21,22] for the solution of 2D SGE. Dehghan and Ghesmati [23] proposed a radial point interpolation method (RPIM)-based local weak meshless method. Jiwari et al. [24] introduced polynomial differential quadrature method (PDQM) to solve damped and undamped 2D SGE.

The differential quadrature method (DQM) has been presented in references [25–30] for solving differential equations. Korkmaz and Dag [31] presented cubic B-spline DQM for Burgers' equation. Gottlieb [32] developed an "optimal strong stability preserving high order time discretization scheme." Abbas et al. [33] proposed "cubic trigonometric B-spline (CTB) collocation method" for the wave equation. Recently, Tamsir et al. [34] presented a cubic trigonometric B-spline DQM for Fisher's reaction–diffusion equations. The CTB functions work as a basis function in DQM for computing the weight coefficients of derivatives with respect to space variables.

In this chapter, we simulate2D SGE numerically for a large number of cases together with ring and line solitons of the elliptic and circular shapes. Both cases of damped and undamped are chosen for numerical simulation. First, the accuracy and convergence rate of the method have been tested by taking a test problem. In rest of the problems, the 2D damped and undamped SGEs have been simulated. The results have been compared with the results existing in the literature.

7.2 DESCRIPTION OF THE METHOD

It is supposed that the domain R is partitioned uniformly by N_x and N_y grid points in x and y directions with step size $h = 1/(N_x - 1)$ and $k = 1/(N_y - 1)$, respectively.

The rth-order partial derivative of u with respect to x and y is given by

$$\frac{\partial^r u_{ij}}{\partial x^r} = \sum_{s=1}^{N_x} w_{is}^{(r)} u_{sj}, \tag{7.4}$$

$$\frac{\partial^r u_{ij}}{\partial y^r} = \sum_{s=1}^{N_y} \tilde{w}_{js}^{(r)} u_{is}, \tag{7.5}$$

where $i = 1, 2, \ldots, N_x$, $j = 1, 2, \ldots, N_y$ and $u_{ij} = u(x_i, y_j, t)$. The terms $w_{ij}^{(r)}$ and $\tilde{w}_{ij}^{(r)}$, $r = 1, 2$ represent the rth-order weight coefficients of the partial derivatives with respect to space variables x and y.

The CTB basis functions [33,34] at the knots are given as

$$T_k(x) = \frac{1}{\varsigma}
\begin{cases}
\delta^3(x_{k-2}), & x \in [x_{k-2}, x_{k-1}) \\
\begin{aligned}
&\{\eta(x_k)\delta(x_{k-2}) + \eta(x_{k+1})\delta(x_{k-1})\}\delta(x_{k-2}) \\
&\quad + \eta(x_{k+2})\delta^2(x_{k-1}),
\end{aligned} & x \in [x_{k-1}, x_k) \\
\begin{aligned}
&\eta^2(x_{k+1})\delta(x_{k-2}) + \{\delta(x_{k-1})\eta(x_{k+1}) \\
&\quad + \delta(x_k)\eta(x_{k+2})\}\eta(x_{k+2}),
\end{aligned} & x \in [x_k, x_{k+1}) \\
\eta^3(x_{k+2}), & x \in [x_{k+1}, x_{k+2}) \\
0 & \text{otherwise}
\end{cases} \tag{7.6}$$

where $\delta = \sin(0.5(x - x_k))$, $\eta = \sin(0.5(x_k - x))$, $\varsigma = \sin(0.5h)\sin(h)\sin(1.5h)$.

The values of s at the knots are given as

$$T_k(x_j) = \begin{cases} b, & \text{if } k-j=0 \\ a, & \text{if } k-j=\pm 1 \\ 0, & \text{else} \end{cases}; \quad T'_j(x_i) = \begin{cases} c, & \text{if } k-j=1 \\ -c, & \text{if } k-j=-1 \\ 0, & \text{else} \end{cases};$$

$$T''_j(x_i) = \begin{cases} e, & \text{if } k-j=0 \\ d, & \text{if } k-j=\pm 1 \\ 0, & \text{else} \end{cases},$$

where $a = \sin^2(0.5h)\csc(h)\csc(1.5h)$, $b = 2/(2\cos(h)+1)$, $c = 0.75\csc(1.5h)$,
$d = \dfrac{3\csc^2(0.5h)\{1+3\cos(h)\}}{16+(\cos(1.5h)+2\cos(0.5h))}$, $e = \dfrac{-3\cot^2(1.5h)}{4\cos(h)+2}$.

The CTB basis functions have been modified as

$$\begin{rcases}
\overset{\leftrightarrow}{T}_1 = T_1(x) + 2T_0(x) \\
\overset{\leftrightarrow}{T}_2(x) = T_2(x) - T_0(x) \\
\overset{\leftrightarrow}{T}_p(x) = T_p(x) \text{ for } p = 3, \ldots, N_x - 2 \\
\overset{\leftrightarrow}{T}_{N_x-1}(x) = T_{N_x-1}(x) - T_{N_x+1}(x) \\
\overset{\leftrightarrow}{T}_{N_x}(x) = T_{N_x}(x) + 2T_{N_x+1}(x)
\end{rcases}, \tag{7.7}$$

where $\left\{ \overset{\leftrightarrow}{T}_1, \overset{\leftrightarrow}{T}_2, \ldots, \overset{\leftrightarrow}{T}_{N_x} \right\}$ forms a basis in the domain R.

For finding the weighting coefficients $w_{ij}^{(1)}$, we take $r = 1$ and use the values of $\overset{\leftrightarrow}{T}_k(x)(k = 1, 2, \ldots, N_x)$ in Equation (7.4) which implies

$$\overset{\leftrightarrow}{T}'_k(x_i) = \sum_{s=1}^{N_x} w_{is}^{(1)} \overset{\leftrightarrow}{T}_k(x_s), \text{ for } i = 1, 2, \ldots, N_x; k = 1, 2, \ldots, N_x \tag{7.8}$$

By using Equation (7.6) and Lemma 1 in Equation (7.8), we get

$$A w_{is}^{(1)} = \vec{R}[i], \text{ for } i = 1, 2, \ldots, N_x; s = 1, 2, \ldots, N_x, \tag{7.9}$$

where A is tri-diagonal matrix and given by

$$A = \begin{bmatrix}
2a+b & b & & & & & \\
0 & b & a & & & & \\
 & a & b & a & & & \\
 & & \ddots & \ddots & \ddots & & \\
 & & & a & b & a & \\
 & & & & a & b & 0 \\
 & & & & & a & 2a+b
\end{bmatrix}_{N \times N}$$

where the vectors $\vec{R}[i]$ are given by

$$\vec{R}[1] = \begin{bmatrix} -2c & 2c & 0 & 0 & \cdots & 0 & 0 \end{bmatrix}^T,$$
$$\vec{R}[2] = \begin{bmatrix} -c & 0 & c & 0 & \cdots & 0 & 0 \end{bmatrix}^T,$$
$$\vdots$$
$$\vec{R}[N_x - 1] = \begin{bmatrix} 0 & 0 & \cdots & 0 & -c & 0 & c \end{bmatrix}^T,$$
$$\vec{R}[N_x] = \begin{bmatrix} 0 & 0 & \cdots & 0 & 0 & -2c & 2c \end{bmatrix}^T.$$

The system Equation (7.9) is solved by using "Thomas algorithm" in order to get $w_{is}^{(1)}$. The weighting coefficients $w_{is}^{(2)}$, $i, s = 1, 2, \ldots, N_x$ are calculated by using the formula of Shu [30] which is given by

$$\begin{cases} a_{is}^{(r)} = r\left(a_{is}^{(1)} a_{ii}^{(r-1)} - \dfrac{a_{is}^{(r-1)}}{x_i - x_s}\right), & \text{for } s \neq i \text{ and } i = 1, 2, \ldots, N_x; \ r = 2, 3, \ldots, N_x - 1, \\ a_{ii}^{(r)} = -\sum_{s=1, j\neq i}^{N_x} a_{is}^{(r)}, & \text{for } s = i. \end{cases}$$

$$(7.10)$$

7.3 IMPLEMENTATION OF METHOD TO 2D SGE

The space derivatives in the 2D SGE Equation (7.1) are estimated by modified cubic trigonometric B-spline (MCTB)-DQM. Consequently, the Equation (7.1) transformed into the following form:

$$\frac{d^2 u_{ij}}{dt^2} + \beta \frac{du_{ij}}{dt} = \sum_{s=1}^{N_x} w_{is}^{(2)} u_{sj} + \sum_{s=1}^{N_y} \tilde{w}_{js}^{(2)} u_{is} - f(x_i, y_j) \sin u_{ij}$$

$$i = 1, 2, \ldots, N_x, \ j = 1, 2, \ldots, N_y, \tag{7.11}$$

together with ICs:

$$u_{ij} = \varphi_1(x_i, y_j) \text{ and } \frac{du_{ij}}{dt} = \varphi_2(x_i, y_j) \text{ at } t = 0s. \tag{7.12}$$

Now let

$$\frac{du_{ij}}{dt} = v_{ij}. \tag{7.13}$$

Using approximation Equation (7.13), Equation (7.11) reduces into first order

$$\frac{dv_{ij}}{dt} + \beta v_{ij} = \sum_{s=1}^{N_x} w_{is}^{(2)} u_{sj} + \sum_{s=1}^{N_y} \tilde{w}_{js}^{(2)} u_{is} - f(x_i, y_j) \sin u_{ij}. \tag{7.14}$$

Now, the estimation of Neumann BCs Equation (7.3) at $x = a_1$ and $x = a_2$ is given as

$$\sum_{s=1}^{N_x} w_{1s}^{(1)} u_{sj} = g_1(a_1, y, t), \tag{7.15}$$

$$\sum_{s=1}^{N_x} w_{N_x s}^{(1)} u_{sj} = g_2(a_2, y, t) \tag{7.16}$$

By solving Equations (7.15) and (7.16) for u_{1j} and $u_{N_x j}$, we have

$$u_{1j} = \frac{w_{1N_x}^{(1)}(g_2 - A_2) - w_{N_x N_x}^{(1)}(g_1 - A_1)}{\left(w_{1N_x}^{(1)} w_{N_x 1}^{(1)} - w_{11}^{(1)} w_{N_x N_x}^{(1)}\right)}, \tag{7.17}$$

$$u_{N_x j} = \frac{w_{N_x 1}^{(1)}(g_1 - A_1) - w_{11}^{(1)}(g_2 - A_2)}{\left(w_{1N_x}^{(1)} w_{N_x 1}^{(1)} - w_{11}^{(1)} w_{N_x N_x}^{(1)}\right)}, \tag{7.18}$$

where $A_1 = \sum_{s=2}^{N_x-2} w_{1s}^{(1)} u_{sj}$, $A_2 = \sum_{s=2}^{N_x-2} w_{N_x s}^{(1)} u_{sj}$, $g_1 = g_1(a_1, y, t)$, $g_2 = g_2(a_2, y, t)$, and $j = 1, 2, \ldots, N_y$.

Similarly, the estimation of Neumann BCs Equation (7.3) at $y = b_1$ and $y = b_2$ is given by

$$\sum_{s=1}^{N_y} \tilde{w}_{1s}^{(1)} u_{is} = g_3, \tag{7.19}$$

$$\sum_{s=1}^{N_y} \tilde{w}_{N_y s}^{(1)} u_{is} = g_4, \tag{7.20}$$

By solving the Equations (7.19) and (7.20) for u_{i1} and u_{iN_y}, we have

$$u_{i1} = \frac{\tilde{w}_{1N_y}^{(1)} (g_4 - A_4) - \tilde{w}_{N_y N_y}^{(1)} (g_3 - A_3)}{\left(\tilde{w}_{1N_y}^{(1)} w_{N_y 1}^{(1)} - \tilde{w}_{11}^{(1)} \tilde{w}_{N_y N_y}^{(1)} \right)}, \tag{7.21}$$

$$u_{iN_y} = \frac{\tilde{w}_{N_y 1}^{(1)} (g_3 - A_3) - \tilde{w}_{11}^{(1)} (g_4 - A_4)}{\left(\tilde{w}_{1N_y}^{(1)} \overline{w}_{N_y 1}^{(1)} - \tilde{w}_{11}^{(1)} \tilde{w}_{N_y N_y}^{(1)} \right)}, \tag{7.22}$$

where $A_3 = \sum_{s=2}^{N_y - 2} \tilde{w}_{1s}^{(1)} u_{is}$, $A_4 = \sum_{s=2}^{N_y - 2} \tilde{w}_{N_y s}^{(1)} u_{is}$, $g_3 = g_3(x, b_1, t)$, $g_4 = g_4(x, b_2, t)$, and $i = 1, 2, \ldots, N_x$

The system of ODEs Equation (7.14) together with ICs Equation (7.12) and Neumann BCs. Equations (7.17), (7.18), (7.21), and (7.22) is solved by strong-stability preserving Runge-Kutta (SSP-RK) (5,4) method.

7.4   RESULTS AND DISCUSSION

The six examples have been considered in this section for numerical simulation. First of all, the accuracy and efficiency of the method are tested in terms of RMS and L_∞ error norms through test problem. The RMS and L_∞ error norms have been evaluated by the formulae:

$$L_\infty = \max \left| u_{ij}^e - u_{ij}^{num} \right| \text{ and RMS} = \frac{1}{N_x \times N_y} \sqrt{\sum_{i=1}^{N_x} \sum_{j=1}^{N_y} \left| u_{ij}^e - u_{ij}^{num} \right|^2},$$

where $i = 1, 2, \ldots, N_x$ and $j = 1, 2, \ldots, N_y$. The terms u_{ij}^e and u_{ij}^{num} represent exact and numerical solutions. For rest of problems, the obtained results have been simulated through surface plots.

Test Problem: The SGE Equation (7.1) is considered for $f(x, y) = -1$ and $\beta = 0$ in $[-7, 7] \times [-7, 7]$ with ICs:

$$u(x, y, 0) = 4 \tan^{-1} \left(e^{x+y} \right) \text{ and } u_t(x, y, 0) = -\frac{4e^{x+y}}{1 + e^{2(x+y)}} \tag{7.23}$$

and Neumann BCs:

$$\left. \begin{array}{l} u_x = -\frac{4e^{x+y+t}}{\exp(2t) + e^{2(x+y)}}, \ x = -7 \text{ and } x = 7, y \in [-7, 7], \\[2mm] u_y = -\frac{4e^{x+y+t}}{\exp(2t) + e^{2(x+y)}}, \ y = -7 \text{ and } y = 7, x \in [-7, 7], \end{array} \right\} \tag{7.24}$$

Table 7.1

Comparison of Error Norms for Test Problem

t	L_∞ – Error					RMS-Error			
	[6]	[18]	[24]	[35]	Present	[18]	[24]	[35]	Present
1	3.50e-02	6.70e-02	2.7e-03	3.0e-04	2.52e-04	5.0e-03	5.0e-04	2.0e-04	1.53e-04
3	4.31e-02	8.34e-02	2.0e-03	6.0e-04	5.79e-04	1.03e-02	5.0e-05	4.0e-04	3.79e-04
5	4.04e-02	1.01e-01	3.3e-03	8.0e-04	7.74e-04	1.45e-02	7.0e-04	7.0e-04	6.63e-04
7	3.50e-02	1.516e-01	5.9e-03	1.2e-03	1.15e-03	1.87e-02	1.1e-03	1.0e-03	9.94e-4

and with the exact solution

$$u(x,y,t) = 4\ \tan^{-1}\left(e^{x+y-t}\right).$$

The solution of test problem is solved using the parameter values: $\Delta t = 0.001$ and $h = \Delta x = \Delta y = 0.25$ at different t and presented in Table 7.1 which shows that obtained results are finer than the results presented in [13,14,17,32].

7.4.1 CIRCULAR RING SOLITONS

The circular ring solitons of SGE Equation (7.1) have been obtained for $f(x,y) = -1$ and $\beta = 0$ in $[-7, 7] \times [-7, 7]$ with ICs:

$$\left.\begin{array}{l} u(x,y,0) = 4\ \tan^{-1}\left(e^{\left(3-\sqrt{x^2+y^2}\right)}\right), \\ u_t(x,y,0) = 0, \end{array}\right\} \tag{7.25}$$

and Neumann B.C.'s:

$$\left.\begin{array}{l} u_x = 0,\ \text{for } x = -7 \text{ and } x = 7, y \in [-7,7], t > 0 \\ u_y = 0,\ \text{for } y = -7 \text{ and } y = 7, x \in [-7,7], t > 0 \end{array}\right\}. \tag{7.26}$$

The surface and contour curves of the circular ring soliton waves are depicted in Figure 7.1 at $t = 0, 2.8, 5.6, 8.4, 11.2, 12.6$ with $\Delta t = 0.2$ and $\Delta x = \Delta y = 0.4$ in terms of $\sin(u/2)$. The ring soliton is shrinking from its initial position until $t = 2.8$ and comes into view of a single ring soliton. A radiation comes into view from $t = 5.6$ which is an expansion stage and continues until $t = 11.2$ where the ring soliton is almost rehabilitated. The graphs shown in Figure 7.1 are in excellent agreement with the plots given in Refs. [6–8,14,15,23,24,35].

7.4.2 ELLIPTICAL RING SOLITONS

The elliptical ring solitons of SGE Equation (7.1) have been obtained for $f(x,y) = -1$ and $\beta = 0$ in $[-7, 7] \times [-7, 7]$ with ICs:

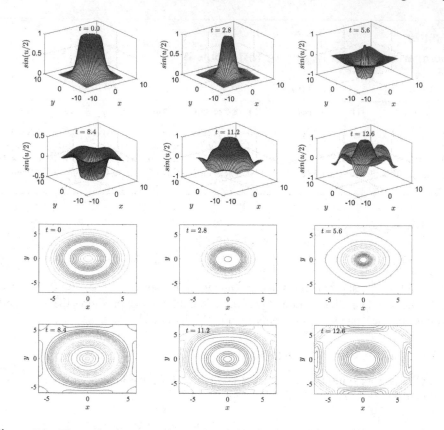

Figure 7.1 The surface curves and contours of circular ring solitons at different times.

$$u(x,y,0) = 4\tan^{-1}\left(e^{\left(3-\sqrt{\frac{(x-y)^2}{3}+\frac{(x+y)^2}{2}}\right)}\right),\left.\begin{array}{c}\\\\\end{array}\right\}$$

$$u_t(x,y,0) = 0.$$
(7.27)

The Neumann boundary conditions in Equation (7.26) have been used. The surface and contour curves of the elliptical ring solitons are depicted in Figure 7.2 at $t = 0$, 1.6, 3.2, 4.8, 6.4, 8.0 with $\Delta t = 0.2$ and $\Delta x = \Delta y = 0.4$ in terms of $\sin(u/2)$. It is observed that the soliton of the elliptical ring contracts from its early position up to $t = 3.2$. The temporary behavior of the soliton waves consists of contraction and explosion phases. The formation of a circular ring soliton has begun from $t = 8.0$ as given in references [8,16,19,20,23,24,35].

7.4.3 ELLIPTICAL BREATHER

The elliptical breather for SGE Equation (7.1) has been obtained for $f(x,y) = -1$ and $\beta = 0$ in the $[-7, 7] \times [-7, 7]$ with ICs:

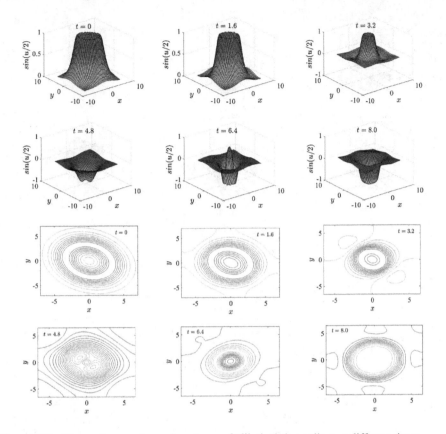

Figure 7.2 The surface curves and contours of elliptical ring soliton at different times.

$$u(x,y,0) = 4\tan^{-1}\left(2.0\operatorname{sech}\left(0.866\sqrt{\frac{(x-y)^2}{3} + \frac{(x+y)^2}{2}}\right)\right), \right\}$$

$$u_t(x,y,0) = 0. \tag{7.28}$$

The Neumann boundary conditions in Equation (7.26) have been used. Figure 7.3 shows the surface curves and contours at $t = 0, 1.6, 8.0, 9.6, 11.2, 12.8$ with $\Delta t = 0.2$ and $\Delta x = \Delta y = 0.4$ for the elliptical breather. The graphs disclose that the elliptical breather seems rotating in a clockwise direction around the major axis $y = -x$ from its early position and seems shrinking until $t = 1.6$ a reflection stage is observed at $t = 8.0$ and 9.6. At $t = 11.2$, the major axis almost recaptures its early position ($y = -x$) but observes a strong oscillation. From $t = 12.8$, an expansion stage is seen. The graphs presented in this solution wave are in excellent correspondence with the plots given in references [7,8,16,17,23,24,35].

7.4.4 SUPERPOSITION OF TWO ORTHOGONAL LINE SOLITONS

The superposition of two orthogonal line solitons has been studied for $f(x,y) = -1$ and $\beta = 0.05$ in $[-6, 6] \times [-6, 6]$ with ICs:

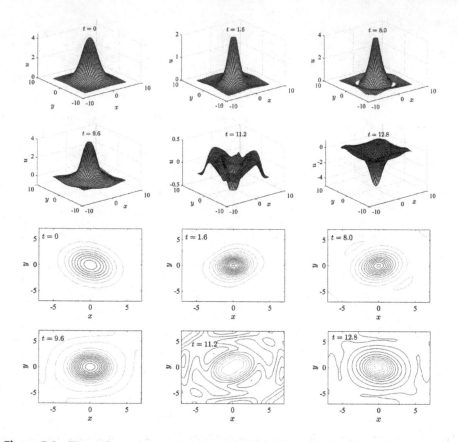

Figure 7.3 The surface curves and contours of elliptical breather at different times.

$$u(x,y,0) = 4\tan^{-1}(e^x + e^y), \\ u_t(x,y,0) = 0. \qquad \left.\right\} \qquad (7.29)$$

This problem has been solved numerically with parameters: $t = 1, 3, 7$, $\Delta t = 0.001$ and size of grid 31×31. Figure 7.4 shows the surface curves and contours. We can see on the graphs that we find the rupture of two orthogonal line solitons moving away and that separations occur between them without any deformation from $t = 1$ to $t = 3$. At $t = 7$, appearance of a deformation has been noticed. A fine agreement has been found with those given in references [24,35].

7.4.5   LINE SOLITONS IN AN INHOMOGENOUS MEDIUM

The line solitons have been studied in an inhomogenous medium for $f(x,y) = -1$ and $\beta = 0.05$ in $[-7, 7] \times [-7, 7]$ with ICs:

$$u(x,y,0) = 4\tan^{-1}\left(e^{\left(\frac{x-3.5}{0.954}\right)}\right) \text{ and } u_t(x,y,0) = 0.629\,\mathrm{sech}\left(e^{\left(\frac{x-3.5}{0.954}\right)}\right). \qquad (7.30)$$

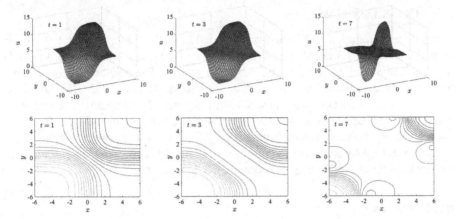

Figure 7.4 The surface curves and contours of superposition of two orthogonal lines at different times.

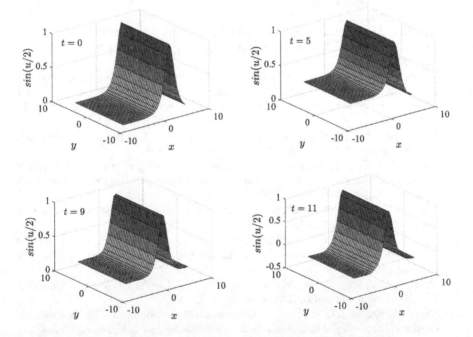

Figure 7.5 The surface curves and contours of line solitons at different times.

This problem has been solved numerically with parameters: $t = 0, 5, 9, 11$, $\Delta t = 0.001$ and size of grid 31×31. Figure 7.5 shows the surface curves and contours. It is clear that the soliton line goes into straight line to a slight extent during the transmission period. A twist has been noticed in its straightness on increasing t. The movement seems to be prevented due to the medium's inhomogeneity. A fine agreement has been found with those of the references [18,24,35].

7.5 CONCLUSIONS

A numerical simulation of 2D nonlinear damped as well as undamped SGE has been carried out via MCTB-DQM. The different cases of the ring and line solitons including elliptical and circular shapes have been discussed. The results and graphs have been compared to those available in references [6–8,14,16–20,23,24,35]. Obtained results are finer than the results available in references [6,18,24,35]. The excellent numerical approximations to the exact solutions have been attained. The surface and contour curves showed that the results are very similar to those available in the aforesaid references. Moreover, the method is economically easy to implement and may be extended for higher dimensional wave-related problems.

REFERENCES

1. Barone, A., Esposito, F., Magee, C.J., Scott, A.C. Theory and applications of the sine-Gordon equation. *RivistadelNuovoCimento*, 1971, 1(2): 227–267.
2. Dodd, R.K., Eilbeck, I.C., Gibbon, J.D. *Solitons and Nonlinear Wave Equations*. Academic Press, London, 1982.
3. Josephson, J.D. Supercurrents through barriers. *Advances in Physics*, 1965, 14: 419–451.
4. Perring, J.K., Skyrme, T.H. A model unified field equation. *Nuclear Physics,* 1962, 31: 550–555.
5. Whitham, G.B. *Linear and Nonlinear Waves*. Wiley-Interscience, New York, 1999.
6. Djidjeli, K., Price, W.G., Twizell, E.H. Numerical solutions of a damped sine-Gordon equation in two space variables. *Journal of Engineering Mathematics*, 1995, 29: 347–369.
7. Christiansen, P.L., Lomdahl, P.S. Numerical solution of (2 + 1) dimensional sine-Gordon solitons. *Physica D*, 1981, 2: 482–494.
8. Argyris, J., Haase, M., Heinrich, J.C. Finite element approximation to two dimensional sine-Gordon solitons. *Computer Methods in Applied Mechanics and Engineering*, 1991, 86: 1–26.
9. Xin, J.X. Modeling light bullets with the two-dimensional sine-Gordon equation. *Physica D*, 2000, 135: 345–368.
10. Minzoni, A.A., Smythb, N.F., Worthy, A.L. Evolution of two-dimensional standing and travelling breather solutions for the sine-Gordon equation. *Physica D*, 2004, 189: 167–187.
11. Minzoni, A.A., Smythb, N.F., Worthy, A.L. Pulse evolution for a two-dimensional sine-Gordon equation. *Physica D*, 2001, 159: 101–123.
12. Sheng, Q., Khaliq, A.Q.M., Voss, D.A. Numerical simulation of two-dimensional sine-Gordon solitons via a split cosine scheme. *Mathematics and Computers in Simulation*, 2005, 68: 355–373.
13. Bratsos, A.G. An explicit numerical scheme for the sine-Gordon equation in 2+1 dimensions. *Applied Numerical Analysis and Computational Mathematics*, 2005, 2(2): 189–211.
14. Bratsos, A.G. An improved numerical scheme for the sine-Gordon equation in (2 +1) dimensions. *International Journal for Numerical Methods in Engineering*, 2008, 75: 787–799.
15. Bratsos, A.G. A modified predictor-corrector scheme for the two dimensional sine-Gordon equation. *Numerical Algorithms*, 2006, 43: 295–308.

16. Bratsos, A.G. The solution of two-dimensional Sine-Gordon equation using the method of lines. *Journal of Computational and Applied Mathematics*, 2007, 206: 251–277.

17. Bratsos, A.G. A third order numerical scheme for the two-dimensional sine-Gordon equation. *Mathematics and Computers in Simulation*, 2007, 76: 271–282.

18. Dehghan, M., Shokri, A. A numerical method for solution of the two dimensional sine-Gordon equation using the radial basis functions. *Mathematics and Computers in Simulation*, 2008, 79: 700–715.

19. Dehghan, M., Mirzaei, D. The dual reciprocity boundary element method (DRBEM) for two-dimensional sine-Gordon equation. *Computer Methods in Applied Mechanics and Engineering*, 2008, 197: 476–486.

20. Mirzaei, D., Dehghan, M. Boundary element solution of the two dimensional sine-Gordon equation using continuous linear elements. *Engineering Analysis with Boundary Elements*, 2009, 33: 12–24.

21. Mirzaei, D., Dehghan, M. Implementation of meshless LBIE method to the 2D nonlinear SG problem. *International Journal for Numerical Methods in Engineering*, 2009, 79: 1662–1682.

22. Mirzaei, D., Dehghan, M. Meshless local Petrov–Galerkin (MLPG) approximation to the two-dimensional sine-Gordon equation. *Journal of Computational and Applied Mathematics*, 2010, 233: 2737–2754.

23. Dehghan, M., Ghesmati, A. Numerical simulation of two-dimensional sine-Gordon solitons via a local weak meshless technique based on the radial point interpolation method (RPIM). *Computer Physics Communications*, 2010, 181: 772–786.

24. Jiwari, R., Pandit, S., Mittal, R.C. Numerical simulation of two-dimensional sine-Gordon solitons by differential quadrature method. *Computer Physics Communications*, 2012, 183: 600–616.

25. Bellman, R., Kashef, B.G., Casti, J. Differential quadrature: A technique for the rapid solution of nonlinear differential equations. *Journal of Computational Physics*, 1972, 10: 40–52.

26. Shu, C., Richards, B.E. Application of generalized differential quadrature to solve two dimensional incompressible navier-Stokes equations. *International Journal for Numerical Methods in Fluids*, 1992, 15: 791–798.

27. Quan, J.R., Chang, C.T. New insights in solving distributed system equations by the quadrature methods-II. *Computers and Chemical Engineering*, 1989, 13: 1017–1024.

28. Shu, C., Xue, H. Explicit computation of weighting coefficients in the harmonic differential quadrature. *Journal of Sound and Vibration*,1997, 204(3): 549–555.

29. Shu, C., Chew, Y.T. Fourier expansion-based differential quadrature and its application to Helmholtz eigenvalue problems. *Communications in Numerical Methods in Engineering*, 1997, 13(8): 643–653.

30. Shu, C. *Differential Quadrature and its Application in Engineering*. Athenaeum Press Ltd., Great Britain, 2000.

31. Korkmaz, A., Dag, I. Cubic B-spline differential quadrature methods and stability for Burgers' equation. *International Journal for Computer-Aided Engineering and Software*, 2013, 30(3): 320–344.

32. Gottlieb, S., Ketcheson, D.I., Shu, C.W. High order strong stability preserving time discretizations. *Journal of Scientific Computing*, 2009, 38: 251–289.

33. Abbas, M., Majid, A.A., Ismail, A.I.M., Rashid, A. The application of cubic trigonometric B-spline to the numerical solution of the hyperbolic problems. *Applied Mathematics and Computation*, 2014, 239: 74–88.

34. Tamsir, M., Dhiman, N., Srivastava, V.K. Cubic trigonometric B-spline differential quadrature method for numerical treatment of Fisher's reaction-diffusion equations. *Alexandria Engineering Journal*, 2018, 57: 2019–2026.
35. Shukla, H.S., Tamsir, M., Srivastava, V.K. Numerical simulation of two dimensional sine-Gordon solitons using modified cubic B-spline differential quadrature method. *AIP Advances*, 2015, 5: 1–14.

8 Dynamical Complexity of Patchy Invasion in Prey–Predator Model

Ramu Dubey
J.C. Bose University of Science and Technology

Teekam Singh
Graphic Era (Deemed to be) and Graphic Era Hill
University Dehradun

CONTENTS

8.1 INTRODUCTION

Formation of spatial patterns in nature is ubiquitous, with illustrations like zebra stripe patterns on animals skin, Turing patterns in a coherent quantum field, or diffusive patterns in predator-prey models [2,25,29]. The spatial factors of species interplay have been recognized as a vital component in how ecological communities are created and ecological interplay occurs over a broad limit of temporal and spatial scale [12]. Spatial population distribution is of major importance in the study of ecological systems [17,18,30]. Mechanisms and scenarios characterizing the spatial population distribution of ecological species in spatial habitat are a focus of special interest in population dynamics. The spatial population distribution is affected by the proliferation capacity of the species and interactions between individuals [35]. Spatial effect may be disregarded in a certain extent, particularly when the population of a given species stay fixed in space at any moment of time. Albeit this assumption is not completely realistic. Individuals of an ecological species do not fix at all times in space, and their dispersion in space changes incessantly by the self-movement of individuals [17,19,21,25,26].

Spatiotemporal mathematical model is an appropriate tool for investigating fundamental mechanism of complex spatiotemporal population dynamics. An appropriate mathematical structure to explain the spatial aspect of population dynamics is specified by reaction-diffusion equations. Reaction-diffusion models were initially applied to describe the ecological pattern formation by Segel and Jackson in 1972, based on the primary work of Turing [33]. Over the last several decades, a number of papers have been published on the spatial dynamics of predator-prey model based on reaction-diffusion equations, and different types of patterns have emerged from these models [17,19,21,25,26,29,33].

Idea of diffusion may be considered as the natural propensity for a cluster of particles at the beginning concentrated close to a location in space to spread out in time, slowly occupying an ever sizable area close to the initial point. Here, the word "particles" mentions not only to physical portion of the matter, but to biological populations or to any other recognizable elements as well. Moreover, the word "space" mentions not only to general Euclidean n-space but also to an hypothetical living space (such as ecological space) [14,17,25,26]. Diffusion is a natural phenomenon where physical material moves from an area of high concentration to an area of low concentration; that is, diffusion is a natural process by which the particle cluster as an entire dispersion according to the non-uniform movement of every particle. Diffusion can be defined to be basically an invariant process by which particle clusters, population, etc. diffuse inside a given space according to individual random movement [26].

Diffusion–Reaction partial differential equation systems can be used to represent mathematical models, which describe how the individuals of one or more species distributed in space change under the effect of two procedures, first is local interaction, in which the species interact with each other, and second is the diffusion, which causes the species to spread out over a surface in space. Mathematically, reaction–diffusion systems take the form of semi-linear parabolic partial differential equations [17,21,26].

Through mathematical modeling as a viable tool, complex biological processes are studied. Mathematical modeling can be extremely helpful in analyzing factors that may contribute to the complexity intrinsic in insufficiently understood tumor–immune as well as Prey–Predator interactions. Likewise, the primary objective of the mathematical modeling of tumor–immune and Prey–predator models are, briefly, the analysis of the interplay inside and between biological species and their artificial surrounding, and the examination of the temporal transformation of clusters of individuals of different biological species. It is however true that space and time are indivisible "sibling coordinates" and only when population densities (tumor–immune system or Prey–Predator system) are contemplated in both space and time, actual dynamics can be understood [3,5,13,16–18,20,27,39,40].

Cooperative behavior can stimulate a relation among the population density and per capita population growth rate [10,31]. Ecologists have accepted several mechanisms for stimulating cooperative behavior in prey, namely cooperating reproduction and foraging capacity. The cooperative behavior in prey may be generated by predation or by procedure inborn to the prey lifespan history [32]. Theory has pervasively paid attention to cooperative behavior in preys [9,15,22,28,36,37,41], and cooperative behavior in predators is less studied and poorly understood [7,11,34], in particular when space is considered explicitly. A mathematical model of prey and predator population interplay with cooperative behavior in predators through the system of nonlinear ordinary differential equations has been studied in non-spatial domain by Alves et al. [1]. Motivated from their work, we modify and extend the model in a spatial domain to study its spatial dynamics.

Most of the models in mathematical ecology or tumor–immune interaction deal with non-spatial variant. The rate of change of the number of individuals u in a population may be manifested as the derivative with respect to time "t," du/dt. The model equations of a biological community of interacting individuals and their environment are then founded by equating this derivative to another relation expressing the effect of species interaction on population. Same is the situation with tumor–immune interacting models. This type of straightforward analysis is not practicable when spatial models are considered. Directly connected to species interplay is the net population via an arbitrary infinitesimal piece of space rather than the spatial rate of change of the population itself, and thus, a reasonable manifestation is unreachable without knowledge of the mechanism of motion of the individuals.

8.1.1 PREY–PREDATOR SYSTEM

Prey–predator system represents the functional dependence of one species on another, where the first species depends on the second species for food. Predation is a mode of life in which food is primarily obtained by killing and consuming organisms. The prey is part of the predator's habitat, and if the predators do not get any prey for food, then they become extinct. The functional dependence in general depends on many factors, namely, the various species densities, the efficiency with which the predator can search out and kill the prey, and the handling time [4,14,38].

8.1.1.1 Cooperative Behavior of Hunting

Cooperative behavior of hunting is one of the highly fascinating tactical natural instincts in the animal kingdom. A successful hunt requires a great deal of cooperation and coordination within the group. Hunting cooperation in animal kingdom is very frequent; for example, group hunting enables lionesses to have greater success in capturing preys, and it involves both divisions of work and role specialization. It has been connected to the social system of animal species and the evolution of society and thus provides a unique approach to study cooperative behavior [1,7,9,11,15,22,28,34,36,41].

8.2 NONLINEAR DYNAMICS PRELIMINARIES

Some mathematical methods and ideas have been depicted in this segment which are used to examine the nonlinear dynamics and pattern formation (spatiotemporal models), introduced in this thesis. The details of the ideas are depicted for non-spatial (ordinary differential equations) as well as spatial (reaction–diffusion partial differential equations) systems.

8.2.1 BASICS OF STABILITY ANALYSIS

Let $X(t) \in R^n$ depicts the states of a system at time "t. The dynamics of the system is ruled by a system of first-order nonlinear ordinary differential equations:

$$\dot{X}(t) = G(X(t), \theta), \ X(0) = X_0, \tag{8.1}$$

where $X = [x_1, x_2, ..., x_n]^T$ stands for the n state variables, θ holds the parameter values, and G is a nonlinear function of the state variables and parameter values. If $G(X_*) = 0$, then X_* is an equilibrium solution of the system. Consider X_0 to be its neighboring point. The equilibrium solution X_* is stable if for all $\varepsilon > 0$, there is a $\delta > 0$ such that

$$\| X(t) - X_* \| < \varepsilon, \quad \text{whenever } \| X_* - X_0 \| < \delta.$$

That is, X_* is stable if the equilibrium solutions go ahead to X_* at a said time stay close to X_* for each future time. X_* is asymptotically stable if neighboring solutions not only stay close, but also approach to X_* as t goes to infinity, for each future time. That is, X_* is stable and

$$\lim_{t \to \infty} X(t) = X_*,$$

then the solution X_* is asymptotically stable.

$$\text{Asymptotic Stability} \Rightarrow \text{Stability.}$$

Stability of an equilibrium solution is a local property. An equilibrium solution X_* which is not stable is called unstable.

8.2.1.1 Local Stability Analysis

The system of interacting populations

$$\frac{dX_i(t)}{dt} = G_i(X_1, X_2, ..., X_n), \tag{8.2}$$

with initial conditions

$$X_i(0) = X_{i0} \geq 0, \qquad i = 1, 2, ..., n. \tag{8.3}$$

Let us suppose that the function G_i is such that the solution of above system is unique. Let $X(t)$ be any other solution in the vicinity of equilibrium solution X_*, then

$$X = X_* + \eta, \tag{8.4}$$

where $\eta = (\eta_1, \eta_2, ..., \eta_n)$ is a perturbation from the equilibrium solution. Then, the perturbation vector can be written as

$$\frac{d\eta}{dt} = \left. \frac{\partial G}{\partial X} \right|_{X=X_*} \eta \equiv A\eta, \tag{8.5}$$

where $A = (a_{ij})_{n \times n}$ is the variational matrix at the equilibrium solution X_*. Let $\eta(0)$ be the initial perturbation from the equilibrium solution X_*, then the formal matrix solution to Equation (8.5) can be given by

$$\eta(t) = e^{At}\eta(0). \tag{8.6}$$

The system is stable about the equilibrium solution X_* if the perturbation $\eta(t)$ goes to zero as t tends to ∞. This is feasible only if the real parts of the characteristic values of the variational matrix A, namely, $\text{Re}\{\lambda_i\}$, are negative for each i. If $\text{Re}\{\lambda_i\} > 0$ for at least one value of i, then the equilibrium solution is unstable. Since $\eta(t)$ is the solution of linearized system Equation (8.5), which is a close to actual nonlinear system, the stability is referred to local/linear stability only.

Therefore, the characteristic values of the variational matrix decide whether the equilibrium solution is linearly stable or unstable. The characteristic equation for the variational matrix can be written as

$$\det(A - \lambda I) = a_0 \lambda^n + a_1 \lambda^{n-1} + ... + a_n = 0, \quad a_0 \neq 0. \tag{8.7}$$

The coefficients $a_i, i = 1, 2, ..., n$ of characteristic equation are all real. The system Equation (8.2) is locally stable about the equilibrium point if all of the eigenvalues have negative real parts. On the other hand, the system is unstable if at least one of the eigenvalues has positive real part. In other words, all the eigenvalues of Jacobian matrix must lie in the left half of the complex plane. Accordingly, the necessary condition (not sufficient) for all eigenvalues to have negative real part is

$$\text{Trace } (A) < 0.$$

In the special case, if Trace $(A) = 0$, then either at least one eigenvalue must lie in the right half plane, or all eigenvalues must be purely imaginary (the pathological case of neutral stability). Another necessary, but not sufficient, condition is

$$(-1)^n \det |A| > 0.$$

Routh–Hurwitz Criterion gives necessary and sufficient conditions to make certain that the real part of all characteristic roots is negative that belongs to left half complex plane. These conditions, collectively $a_n > 0$, are [24]

$$H_1 = a_1 > 0, \ H_2 = \begin{vmatrix} a_1 & a_3 \\ 1 & a_2 \end{vmatrix} > 0, \ H_3 = \begin{vmatrix} a_1 & a_3 & a_5 \\ 1 & a_2 & a_4 \\ 0 & a_1 & a_3 \end{vmatrix} > 0,$$

$$H_k = \begin{vmatrix} a_1 & a_3 & \cdot & \cdot & \cdot & \cdot \\ 1 & a_2 & a_4 & \cdot & \cdot & \cdot \\ 0 & a_1 & a_3 & \cdot & \cdot & \cdot \\ 0 & 1 & a_2 & \cdot & \cdot & \cdot \\ \cdot & \cdot & \cdot & \cdot & \cdot & \cdot \\ 0 & 0 & \cdot & \cdot & \cdot & a_k \end{vmatrix} > 0, \quad k = 1, 2, 3, ..., n \tag{8.8}$$

If the values of parameters are such that the above restrictions are simultaneously satisfied, then the given system will be locally asymptotically stable at X_*.

8.2.2   TYPES OF BIFURCATIONS

The theory of bifurcation is the mathematical study of sudden changes in the qualitative behavior of the solutions of a nonlinear dynamical system. Bifurcation analysis shows the long-term dynamics of the interacting population depending on the system parameters. In particular, equilibrium point(s) can be created and destroyed, or their stability can change due to change in parameter values. The parameter values for which the bifurcation occurs are called bifurcating points. In this thesis, we particularly focused on local bifurcations, which occur when a small change in the parameter value of a given dynamical system causes a sudden change in the qualitative behavior of the system in the neighborhood of a critical point of the system. Scientifically, they are important since they provide models of transitions and stabilities as the control parameter is varied. Some different types of local bifurcations are as follows:

- **Hopf bifurcation**: Hopf bifurcation is that type of bifurcation at which a stable equilibrium point loses its stability at a threshold value and gives birth to a limit cycle with the variation of the bifurcation parameter. The system experiences Hopf bifurcation when a purely complex conjugate crosses the boundary of stability. The Hopf bifurcation destroys the temporal symmetry of a system and gives rise to oscillations, which are uniform in space and periodic in time. Two types of Hopf bifurcation are observed: one is supercritical, and other is subcritical. Supercritical Hopf bifurcation is a phenomenon in which the unstable limit cycle becomes stable at the bifurcation point. Subcritical Hopf bifurcation is a phenomenon in which the stable limit cycle becomes unstable at the bifurcation point.
- **Turing bifurcation**: Turing bifurcation is the primary bifurcation which gives rise to spatiotemporal patterns and is crucial for almost all reaction–diffusion-type mathematical systems for pattern formation in embryology,

ecology, epidemiology, and to some other areas of biology, physics, and chemistry [2,19,23,25,33]. The primary concept of the Turing bifurcation is that a uniform steady-state solution can be stable to uniform spatiotemporal perturbations, but unstable to definite spatiotemporally changing perturbations, leading to the formation of patterns, that is, a spatial pattern. The straightforward model to contemplate mathematically is fundamentally treated by Turing in 1952, namely, two reaction–diffusion-type partial differential equations, the interacting chemicals having distinct coefficients of diffusion. For appropriate reaction kinetics, as the proportion of diffusivity increases (or decreases) from unity, e.g., there is a critical value at which the homogenous equilibrium solution becomes unstable to a particular spatiotemporal mode. Such kind of bifurcation is called the Turing bifurcation.

8.3 TURING (DIFFUSIVE) INSTABILITY

Spatiotemporal patterns are formed via the diffusive instability of the uniform equilibrium solution to small spatiotemporal perturbations. If the uniform equilibrium solution is stable, then small spatiotemporal perturbations from the equilibrium state will move towards back to the equilibrium state. In 1952, Alan Mathison Turing pointed out how a reaction–diffusion system, showing such instabilities can form diffusive patterns [29,33].

Alan Turing, in 1952, demonstrated that the reaction–diffusion system may form the spatial pattern, if the following two conditions holds:

- the coexistence steady state is linearly stable in the non-spatial (without diffusion) system
- after adding the diffusion term in system, the coexistence steady state is linearly unstable.

Proper mathematical analysis demonstrates that, on the beginning of instability, the model initially becomes unstable with regard to a spatiotemporally nonhomogenous perturbation with a definite wave number. Such type of instability is called a diffusive instability (Turing instability).

The mathematical foundation of diffusive instability by considering two state variables, Y_1 and Y_2, which are subject to one-dimensional space:

$$\frac{\partial Y_1}{\partial t} = H_1(Y_1, Y_2) + D_1 \frac{\partial^2 Y_1}{\partial x^2},$$

$$\frac{\partial Y_2}{\partial t} = H_2(Y_1, Y_2) + D_2 \frac{\partial^2 Y_2}{\partial x^2}, \tag{8.9}$$

where "x is the space coordinate, and "t is the time. D_1 and D_2 are diffusion coefficients of Y_1 and Y_2, respectively. $H_1(Y_1, Y_2)$ and $H_2(Y_1, Y_2)$ are the arbitrary interaction terms of Y_1 and Y_2, respectively.

To understand the effect of diffusion in pattern formation, we assume that in absence of diffusion, that is, when solutions are well mixed, the system has some positive spatially homogeneous steady state, (Y_1^*, Y_2^*). Mathematically, this means that

$$\frac{\partial Y_1^*}{\partial t} = 0 = \frac{\partial Y_2^*}{\partial t},$$

$$\frac{\partial^2 Y_1^*}{\partial x^2} = 0 = \frac{\partial^2 Y_2^*}{\partial x^2}, \tag{8.10}$$

$$\Rightarrow \quad H_1\left(Y_1^*, Y_2^*\right) = 0 = H_2\left(Y_1^*, Y_2^*\right). \tag{8.11}$$

Additionally, suppose that (Y_1^*, Y_2^*) is stable with respect to spatially uniform perturbations, system is stable without diffusion.

To examine the effects of small nonhomogeneous perturbation on the stability of the system with respect to homogeneous steady state, we write

$$Y_1(t,x) = Y_1^* + Y_1'(t,x),$$

$$Y_2(t,x) = Y_2^* + Y_2'(t,x). \tag{8.12}$$

It is assumed that the perturbations are sufficiently small; that is, we analyze the local stability of the system. Substituting Equation (8.12) into (8.9), using Equation (8.10), and linearizing the equations, we obtain

$$\frac{\partial Y_1'}{\partial t} = a_{11}Y_1' + a_{12}Y_2' + D_1\frac{\partial^2 Y_1'}{\partial x^2},$$

$$\frac{\partial Y_2'}{\partial t} = a_{21}Y_1' + a_{22}Y_2' + D_2\frac{\partial^2 Y_2'}{\partial x^2}, \tag{8.13}$$

where

$$a_{11} = \frac{\partial H_1}{\partial Y_1}\bigg|_{\left(Y_1^*, Y_2^*\right)}, \quad a_{12} = \frac{\partial H_1}{\partial Y_2}\bigg|_{\left(Y_1^*, Y_2^*\right)},$$

$$a_{21} = \frac{\partial H_2}{\partial Y_1}\bigg|_{\left(Y_1^*, Y_2^*\right)}, \quad a_{22} = \frac{\partial H_2}{\partial Y_2}\bigg|_{\left(Y_1^*, Y_2^*\right)}, \tag{8.14}$$

and Y_1' and Y_2' are perturbations from Y_1^* and Y_2^*. Equations (8.13) can be written in the compact matrix form:

$$Y_t' = AY' + DY_{xx}', \tag{8.15}$$

where

$$Y' = \begin{pmatrix} Y_1'(t,x) \\ Y_2'(t,x) \end{pmatrix} = \begin{pmatrix} Y_1(t,x) - Y_1^* \\ Y_2(t,x) - Y_2^* \end{pmatrix},$$

$$A = \begin{pmatrix} a_{11} & a_{12} \\ a_{21} & a_{22} \end{pmatrix},$$

$$D = \begin{pmatrix} D_1 & 0 \\ 0 & D_2 \end{pmatrix}.$$

For linear stability analysis, it is sufficient to assume solution of Equation (8.13) in the form

$$Y_1' = \exp(\mu t + ikx),$$
$$Y_2' = \exp(\mu t + ikx), \tag{8.16}$$

where k and μ are the wave number and frequency, respectively. Corresponding characteristic equation is

$$\begin{vmatrix} a_{11} - D_1 k^2 - \mu & a_{12} \\ a_{21} & a_{22} - D_2 k^2 - \mu \end{vmatrix} = 0. \tag{8.17}$$

Solving for μ, we obtain

$$\mu = \frac{1}{2}\left(a_{11} + a_{22} - k^2(D_1 + D_2) \pm \right.$$

$$\left. \sqrt{\left(a_{11} + a_{22} - k^2(D_1 + D_2) \right)^2 - 4\left((a_{11} - D_1 k^2)(a_{22} - D_2 k^2) - a_{12} a_{21} \right)} \right).$$

The condition $k = 0$ corresponds to the neglect of diffusion, and by definition, perturbations of zero wave number are stable when diffusive instability sets in. It is thus required that

$$a_{11} + a_{22} < 0,$$
$$a_{11} a_{22} - a_{12} a_{21} > 0. \tag{8.18}$$

Diffusive instability sets in when at least one of the following conditions is violated subject to the conditions (8.18):

$$\tilde{a}_{11} + \tilde{a}_{22} < 0,$$
$$\tilde{a}_{11} \tilde{a}_{11} - a_{12} a_{21} > 0. \tag{8.19}$$

It is seen that the first condition $\tilde{a}_{11} + \tilde{a}_{22} < 0$ is not violated when the requirement $a_{11} + a_{22} < 0$ is met. Hence, only violation of the second condition $\tilde{a}_{11} \tilde{a}_{11} - a_{12} a_{21} > 0$ gives rise to diffusive instability. Reversal of the second inequality of Equation (8.19) yields

$$Q(k^2) = D_1 D_2 k^4 - (D_1 a_{22} + D_2 a_{11}) k^2 + a_{11} a_{22} - a_{12} a_{21} < 0. \tag{8.20}$$

The minimum of $Q(k^2)$ occurs at $k^2 = k_m^2$, where

$$k_m^2 = \frac{D_1 a_{22} + D_2 a_{11}}{2 D_1 D_2} > 0. \tag{8.21}$$

Thus, a sufficient condition for instability is that $Q(k_m^2)$ be negative. Therefore,

$$(a_{11} a_{22} - a_{12} a_{21}) - \frac{(D_1 a_{22} + D_2 a_{11})^2}{4 D_1 D_2} < 0. \tag{8.22}$$

Combination of Equations (8.18), (8.21), and (8.22) leads to the following final criterion for diffusive instability:

$$D_1 a_{22} + D_2 a_{11} > 2(a_{11}a_{22} - a_{12}a_{21})^{\frac{1}{2}} (D_1 D_2)^{\frac{1}{2}} > 0. \tag{8.23}$$

The critical conditions for the occurrence of the instability are obtained when the first inequality of Equation (8.23) is an equality.

8.4 MODELS DESCRIPTION

By incorporating the diffusion and Holling type III functional response in the general prey–predator system with hunting cooperation [1,6], we obtain the following predator–prey model:

$$\frac{dX}{dt'} = rX\left(1 - \frac{X}{K}\right) - \frac{(\lambda + aY)X^2 Y}{1 + H_1(\lambda + aY)X^2}$$
$$\frac{dY}{dt'} = e\frac{(\lambda + aY)X^2 Y}{1 + H_1(\lambda + aY)X^2} - mY \tag{8.24}$$

where $X(t')$ and $Y(t')$ are the densities of prey and predator population at time t', respectively. The parameter r is rate of growth of prey, K is its holding efficiency, λ is the constant invasion rate, and a is the rate of predator hunting cooperation. The parameter e is the efficiency of conversion, and m is the natural death rate of predator. H_1 is the predator's handling time. All parameters are positive.

On the other hand, we assume that the prey and predator populace densities spread without any order, and this random distribution of species is depicted by diffusion. Then, we propose a spatial prey–predator model with hunting cooperation and Allee effects in predators corresponding to Equation (8.24) as follows:

$$\frac{\partial X}{\partial t'} = rX\left(1 - \frac{X}{K}\right) - \frac{(\lambda + aY)X^2 Y}{1 + H_1(\lambda + aY)X^2} + d_1 \nabla^2 X$$
$$\frac{\partial Y}{\partial t'} = e\frac{(\lambda + aY)X^2 Y}{1 + H_1(\lambda + aY)X^2} - mY + d_2 \nabla^2 Y \tag{8.25}$$

where the non-negative constants d_1 and d_2 are the diffusion coefficients of prey and predator, respectively. ∇^2 is the usual Laplacian operator in $d \leq 3$ space dimensions.

The great number of the research work is based upon the nondimensional models of the nonlinear coupled partial differential equations as they have less number of parameters. Following [1], the variables are scaled as

$$u = \sqrt{\frac{e\lambda}{m}}X, \quad v = \sqrt{\frac{\lambda}{em}}Y, \quad t = mt',$$

and dimensionless values of other parameters are given by

$$\sigma = \frac{r}{m}, \quad N = \sqrt{\frac{e\lambda}{m}}K, \quad \alpha = \frac{a}{\lambda}\sqrt{\frac{em}{\lambda}}, \quad h_1 = \frac{m}{e}H_1.$$

With these changes, Equations (8.25) becomes

$$\frac{\partial u}{\partial t} = \sigma u \left(1 - \frac{u}{N}\right) - \frac{(1 + \alpha v) u^2 v}{1 + h_1 (1 + \alpha v) u^2} + d_1 \nabla^2 u$$

$$\frac{\partial v}{\partial t} = \frac{(1 + \alpha v) u^2 v}{1 + h_1 (1 + \alpha v) u^2} - v + d_2 \nabla^2 v. \tag{8.26}$$

This is the working spatial prey–predator model with hunting cooperation and Allee effects in predators. Generally, to make certain that spatial patterns are governed by reaction–diffusion method, model (8.26) is to be analyzed with the following initial conditions:

$$2D : u(x,y,0) > 0, \quad v(x,y,0) > 0, \quad (x,y) \in \Omega = [0,L] \times [0,L] \tag{8.27}$$

and Neumann's boundary condition

$$\frac{\partial u}{\partial v} = \frac{\partial v}{\partial v} = 0, \tag{8.28}$$

where L is the size of the homogeneous spatial domain, and v is the outward unit normal on the boundary $\partial \Omega$. The aim of our study in this chapter is to investigate the phenomena of diffusion-driven instability (spatial pattern) and higher order instability analysis outside the diffusion Driven instability domain, in the predator–prey system with hunting cooperation and Allee effects in predator.

8.5 SPATIOTEMPORAL MODEL

8.5.1 INITIAL DENSITY DISTRIBUTION

The spatiotemporal pattern formation generally onsets with a community interplays of species. The initial conditions for model (8.26) should be stated by the mathematical compact support function that is within a definite domain the initial distribution of prey & predator is non-zero and elsewhere zero. The structure of the realm and the outlines of the species densities can be dissimilar in different prey-predator models. In this study, we have employed the initial distribution of species such as statistically uncorrelated Gaussian white noise perturbation in space:

$$u(x_i, y_j, 0) = u^* + \gamma_1 \varepsilon_{ij},$$
$$v(x_i, y_j, 0) = v^* + \gamma_2 \eta_{ij}, \tag{8.29}$$

where γ_1 and γ_2 are very small real numbers, and ε_{ij} and η_{ij} are statistically uncorrelated Gaussian white noise perturbations with zero mean and fixed variance in two-dimensional space.

8.5.2 EQUILIBRIA OF SYSTEM

In absence of diffusion, the equilibrium points of the system are given by

$$\sigma u\left(1 - \frac{u}{N}\right) - \frac{(1+\alpha v)u^2 v}{1 + h_1(1+\alpha v)u^2} = 0,$$

$$\frac{(1+\alpha v)u^2 v}{1 + h_1(1+\alpha v)u^2} - v = 0. \tag{8.30}$$

Clearly, the above system has the following meaningful steady-state points: (i) $E_0(0,0)$ (prey and predator both extinct), (ii) $E_1(N,0)$ (prey only survive), and (iii) two positive coexistence equilibrium points $E_2(u_2,v_2)$ and $E_3(u_3,v_3)$ (interior equilibrium solutions), where $v_2 = \frac{1}{\alpha}\left(\frac{1}{(1-h_1 u_2)u_2} - 1\right)$ and $v_3 = \frac{1}{\alpha}\left(\frac{1}{(1-h_1 u_3)u_3} - 1\right)$, and u_2, u_3 be the positive solution of

$$A_0 u^4 + A_1 u^3 + A_2 u^2 + A_3 u + A_4 = 0, \tag{8.31}$$

where
$A_0 = h_1 \alpha \sigma$, $A_1 = (h_1 N - 1)\alpha \sigma$, $A_2 = ((1 - h_1)\alpha \sigma - h_1)N$, $A_3 = (1 - h_1)N$, $A_4 = -N$.

The number of coexistence equilibrium points and their stability depends upon the parameter values for the model (8.26).

8.5.3 STABILITY ANALYSIS OF SYSTEM

The Jacobian matrix (**J**) of the model of Equation (8.26) is as follows:

$$\begin{bmatrix} \sigma(1 - \frac{2u}{N}) + \frac{v(1+v\alpha)[(1+v\alpha)h_1 u^2 - 1]}{[1+(1+v\alpha)h_1 u + (1+v\alpha)h_1 u^2]^2} & -\frac{u[1+2\alpha v + (1+\alpha v)^2 h_1 u + (1+\alpha v)^2 h_1 u^2]}{[1+(1+\alpha v)h_1 u + (1+\alpha v)h_1 u^2]^2} \\ -\frac{v(1+v\alpha)[(1+v\alpha)h_1 u^2 - 1]}{[1+(1+v\alpha)h_1 u + (1+v\alpha)h_1 u^2]^2} & \frac{u[1+2\alpha v + (1+v\alpha)^2 h_1 u + (1+v\alpha)^2 h_1 u^2]}{[1+(1+v\alpha)h_1 u + (1+v\alpha)h_1 u^2]^2} - 1 \end{bmatrix},$$

$$= \begin{bmatrix} \Delta_{11} & \Delta_{12} \\ \Delta_{21} & \Delta_{22} \end{bmatrix} = (\Delta_{ij})_{2\times 2}. \tag{8.32}$$

1. At $E_0(0,0)$, Jacobian matrix is

$$\mathbf{J}_0 = \begin{bmatrix} \sigma & 0 \\ 0 & -1 \end{bmatrix},$$

whose eigenvalues are -1 and σ (which is a positive parameter). Hence, the system is unstable at the origin.

2. At $E_1(N,0)$, Jacobian matrix is

$$\mathbf{J}_1 = \begin{bmatrix} -\sigma & -\frac{N}{1+h_1 N + h_1 N^2} \\ 0 & \frac{N}{1+h_1 N + h_1 N^2} - 1 \end{bmatrix},$$

whose eigenvalues are $-\sigma$ and $\frac{N}{1+h_1 N + h_1 N^2} - 1$. Hence, the system is locally asymptotically stable if $\frac{N}{1+h_1 N + h_1 N^2} < 1$.

3. At $E_2(u_2,v_2)$

Lemma 8.5.1 The equilibrium solution $E_2(u_2,v_2)$ is locally asymptotically stable if and only if

$$(N-N\sigma+2\sigma u_2)(1+h_1u_2(1+v_2\alpha)+h_1u_2{}^2(1+v_2\alpha))^2-N(u_2(1+2\alpha v_2)-v_2(1+\alpha v_2)+(h_1u_2+h_1v_2+h_1)(1+\alpha v_2)^2u_2{}^2)>0,$$

and

$$u_2v_2(1+\alpha v_2)(1-h_1u_2{}^2(1+\alpha v_2))N(1+2\alpha v_2+h_1u_2(1+\alpha v_2)^2+h_1u_2(1+\alpha v_2)^2)+(-(1+h_1u_2{}^2(1+\alpha v_2)+h_1u_2(1+\alpha v_2))^2+u_2(1+2\alpha v_2+h_1u_2(1+\alpha v_2)^2+h_1u_2{}^2(1+\alpha v_2)^2))(v_2(1+\alpha v_2)N(-1+h_1u_2{}^2(1+\alpha v_2))-2\sigma u_2(1+h_1u_2(1+\alpha v_2)+h_1u_2{}^2(1+\alpha v_2))^2+\sigma N(1+h_1u_2(1+\alpha v_2)+h_1u_2{}^2(1+\alpha v_2))^2)>0.$$

Proof. The eigenvalues of the corresponding Jacobian matrix $\mathbf{J}$ at equilibrium solution $E_2(u_2,v_2)$ are given by $\frac{1}{2}(\mu_1\pm\mu_2)$, where

$$\mu_1=\sigma(1-\frac{2u_2}{N})-1+\frac{v_2(1+\alpha v_2)(h_1u_2{}^2(1+\alpha v_2)-1)}{(1+h_1u_2(1+\alpha v_2)+h_1u_2{}^2(1+\alpha v_2))^2}$$
$$+\frac{u_2(1+2\alpha v_2+h_1u_2(1+\alpha v_2)^2+h_1u_2{}^2(1+\alpha v_2)^2)}{(1+h_1u_2(1+\alpha v_2)+h_1u_2{}^2(1+\alpha v_2))^2},$$

$\mu_2=\sqrt{\mu_1{}^2-4\mu_3}$ and μ_3 is the determinant of Jacobian $\mathbf{J}$ at $E_2(u_2,v_2)$. Hereby, the coexistence equilibrium solution $E_2(u_2,v_2)$ is locally asymptotically stable if and only if $\mu_1<0$ and $\mu_3<0$.

Lemma 8.5.2 The reaction–diffusion system Equation (8.26) enters into a Hopf bifurcation around $E_2(u_2,v_2)$ at $\sigma=\sigma_{hb}$, where σ_{hb} satisfies the equality $\sigma_{hb}=\frac{N((1+h_1u_2(1+\alpha v_2)+h_1u_2{}^2(1+\alpha v_2))^2-u_2(1+2\alpha v_2)+v_2(1+\alpha v_2)-(h_1+v_2h_1+u_2h_1)(1+\alpha v_2)^2u_2{}^2)}{(N-2u_2)(1+h_1u_2(1+\alpha v_2)+h_1u_2{}^2(1+\alpha v_2))^2}.$

Proof. The characteristic equation corresponding to the equilibrium solution $E_2(u_2,v_2)$ is given by $\lambda_{11}{}^2-\mu_1\lambda_{11}+\mu_3=0$. Let us assume that

$$u\approx\exp(\lambda_{11}t),\quad v\approx\exp(\lambda_{11}t).$$

If $\mu_1=0$, then both the eigenvalues will be purely imaginary provided μ_3 is positive and there are no other eigenvalues with negative real part. Presently, $\mu_1=0$ gives $\sigma=\sigma_{hb}$. Substituting $\lambda_{11}=a_1+ib_1$ into the equation $\lambda_{11}{}^2-\mu_1\lambda_{11}+\mu_3=0$ and separating real and imaginary parts, we obtain $(a_1{}^2-b_1{}^2)-\mu_1a_1+\mu_3=0$ and $2a_1b_1-\mu_1b_1=0$. Differentiating $2a_1b_1-\mu_1b_1=0$ both sides with respect to σ at $\sigma=\sigma_{hb}$ and considering $a_1=0$, we get

$$\left.\frac{da_1}{d\sigma}\right|_{\sigma=\sigma_{hb}}=\frac{1}{2}\left(1-\frac{2u_2}{N}\right)\neq0.$$

For a change of stability about $E_2(u_2, v_2)$, we should have the real part of λ_{11}, that is, $a_1 = 0$. Hence, the system undergoes a Hopf bifurcation at $E_2(u_2, v_2)$ as σ passes through the value σ_{hb}.

8.6 ANALYSIS OF THE SPATIOTEMPORAL MODEL

Interior equilibrium point $E_2(u_2, v_2)$ of non-spatial system is spatially homogenous steady state, that is, constant in space and time for the reaction–diffusion system (spatiotemporal model). We assume that $E_2(u_2, v_2)$ is stable in non-spatial system which means the spatially homogenous steady state is stable with respect to spatially homogenous perturbations. Though the diffusion is often consider as a stabilizing process, it is a well-known fact that diffusion can make a spatially homogenous steady-state linearly unstable (Turing instability) with respect to heterogenous perturbations in a system of two interacting species [29,33]. The condition for Turing instability may be obtained by introducing a small heterogenous perturbation of the homogenous steady state as follows:

$$u(x, y, t) = u_2 + \varepsilon_1 \exp(\lambda_k t) \cos(k_x x) \cos(k_y y),$$
$$v(x, y, t) = v_2 + \varepsilon_2 \exp(\lambda_k t) \cos(k_x x) \cos(k_y y), \tag{8.33}$$

where ε_1 and ε_2 are two non-zero reals and $k = (k_x, k_y)$, such that $k^2 = (k_x^2 + k_y^2)$, is the wave number.

Substituting Equation (8.33) into (8.26) and then linearizing it about interior equilibrium point $E_2(u_2, v_2)$, we obtain the variational matrix as

$$
\begin{bmatrix}
\sigma\left(1 - \frac{2u_2}{N}\right) + \frac{v_2(1+v_2\alpha)[(1+v_2\alpha)h_2u_2^2-1]}{[1+(1+v_2\alpha)h_1u_2+(1+v_2\alpha)h_2u_2^2]^2} - k^2 & -\frac{u_2[1+2\alpha v_2+(1+\alpha v_2)^2h_1u_2+(1+\alpha v_2)^2h_1u_2^2]}{[1+(1+\alpha v_2)h_1u_2+(1+\alpha v_2)h_1u_2^2]^2} \\
-\frac{v_2(1+v_2\alpha)[(1+v_2\alpha)h_1u_2^2-1]}{[1+(1+v_2\alpha)h_1u_2+(1+v_2\alpha)h_1u_2^2]^2} & \frac{u_2[1+2\alpha v_2+(1+v_2\alpha)^2h_1u_2+(1+\alpha v_2)^2h_1u_2^2]}{[1+(1+v_2\alpha)h_1u_2+(1+v_2\alpha)h_1u_2^2]^2} - 1 - dk^2
\end{bmatrix}
$$

The corresponding characteristic equation is

$$\lambda^2 + C_1(k^2)\lambda + C_2(k^2) = 0, \tag{8.34}$$

where

$$C_1(k^2) = (1+d)k^2 + 1 + \sigma\left(\frac{2u_2}{N} - 1\right) - \frac{\alpha u_2 v_2 + (1+\alpha v_2)(u_2 - v_2 + h_1 u_2^3(1+\alpha v_2) + (h_1 + h_1 v_2)u_2^2(1+\alpha v_2))}{(1 + h_1 u_2^2(1+\alpha v_2) + h_1 u_2(1+\alpha v_2))^2},$$

$$C_2(k^2) = \frac{2(1+dk^2)u_2(1+u_2^4h_1^2(1+\alpha v_2)^2)\sigma + (1+dk^2)v_2N(1+\alpha v_2) + (k^2-\sigma)((1+dk^2)N + u_2(-1-2v_2\alpha + 2h_1(1+dk^2)(1+\alpha v_2))N)}{(1+h_1u_2^2(1+v_2\alpha) + h_1u_2(1+v_2\alpha))^2N} +$$
$$\frac{h_1u_2^4(1+v_2\alpha)^2(2(-1+2h_1(1+dk^2))\sigma + h_1(1+dk^2)(k^2-\sigma)N) + 2u_2^3(1+\alpha v_2)(2h_1(1+dk^2) - h_1(1+\alpha v_2) + h_1^2(1+dk^2)(1+\alpha v_2))\sigma}{(1+h_1u_2^2(1+v_2\alpha) + h_1u_2(1+v_2\alpha))^2N} +$$
$$\frac{u_2^3(1+\alpha v_2)h_1(-1+2h_1(1+dk^2))(1+\alpha v_2)(k^2-\sigma)N - 2\sigma u_2^2(-1-2v_2\alpha + 2h_1(1+dk^2)(1+\alpha v_2)) + u_2^2(1+\alpha v_2)h_1(1+\alpha v_2)(k^2-\sigma)}{(1+h_1u_2^2(1+v_2\alpha) + h_1u_2(1+v_2\alpha))^2N} +$$
$$\frac{u_2^2h_1^2(1+dk^2)(1+\alpha v_2)(k^2-\sigma) + h_1(1+dk^2)(-2k^2 + v_2 + \alpha v_2^2 + 2\sigma)}{(1+h_1u_2^2(1+v_2\alpha) + h_1u_2(1+v_2\alpha))^2N}.$$

By Routh–Hurwitz criterion, the system Equation (8.26) will be stable about $E_2(u_2, v_2)$ if $C_1(k^2) > 0$ and $C_2(k^2) > 0$. As the parameters D and k^2 are all positive and by the stability of the non-spatial model, $C_1(k^2) > 0$ is always positive.

Therefore, the condition for diffusive instability is $C_2(k^2) < 0$.

The polynomial function $C_2(k^2)$ has a minimum for some value of k, say k_{min}, where

$$k_{min}^2 = \frac{1}{2Nd(h_1 + h_1 u_2)(1 + u_2(h_1 + h_1 u_2)(1 + \alpha v_2))^2}$$

$$\times \left(\left(-1 - u_2(h_1 + h_1 u_2) - (-1 + h_1 + h_1 u_2)(1 + u_2(h_1 + h_1 u_2)(1 + \alpha v_2))^2 \right)N \right.$$

$$+ d(h_1 + h_1 u_2)(v_2(1 + \alpha v_2)(-1 + h_1 u_2{}^2(1 + \alpha v_2)) - 2\sigma u_2$$

$$\left. (1 + u_2(h_1 + h_1 u_2)(1 + \alpha v_2))^2 + \sigma N(1 + u_2(h_1 + h_1 u_2)(1 + \alpha v_2))^2 \right).$$

For this minimum value of k, Turing instability will occur when $C_2(k_{min}^2) < 0$. Therefore, substituting k_{min}^2 in $C_2(k^2)$, we get the sufficient condition for Turing instability as

$$d\Delta_{11} + \Delta_{22} - 2\sqrt{d}\sqrt{\Delta_{11}\Delta_{22} - \Delta_{12}\Delta_{21}} > 0 \qquad (8.35)$$

The interval of the wave number for which Turing instability takes place is (k_-, k_+), and in this interval, we have $C_2(k^2) < 0$, where

$$k_- = \frac{d\Delta_{11} + \Delta_{22} - \sqrt{(d\Delta_{11} + \Delta_{22})^2 - 4d(\Delta_{11}\Delta_{22} - \Delta_{12}\Delta_{21})}}{2d},$$

$$k_+ = \frac{d\Delta_{11} + \Delta_{22} + \sqrt{(d\Delta_{11} + \Delta_{22})^2 - 4d(\Delta_{11}\Delta_{22} - \Delta_{12}\Delta_{21})}}{2d},$$

where the values of Δ_{ij}, $i, j = 1, 2$ are obtain from Equation (8.32) about $E_2(u_2, v_2)$.

8.7 NUMERICAL SIMULATIONS

We will now investigate the numerical results of both spatiotemporal as well as non-spatial models. For numerical simulation, we set σ, N, h_1 as $\sigma = 10.0$, $N = 1.2$, $h_1 = 0.01$ and consider cooperation rate (α), as a controlling parameter. For these values of parameters, the positive equilibrium points are $(0, 0)$, $(1.2, 0)$, $(0.6875, 2.9362)$, and $(0.7638, 2.7763)$. The steady state $(0.6875, 2.9362)$ is stable and $(0.7638, 2.7763)$ is unstable. Hence, throughout our study in the spatiotemporal domain, we have considered the stable steady state $(0.6875, 2.9362)$. Figure 8.1 shows the time evolution of prey and predator in the non-spatial domain. Please note that the nondimensional parameter $N = \frac{e\lambda}{m}K$, comprising of the dimensional carrying capacity, attack rate, per capita mortality rate of predators and the conversion efficiency, which is defined as the average number of offspring produced by a single predator during its life time, when introduced into the prey population at carrying capacity. If $N = 1.2$ and the cooperation coefficient α ($= 0.1, 0.5$) is small, then the predator population goes to extinct as the prey population is too small to sustain them (see Figure 8.1a,b). However, for large value of α (= 0.6, 0.7, 0.8), the predator survives due to hunting cooperation behavior in predators (see Figure 8.1a,b). Figure 8.2 shows the bifurcation diagram for prey and predator species density with α as the bifurcation parameter. We begin by varying the rate of hunting cooperation α (see Figure 8.2). The prey only equilibrium solution ($u = N$, $v = 0$) is always stable for all values of cooperation rate α (solid red lines). For some value of cooperation rate α, a Hopf bifurcation occurs ($\alpha = 0.7671$): the stable coexistence equilibrium loses stability so that stable limit cycle oscillations emerge, and their amplitudes quickly increase with α.

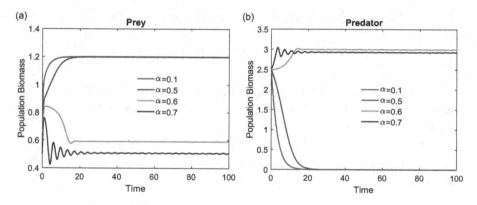

Figure 8.1 Time evolution of (a) prey and (b) predator in the non-spatial domain of the model for fixed parameters $\sigma = 10.0$, $N = 1.2$, $h_1 = 0.01$ and different parameter values of hunting cooperation rate (α) which are mentioned in figures.

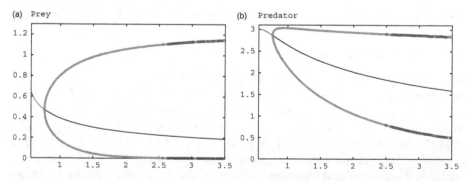

Figure 8.2 The bifurcation diagram with the rate of hunting cooperation (α) as the bifurcation parameter, and the parameters N, σ, h_1 are fixed at $N = 1.2$, $\sigma = 10$, $h_1 = 0.01$, respectively. 8.2(a) The prey density (y-axis) is plotted verses cooperation rate α (x-axis), and in 8.2(b), the predator density (y-axis) is plotted verses cooperation rate α (x-axis). The solid red part of curve is where the steady state is stable, the solid gray part of curve, unstable. The filled solid green circles denote stable periodic solution branch, and blue open circles are unstable, which begins at the Hopf bifurcation value $\alpha = 0.7671$.

For spatiotemporal model, we perform all the numerical simulations of the system (8.26) over the non-zero initial condition and zero-flux boundary conditions, in two-dimensional spatial domain. The domain size is 50×50 with time-step $\Delta t = 0.001$ and space-step $\Delta x = \Delta y = 0.5$. The parameter values of σ, N, and h_1 remain same ($\sigma = 10$, $N = 1.2$, $h_1 = 0.01$), and α is used as the controlling parameter (just like the non-spatial case).

Note: The Neumann zero-flux conditions are placed at boundary of the numerical domain in two-dimensional problems. The size of the domain is chosen large enough so that the impact of the boundaries has been kept as small as possible during the simulation time.

We now demonstrate diffusive induced instability (Turing instability) and the corresponding pattern formation for the system Equation (8.26). Although the sufficient conditions for Turing instability were obtained analytically in the previous section, whether they are satisfied with our corresponding set of parameter values, they are yet to be tested. In order to do so, we sketch the Turing instability condition (8.35) for distinct values of d (other parameter values are fixed, namely, $\sigma = 10.0$, $\alpha = 0.55$, $N = 1.2$, $h_1 = 0.01$). Figure 8.3a shows the zone for the emergence of spatial patterns corresponding to Turing instability condition against the ratio of diffusion coefficients (d). We observe that the sufficient condition of the diffusive instability, that is, Equation (8.35) holds, when d is adequately small, ending on $d = 0.039$ (see Figure 8.3a). The spatial dispersion curve for this particular model is shown in Figure 8.3b, and the dispersion relation is represented by the real part of the largest eigenvalues of the spatial model. The corresponding plot of real part of largest eigenvalue $Re(\lambda)$ against the wave number (k) is shown in Figure 8.3b. The real part of largest eigenvalue $Re(\lambda) > 0$ holds, the wave number (k) fits in the interval (k_-, k_+), that is, $(0.5635, 2.0160)$. Also, we obtain the controlling parameter space for Turing instability via sufficient condition, which is shown in Figure 8.4. In Figure 8.5, we have illustrated the density distributions of prey and predator which covers three kinds of spatial pattern, namely spots, mixed (spots-stripes) and stripes. Figure 8.5a,d shows the two-dimensional stationary diffusive patterns of the model (8.26) at time $t = 2,000$ (2,000,000 iterations) and $\alpha = 0.55$ with diffusion coefficient

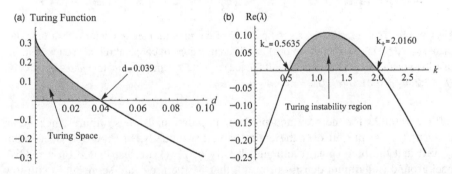

Figure 8.3 (a) Zone for the emergence of spatial pattern corresponding to Turing instability condition and (b) characterization of the dispersal relation for $d = 30$.

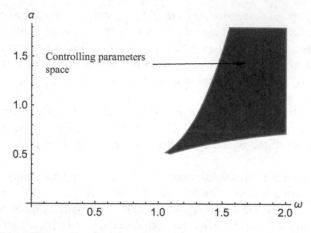

Figure 8.4 Controlling parameters space for Turing patterns corresponding to Turing instability condition in the region.

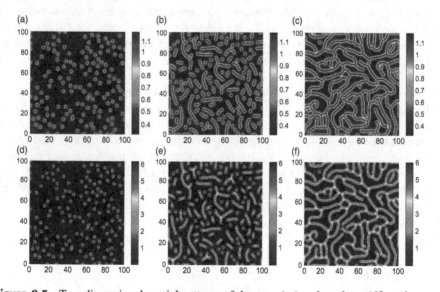

Figure 8.5 Two-dimensional spatial patterns of the prey (ac) and predator (df) at time moment $t = 2,000$ (2,000,000 iterations) for different values of cooperation rate (α) with initial distribution. (a,d) $\alpha = 0.55$; (b,e) $\alpha = 0.56$; (c,f) $\alpha = 0.57$. Other parameter values are $\sigma = 10$, $N = 1.2$, $h_1 = 0.01$, $d = 0.023$.

ratio $d = 0.023$ for the prey and predator population. In these figures, hexagonal patterns (spots) prevail over the entire habitat eventually. In Figure 8.5a, it is observed that the blue spots (minimum density of prey) are distributed on a reddish background (maximum density of prey); that is, the preys are segregated with low population density. On the other hand, Figure 8.5d consists of red spots on a blue background; that is, the predators are isolated with high population density. As the

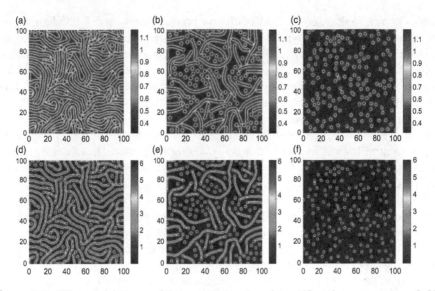

Figure 8.6 2D spatial patterns of the prey (ac) and predator (df) at time moment $t = 2,000$ (2,000,000 iterations) for different values of diffusion coefficient (d) with initial distribution. (a,d) $d = 0.0175$; (b,e) $d = 0.02$; (c,f) $d = 0.028$. Other parameter values are $\sigma = 10$, $\alpha = 0.55$, $N = 1.2$, $h_1 = 0.01$.

α is increased to 0.56, some patches split into stripes resulting in spots-stripes patterns in both prey and predator population (see Figure 8.5b,e). When α is increased to 0.57, the dynamics of the model exhibits a decay in the spot and emergence in stripes pattern only (Figure 8.5c,f). Thus, by increasing the control parameter α, a sequence spots $\rightarrow$ spot-stripes $\rightarrow$ stripes is observed.

Figure 8.6 demonstrates the spatial patterns of prey and predator with respect to different diffusion rates d. Figure 8.6a,d shows the two-dimensional spatial patterns of the model (8.26) at time $t = 2,000$ (2,000,000 iterations) with diffusion rate $d = 0.0175$ for the prey and predator population. In these figures, the stripes pattern prevails over the entire habitat. In Figure 8.6a, it is observed that blue stripes (minimum density of prey) are distributed on a reddish background (maximum density of prey). On the other hand, Figure 8.6d consists of red stripes on a blue background. As the diffusion rate d is increased to 0.02, some stripes split into spots resulting in spots-stripes patterns in both prey and predator population (Figure 8.6b,e). When d is increases to 0.028, the dynamics of the model exhibits a decay in the stripes and emergence in spots pattern (see Figure 8.6c,f). Thus, by increasing the d (rate of diffusion), a sequence stripes $\rightarrow$ spot-stripes $\rightarrow$ spots is observed.

8.8 DISCUSSION AND CONCLUSION

In theoretical ecology, intensive studies of the mechanisms and scenarios of pattern formation, in models of interacting populations, have always been an attraction, as their perception helps to enhance the understanding of real-world ecological systems.

In this chapter, we have considered a diffusive predator–prey model with hunting cooperation in predators and type III functional response under non-zero initial conditions and zero-flux boundary conditions. We have provided elaborate analysis of both non-spatial and spatiotemporal models, and studied possible scenarios of pattern formation in the diffusive predator–prey model with hunting cooperation in predators. While studying the spatiotemporal model, we first obtain the condition for diffusive instability and identified the corresponding domain in the space of controlling parameters. The hunting cooperation coefficient and the ratio of diffusion coefficient are the controlling parameters in our study. Using the parameter values from Turing domain, we investigate the properties of the system using extensive numerical simulations.

Our model simulation has been categorized in two separate domains, namely, the non-spatial and the spatial domains. We have highlighted the effect of hunting cooperation in predators along with carrying capacity of the predators. After simulated numerically, we confirmed that in the non-spatial domain, for fixed N, the increase in the hunting cooperation in the predators helps them to survive. In the spatial domain, for fixed N, the hunting cooperation in predators plays a crucial role in the coexistence. By varying the values of cooperation coefficient, we get dissimilar types of diffusive patterns, namely, patchy pattern (spots), stripe pattern, and mixed pattern (spot-stripe). From the point of view of population dynamics, one can observe that there exists the pattern formation (spot) for preys implying that the preys are scattered with low density and the remaining region is high dense, which means that the preys have segregated in very small groups over the large area and are safe. Similarly, spot formation in predators conveys that with hunting cooperation, the predators are scattered and isolated but still survived. Large African predators such as cheetah (Acinonyx jubatus), leopard (Panthera pardus), and lion (Pathera leo) regularly predate ungulates and double their mass with the possibility of injury or death to the predator during prey capture but can easily be overcome by cooperative hunting that may improve hunting success rate [8].

The methods and consequences in the study may amplify the systematic investigation of spatial pattern formation in the predator–prey systems, and may nicely enforce in some different research dimensions. Further analysis are important to study the patterns dynamics of some more diffusive ecological models. It would be interesting to study the traveling waves in the spatial predator–prey models with hunting cooperation in predators. This work highlights a number of research areas for future consideration in spatial pattern formation.

REFERENCES

1. M.T. Alves and F.M. Hilker. Hunting cooperation and allee effects in predators. *Journal of Theoretical Biology*, 419(1):13–22, 2017.
2. V. Ardizzone, P. Lewandowski, M.H. Luk, Y.C. Tse, N.H. Kwong, A. Lücke, M. Abbarchi, E. Baudin, E. Galopin, J. Bloch, A. Lemaitre, P.T. Leung, P. Roussignol, R. Binder, J. Tignon, and S. Schumacher. Formation and control of turing patterns in a coherent quantum fluid. *Scientific Reports*, 3(1):1–6, 2013.

3. M. Banerjee and S. Banerjee. Turing instabilities and spatiotemporal chaos in ratio-dependent holling–tanner model. *Mathematical Biosciences*, 236(1):64–76, 2012.

4. M. Banerjee and S.V Petrovskii. Self-organised spatial patterns and chaos in a ratio-dependent predator–prey system. *Theoretical Ecology*, 4(1):37–53, 2011.

5. R.A. Barrio, C. Varea, J.L. Aragón, and P.K. Maini. A two–dimensional numerical study of spatial pattern formation in interacting turing systems. *Bulletin of Mathematical Biology*, 61(3):483–505, 1999.

6. L. Berec. Impacts of foraging facilitation among predators on predator–prey dynamics. *Bulletin of Mathematical Biology*, 72(1):94–121, 2010.

7. A. Bompard, I. Amat, X. Fauvergue, and T. Spataro. Host-parasitoid dynamics and the success of biological control when parasitoids are prone to allee effects. *Bulletin of Mathematical Biology*, 8(10):076768–076779, 2013.

8. H.S. Clements, C.J. Tambling, and G.I.H. Kerley. Prey morphology and predator sociality drive predator–prey preferences. *Journal of Mammalogy*, 97(3):919–927, 2016.

9. F. Courchamp, T. Clutton-Brock, and B. Grenfell. Inverse density dependence and the allee effect. *Trends in Ecology and Evolution*, 14(10):405–410, 1999.

10. B. Dennis and J. Couchamp. Allee effects in ecology and conservation. *Environmental Conservation*, 36(1):1–80, 2009.

11. A. DeRoss, L. Persson, and H.R. Thieme. Emergent allee effects in top predators feeding on structured prey populations. *Proceedings of the Royal Society of London B: Biological Sciences*, 270(1515):611–618, 2003.

12. U. Dieckmann, R. Law, and J.A.J. Metz. *The Geometry of Ecological Interactions: Simplifying Spatial Complexity*. Cambridge University Press, Cambridge, 1988.

13. B. Dubey, B. Das, and J. Hussain. A predator–prey interaction model with self and cross-diffusion. *Ecological Modelling*, 141(1):67–76, 2001.

14. L. Edelstein-Keshet. *Mathematical Models in Biology*. SIAM, Philadelphia, PA, 1988.

15. E. González-Olivares, H. Meneses-Alcay, B. González-Yañez, J. Mena-Lorca, A. Rojas-Palma, and R. Ramos-Jiliberto. Multiple stability and uniqueness of the limit cycle in a gause-type predator–prey model considering the allee effect on prey. *Nonlinear Analysis: Real World Applications*, 12(6):2931–2942, 2011.

16. H.P. Greenspan. Models for the growth of a solid tumor by diffusion. *Studies in Applied Mathematics*, 51(4):317–340, 1972.

17. S.A. Levin and L.A. Segel. Hypothesis for origin of planktonic patchiness. *Nature*, 259(5545):659–659, 1976.

18. P.K. Maini, D.L. Benson, and J.A. Sherratt. Pattern formation in reaction–diffusion models with spatially inhomogeneous diffusion coefficients. *Mathematical Medicine and Biology: A Journal of the IMA*, 9(3):197–213, 1992.

19. H. Malchow, S.V. Petrovskii, and E. Venturino. *Spatiotemporal Patterns in Ecology and Epidemiology: Theory, Models, and Simulation*. Chapman and Hall/CRC, Boca Raton, FL, 2008.

20. A. Matzavinos, M.A. Chaplain, and V.A. Kuznetsov. Mathematical modeling of the spatiotemporal response of cytotoxic t–lymphocytes to a solid tumour. *Mathematical Medicine and Biology*, 21(1):1–34, 2004.

21. A.B. Medvinsky, S.V. Petrovskii, I.A. Tikhonova, H. Malchow, and B.L. Li. Spatiotemporal complexity of plankton and fish dynamics. *SIAM Review*, 44(3): 311–370, 2002.

22. A.Y. Morozov, S.V. Petrovskii, and B.L. Li. Bifurcations and chaos in a predator–prey system with the allee effect. *Proceedings of the Royal Society of London B: Biological Sciences*, 271(1546):1407–1414, 2004.

23. J.D. Murray. A pre-pattern formation mechanism for animal coat markings. *Journal of Theoretical Biology*, 88(1):161–199, 1981.

24. J.D. Murray. *Mathematical Biology I. An Introduction.* Springer–Verlag, Berlin, Germany, 2002.

25. J.D. Murray. *Mathematical Biology II. Spatial Models and Biomedical Applications.* Springer–Verlag, Berlin, Germany, 2003.

26. A. Okubo and S.A. Levin. *Diffusion and Ecological Problems: Modern Perspectives.* Springer Science and Business Media, Berlin, Germany, 2013.

27. M. Papadogiorgaki, P. Koliou, X. Kotsiakis, and M.E. Zervakis. Mathematical modelling of spatiotemporal glioma evolution. *Theoretical Biology and Medical Modelling*, 10(1):1–47, 2013.

28. F. Rao and Y. Kang. The complex dynamics of a diffusive prey–predator model with an allee effect in prey. *Ecological Complexity*, 28(1):123–144, 2016.

29. L.A. Segel and J.L. Jackson. Dissipative structure: An explanation and an ecological example. *Journal of Theoretical Biology*, 37(3):545–559, 1972.

30. G.R. Shaver. *Spatial Heterogeneity: Past, Present, and Future, in Ecosystem Function in Heterogeneous Landscapes, Springer, New York.*

31. P.A. Stephens and W.J. Sutherland. Consequences of the allee effect for behaviour, ecology and conservation. *Trends in Ecology and Evolution*, 14(3):401–405, 1999.

32. P.A. Stephens, W.J. Sutherland, and R.P. Freckleton. What is the allee effect? *Oikos*, 87(1):185–190, 1999.

33. A.M. Turing. The chemical basis of morphogenesis. *Philosophical Transactions of the Royal Society of London B: Biological Sciences*, 237(641):37–72, 1952.

34. A. Verdy. Modulation of predator–prey interactions by the allee effect. *Ecological Modelling*, 221(8):1098–1107, 2010.

35. F. Vinatier, P. Tixier, P.F. Duyck, and F. Lescourret. Methods ecol evol. *Factors and Mechanisms Explaining Spatial Heterogeneity: A Review of Methods for Insect Populations*, 2(641):11–22, 2011.

36. J. Wang, J. Shi, and J. Wei. Dynamics and pattern formation in a diffusive predator–prey system with strong allee effect in prey. *Journal of Differential Equations*, 251(5):1276–1304, 2011.

37. J. Wang, J. Shi, and J. Wei. Predator–prey system with strong allee effect in prey. *Journal of Mathematical Biology*, 62(2):291–331, 2011.

38. W. Wang, Q.X. Liu, and Z. Jin. Spatiotemporal complexity of a ratio–dependent predator–prey system. *Physical Review E*, 75(5):051913–051922, 2007.

39. X.C. Zhang, G.Q. Sun, and Z. Jin. Spatial dynamics in a predator–prey model with beddington-deangelis functional response. *Physical Review E*, 85(2):021924–021938, 2012.

40. Q.Q. Zheng and J.W. Shen. Dynamics and pattern formation in a cancer network with diffusion. *Communications in Nonlinear Science and Numerical Simulation*, 27(1):93–109, 2015.

41. S.R. Zhou, Y.F. Liu, and G. Wang. The stability of predator–prey systems subject to the allee effects. *Theoretical Population Biology*, 67(1):23–31, 2005.

9 Developments in Runge–Kutta Method to Solve Ordinary Differential Equations

Geeta Arora and Varun Joshi
Lovely Professional University

Isa Sani Garki
Jigawa State Polytechnic Dutse

CONTENTS

9.1 INTRODUCTION

Numerical methods are well known in the field of science and engineering to solve various linear and nonlinear ordinary differential equations (ODEs). Some of the well-known methods developed to solve the ODEs include but not limited to Euler's method, Picard's method, Taylor's series method, and Runge–Kutta (R-K) method. These methods have gone through various stages of development with the advancement of the programming languages and for the various applications to real life applications. This chapter is an attempt to discuss the development in R-K method since its existence. This method is developed for various orders of convergence, i.e., ranging from order 1 to order 4. In all these available versions of R-K method, the method of fourth order becomes more famous because of its convergent properties. This method is of excessive practical significance with mentioned accuracy and numerical stability in comparison with the well-known Euler's method.

The Taylor's series method [8] which is well known for solving differential equations numerically becomes ineffective for the problem which involves the higher order derivatives. Also, the well-known Euler's method is less effective in the practical problems since it requires the step size (h) to be small for obtaining reasonable accuracy. The merit of R-K methods as compared to the above-discussed methods involves no requirement of the calculations for higher order derivatives, and also they are designed to give greater accuracy with the advantage of requirement of only the function values at some selected points on the sub-interval. These methods agree with Taylor's series solution up to the terms of h^r where r is the order of the R-K method.

According to Arumugan et al. [1], problems in science and engineering can be solved by reducing them to differential equations satisfying certain conditions. Analytic methods can be applied to solve many standard types of differential equations. However, some differential equations generated from physical problems are so complex that they can be solved efficiently by numerical methods. The general solution of a differential equation of the nth order has n arbitrary constants; hence, to compute its numerical solution, n conditions are needed. If these n conditions are specified at the initial point only, then it is called an initial value problem. If on the other hand, the conditions are specified at two or more points, then it is called a boundary value problem.

The initial value problem, $\frac{dy}{dx} = f(x,y)$ with the initial condition $y(x_0) = y_0$, can be solved by any method from the methods categorized in the following groups of methods:

1. **Single-step or pointwise methods:** In these methods, the solution is approximated by a truncated series and each term of the series is a function of x. A solution of this type is called a pointwise solution. The methods by Taylor and Picard belong to this category.
2. **Step by step methods:** Here, the values of solution are computed by short steps ahead for equal intervals h of the independent variable. These values are then iterated till the desired accuracy is attained. The methods of Euler, Milne, Adams–Bashforth, Runge–Kutta, and others constitute this category. Among all these methods, this chapter will focus and discuss on the R-K method of various orders.

9.2 DEVELOPMENT OF RUNGE–KUTTA METHOD

In the last decade, many researchers have devoted their efforts in development of numerical methods to solve the ODEs efficiently with good accuracy. One such method is the R-K method that was developed by Runge [13] and Kutta [9], the two German Mathematicians in early 1900 in context to find the solution to differential equation in field of atomic spectra. This method came into existence from a procedure that was basically developed while working on the numerical solution of algebraic equations. These methods possess the strength that they are easy to program and are capable to compute solution of ODEs as compared to other methods except if the calculations of the function are complicated. The idea of generalizing the Euler's method, by allowing for a number of evaluations of the derivative to take place in a step, was due to Runge's paper of 1895.

In the paper, Runge dealt with an initial value problem of the form $\frac{dy}{dx} = f(x,y), y(x_0) = y_0$. He explored three main schemes. The first of these three methods is the midpoint role adapted to ODEs, whereas the second and third methods are different versions of the trapezoidal rule. The last of these methods suggests iterative computation of the stage values.

Further contributions were made by Heun in 1900 [7] and Kutta 1901 [9]. The latter completely characterized the set of R-K methods of order 4 and proposed the first method of order 5. In his paper, which appeared in 1901, Kutta discussed the analysis of R-K methods as far as order five and systematically obtained the order conditions up to fifth order. However, this was incomplete in two different respects. First, his analysis was for only a first-order differential equation, rather than a system of equations. The other sense in which the work of Kutta was incomplete was that his order 5 methods have slight errors in them. These errors were corrected by Nystrom in 1925 [11]. The first phase in the history of R-K methods ended by the work of Nystrom. He took the analysis of the fifth-order methods to its completion. It was not until the work of Huta in 1957 that the sixth-order methods were introduced. The theory for a system comes out of the work of Gill in 1951 [6] and Merson in 1957 [10] and by Butcher in 1996 [3].

Since the advent of digital computers, most of the researchers' interest has been focused on R-K methods, and a large number of research workers have contributed to recent extensions of the theory, and to the development of extended R-K methods. Although early studies were devoted entirely to explicit R-K methods, interest then moved to include implicit methods, which got recognized as an appropriate method for the solution of stiff differential equations.

9.3 THE RUNGE–KUTTA METHODS

Consider the first-order ODE of the form given by

$$\frac{dy}{dx} = f(x,y), \tag{9.1}$$

with condition $y(x_0) = y_0$ where points of the domain $[x_0, x_n]$ are considered at uniform distance. The solution at the point x_{n+1} obtained by $y(x_{n+1})$ can be obtained by using the R-K method. The method is based upon the concept of weighted average of slopes at various points in the domain. Considering m slopes, in general the formula for the method can be written as

$$y_{n+1} = y_n + \sum_{i=-1}^{m} w_i k_i \tag{9.2}$$

where w_i are the constant weight coefficients and k_i are the coefficients to be calculated. As discussed and presented by Bucher [4], each formula of R-K method can be represented in a tabular form known as Bucher table. This table provides a summary of the values of the constants with respect to k_i inform of steps involved with final row of table as the represented final formula.

Following are the details of R-K methods of different orders:

1. **First-Order R-K method:** The Euler's formula for first approximation to the solution of the above differential Equation (15.2) is given by:

$$y_1 = y(x_0 + h)$$

On expanding by Taylor's series, it can be rewritten as

$$y_1 = y(x_0 + h) = y_0 + \frac{h}{1!}y_0' + \frac{h^2}{2!}y_0'' + \dots$$

Clearly, Euler's method agrees with the Taylor's series solution up to the term in h. Hence, Euler's method is the R-K method of first order given as

$$y_1 = y_0 + hy_0' = y_0 + hf(x_0, y_0)$$

2. **Second-Order R-K method:** The modified Euler's formula for numerical solution ODE given by Equation (15.2) can be obtained from

$$y_1 = y_0 + \frac{h}{2}(f(x_0, y_0) + f(x_0 + h, y_0 + hf(x_0, y_0)))$$

$$= y_0 + \frac{h}{2}(f_0 + f(x_0 + h, y_0 + hf_0)), \tag{9.3}$$

where $f_0 = f(x_0, y_0)$. Expanding the LHS by Taylor's series, we get

$$y_1 = y(x_0 + h) = y_0 + \frac{h}{1!}y_0' + \frac{h^2}{2!}y_0'' + \frac{h^3}{3!}y_0''' + \cdot$$

Expanding $f(x_0 + h, y_0 + hf_0)$ by Taylor's series for a function of two variables, we have

$$f(x_0 + h, y_0 + hf_0) = f(x_0, y_0) + \frac{h}{1!}\left(\left(\frac{df}{dx}\right)(x_0, y_0)\right.$$

$$\left. + h\left(\frac{df}{dx}\right)(x_0, y_0)\right) + O(h^2)$$

where $O(h^2)$ represents all the terms involving second and higher powers of h. Using this expression in Equation (9.3), we get

$$y_1 = y_0 + \frac{h}{2}(f_0 + f(x_0, y_0) + h\left(\frac{df}{dx}\right)(x_0, y_0) + h\left(\frac{df}{dx}\right)(x_0, y_0)) + O(h^2)$$

$$y_1 = y_0 + hy_0' + \frac{h^2}{2!}y_0'' + O(h^3), \tag{9.4}$$

Comparing Equations (9.3) and (9.4), it can be concluded that the modified Euler's method agrees with Taylor's series solution up to the h^2 term. Hence, the modified Euler's method is the R-K method of second order or midpoint method. Therefore, the second-order R-K formula is given by

$$y_1 = y_0 + \frac{1}{2}(k_1 + k_2)$$

where $k_1 = hf(x_0, y_0)$, $k_2 = hf(x_0 + h, y_0 + \frac{k_1}{2})$. In form of Bucher table, the formula can be represented as in Table 9.1.

3. **Third-Order R-K formula:** The formula for the third order R-K method is given by

$$y_1 = y_0 + \frac{1}{6}(k_1 + 4k_2 + k_3) \qquad (9.5)$$

where $k_1 = hf(x_0, y_0)$, $k_2 = hf(x_0 + \frac{h}{2}, y_0 + \frac{k_1}{2})$, $k_3 = hf(x_0 + h, y_0 - k_1 + 2k_2)$. In form of Bucher table, the formula can be represented as in Table 9.2.

4. **Fourth-Order R-K method:** This method is the most commonly used Runge–Kutta (R-K) method. The working rule for solving the initial value problem given by Equation (15.2) using the fourth-order R-K method is as follows:

$$y_1 = y_0 + \frac{1}{6}(k_1 + 2k_2 + 2k_3 + k_4) \qquad (9.6)$$

Table 9.1

Bucher Table for the Second-Order R-K Formula

h	k_1	k_2
0	0	0
$\frac{1}{2}$	$\frac{1}{2}$	0
	$\frac{1}{2}$	$\frac{1}{2}$

Table 9.2

Bucher table for the Third-Order R-K Formula

h	k_1	k_2	k_3
0	0	0	0
$\frac{1}{2}$	$\frac{1}{2}$	0	0
1	-1	2	
	$\frac{1}{6}$	$\frac{4}{6}$	$\frac{1}{6}$

The four required constants are $k_1 = hf(x_0, y_0)$, $k_2 = hf\left(x_0 + \frac{h}{2}, y_0 + \frac{k_1}{2}\right)$, $k_3 = hf\left(x_0 + \frac{h}{2}, y_0 + \frac{k_2}{2}\right)$, $k_4 = hf(x_0 + h, y_0 + k_3)$. Similarly, the value of y_2 in the second interval is obtained by replacing x_0 by x_1 and y_0 by y_1 in the above set of formulae. In general, to find y_n, we substitute x_{n-1}, y_{n-1} in the expression for k_1, k_2, etc. It should be noted that

- The operation is identical for both linear and nonlinear differential equations.
- To evaluate y_{n+1}, we need information only at the point y_n. Information at the points y_{n-1}, y_{n-2}, etc. is not directly required. This indicates that R-K methods are step-based methods.

In form of Bucher table, the formula is represented in Table 9.3.

5. **Fifth-Order R-K method:** The fifth-order R-K method was introduced by Kutta [9], but since there were errors in the presentation of his results, it was then partly corrected by Nystrom [12] and Lawson [10] separately that give rise to two different formulations of the fifth-order R-K method. But the formula by Lawson becomes famous as the fifth-order R-K method given as follows with Bucher table of the formula can be represented as in Table 9.4:

$$y_1 = y_0 + \frac{1}{90}(7k_1 + 32k_3 + 12k_4 + 32k_5 + 7k_6) \tag{9.7}$$

Table 9.3

Bucher table for the Fourth-Order R-K Formula

h	k_1	k_2	k_3	k_4
0	0	0	0	0
$\frac{1}{2}$	$\frac{1}{2}$	0	0	0
$\frac{1}{2}$	0	$\frac{1}{2}$	0	0
1	0	0	1	0
	$\frac{1}{6}$	$\frac{2}{6}$	$\frac{2}{6}$	$\frac{1}{6}$

Table 9.4

Bucher table for the Fourth-Order R-K Formula

h	k_1	k_2	k_3	k_4	k_5	k_6
0	0	0	0	0	0	0
$\frac{1}{2}$	$\frac{1}{2}$	0	0	0	0	0
$\frac{1}{4}$	$\frac{3}{16}$	$\frac{1}{16}$	0	0	0	0
$\frac{1}{2}$	0	0	$\frac{1}{2}$	0	0	0
$\frac{3}{4}$	0	$\frac{-3}{16}$	$\frac{6}{16}$	$\frac{9}{16}$	0	0
1	$\frac{1}{7}$	$\frac{4}{7}$	$\frac{6}{7}$	$\frac{-12}{7}$	$\frac{8}{7}$	0
	$\frac{7}{90}$	0	$\frac{32}{90}$	$\frac{12}{90}$	$\frac{32}{90}$	$\frac{7}{90}$

with $k_1 = hf(x_0, y_0)$, $k_2 = hf\left(x_0 + \frac{h}{2}, y_0 + \frac{k_1}{2}\right)$, $k_3 = hf\left(x_0 + \frac{h}{4}, y_0 + \frac{3k_1}{16} + \frac{k_2}{16}\right)$, $k_4 = hf\left(x_0 + \frac{h}{2}, y_0 + \frac{k_3}{2}\right)$, $k_5 = hf\left(x_0 + \frac{3h}{4}, y_0 - \frac{3k_2}{16} + \frac{6k_3}{16} + \frac{9k_4}{16}\right)$, $k_6 = hf\left(x_0 + h, y_0 + \frac{k_1}{7} + \frac{4k_2}{7} + \frac{6k_3}{7} + \frac{12k_4}{7} - \frac{12k_4}{7} + \frac{8k_5}{7}\right)$.

9.4   EXTENSION OF RUNGE–KUTTA METHODS

Apart from the R-K method of order first to fifth, all of which are due to Runge and Kutta, the original authors, there are other special methods which are a sort of extension to the earlier methods. These modified methods developed on the theme of R-K method are discussed as follows:

1. **RK34 method:** This method was proposed by Abebe [2] and Gustaf [14] as a combination of the R-K third and fourth methods usually called Embedded R-K method and is given as follows:
   RK34 of order four

$$y_1 = y_0 + \frac{1}{6}(k_1 + 2k_2 + 2k_3 + k_4)$$

   RK34 of order three

$$y_1 = y_0 + \frac{1}{6}(k_1 + 4k_2 + l_3)$$

   where $k_1 = hf(x_0, y_0)$, $k_2 = hf(x_0 + \frac{h}{2}, y_0 + \frac{k_1}{2})$, $k_3 = hf(x_0 + \frac{h}{2}, y_0 + \frac{k_2}{2})$, $l_3 = hf(x_0 + h, y_0 - hk_1 + 2hk_2)$, $k_4 = hf(x_0 + h, y_0 + k_3)$

2. **Runge–Kutta–Fehlberg (RKF45) method:** This is a numerical method which was developed by famous mathematician Erwin Fehlberg with the modification of the well-known R-K method in 1969. This [5] method has a procedure to determine if the proper step size h is being used. At each step, two different approximations for the solution are made and compared. If the two answers are in close agreement, the approximation is accepted. If the two answers do not agree to a specified accuracy, the step size is reduced. If the answers agree to more significant digits than required, the step size is increased. The method is given by
   RKF45 of order four

$$y_1 = y_0 + \frac{25}{216}k_1 + \frac{1408}{2565}k_3 + \frac{2197}{4104}k_4 + \frac{1}{5}k_5$$

   RKF45 of order five

$$y_1 = y_0 + \frac{16}{135}k_1 + \frac{6656}{12825}k_3 + \frac{28561}{56430}k_4 - \frac{9}{50}k_5 + \frac{2}{55}k_6$$

   where $k_1 = hf(x_0, y_0)$, $k_2 = hf\left(x_0 + \frac{h}{2}, y_0 + \frac{k_1}{2}\right)$, $k_3 = hf\left(x_0 + \frac{3h}{8}, y_0 + \frac{3k_1}{32} + \frac{9k_2}{32}\right)$, $k_4 = hf\left(x_0 + \frac{12h}{13}, y_0 + \frac{1932k_1}{2197} - \frac{7200k_2}{2197} + \frac{7296k_3}{2197}\right)$, $k_5 = hf\left(x_0 + h, y_0 + \frac{439k_1}{216} - 8k_2 + \frac{3680k_3}{513} - \frac{845k_4}{4104}\right)$, $k_6 = hf\left(x_0 + \frac{h}{2}, y_0 - \frac{8k_1}{27} + 2k_2 - \frac{3544k_3}{2565} + \frac{1859k_4}{4104} - \frac{11k_5}{40}\right)$.

3. **Runge–Kutta–Gill method:** The Runge–Kutta–Gill method [6] is the most widely used single-step method for solving ODEs and is given by

$$y_1 = y_0 + \frac{1}{6}\left(k_1 + 2\left(1 - \frac{1}{\sqrt{2}}\right)k_2 + 2\left(1 + \frac{1}{\sqrt{2}}\right)k_3 + k_4\right) + O(h^5) \quad (9.8)$$

where $k_1 = hf(x_0, y_0)$, $k_2 = hf\left(x_0 + \frac{h}{2}, y_0 + \frac{k_1}{2}\right)$, $k_3 = hf\left(x_0 + \frac{h}{2}, y_0 - \left(-\frac{1}{2} + \frac{1}{\sqrt{2}}\right)k_1 + \left(1 - \frac{1}{\sqrt{2}}\right)k_2\right)$, $k_4 = hf\left(x_0 + h, y_0 - \frac{k_2}{\sqrt{2}} + \left(1 + \frac{1}{\sqrt{2}}\right)k_3\right)$.

4. **Runge–Kutta–Merson method:** This method [10] outlines a process for deciding the step size for better predetermined accuracy. For this method, five functions are evaluated at every step. The algorithm is given by

$$y_1 = y_0 + \frac{1}{6}(k_1 + 4k_4 + k_5) + O(h^5) \quad (9.9)$$

where $k_1 = hf(x_0, y_0)$, $k_2 = hf(x_0 + \frac{h}{3}, y_0 + \frac{k_1}{3})$, $k_3 = hf(x_0 + \frac{h}{3}, y_0 + \frac{k_1}{6} + \frac{k_2}{6})$, $k_4 = hf(x_0 + \frac{h}{2}, y_0 + \frac{k_1}{8} + \frac{3k_2}{8})$, $k_5 = hf(x_0 + h, y_0 + \frac{k_1}{2} - \frac{3k_2}{2} + 2k_4)$.

5. **RK43 method:** This method applied the following four stages to arrive at the solution:

$$y_1 = y_m + \frac{h}{2}f(x_m, y_m)$$

$$y_2 = y_1 + \frac{h}{2}f(x_1, y_1)$$

$$y_3 = \frac{2}{3}y_m + \frac{y_2}{3} + \frac{h}{6}f(x_2, y_2)$$

$$y_{m+1} = y_3 + \frac{h}{2}f(x_3, y_3)$$

6. **RK54 method:** This method applies the following five stages to arrive at the solution:

$$y_1 = y_m + 0.391752226571890hf(x_m, y_m)$$

$$y_2 = 0.444370493651235y_m + 0.555629506348765y_1 + 0.368410593050371hf(x_1, y_1)$$

$$y_3 = 0.620101851488403y_m + 0.379898148511597y_2 + 0.251891774271694hf(x_2, y_2)$$

$$y_4 = 0.178079954393132y_m + 0.821920045606868y_3 + 0.544974750228521hf(x_3, y_3)$$

$$y_{m+1} = 0.517231671970585y_2 + 0.096059710526147y_3 + 0.063692468666290hf(x_3, y_3)$$

$$+ 0.386708617503269y_4 + 0.226007483236906hf(x_4, y_4)$$

Table 9.5

The Results Are Presented in the Form of Maximum Absolute Errors

Methods	Max. Error	Position of Accuracy
Fifth order R-K	9.67E-12	First
ODE45	9.46E-11	Second
Fehlberg RK45 of fourth order	9.48E-10	Third
Fehlberg RK45 of fifth order	9.48E-10	Third
Mersons	9.70E-10	Fifth
Gills	9.87E-9	Sixth
RK34 of third order	8.71E-9	Seventh
Fourth order	8.71E-9	Seventh
Third order	9.99E-7	Ninth
Second order	9.46E-5	Tenth
RK54	9.21E-3	Eleventh
RK43	9.20E-3	Twelfth
RK34 of fourth order	8.27E-3	Thirteenth
First order	7.32E-3	Fourteenth

9.5 NUMERICAL RESULTS

In this section for demonstrating the efficiency of different orders of R-K method in solving the given differential equation, an ODE is solved by the different orders of R-K methods. The obtained solutions are compared with the exact solution of the equation in order to ascertain the level of accuracy with regard to each method.

The Problem: Obtain the approximate value of the solution (y) for the initial value problem

$$y' = -2xy^2, y(1) = 1,$$

using different R-K methods with step $h = 0.05$. The exact solution of the equation is given by $y = \frac{1}{x^2}$. The solution is obtained by R-K method with domain $[1,2]$ with number of partitions as $N = 21$ using a MATLAB program. In Table 9.5, the results are presented in the form of maximum absolute errors.

9.6 CONCLUSION

It may not be clearly identified that which one, out of the discussed different R-K methods, is more accurate by just studying the solution curves and comparing the exact solution with approximated solutions obtained for the differential equation selected for the study. It can be seen from the graphs that the solution curves for exact and approximated solution obtained from almost all different categories coincide throughout the corresponding values. However, the differences among the solutions can be detected by studying the tables displaying both the exact and approximated solutions numerically. On going through observation of all the methods, these methods can be listed in the order of their accuracies from the most accurate down to the least accurate as given in Table 9.5.

REFERENCES

1. S. Arumugan, I. A. Thangapandi, and A. Somasundaram. *Numerical Solutions of Ordinary Differential Equations: Numerical Methods.* Scitech Publications Pvt. Ltd., India, 2015.

2. U. M. Ascherm and L. R. Petzold. *Computer Methods for Ordinary Differential Equations and Differential Algebraic Equations.* SIAM, Philadelphia, PA, 1998.

3. J. C. Butcher. A history of Runge-Kutta methods. *Appl. Num. Math.*, 20(3):247–260, 1996.

4. J. C. Butcher. *Numerical Methods for Ordinary Differential Equations.* John Wiley and Sons, Ltd, Hoboken, NJ, 2008.

5. E. Fehlberg. Classical fifth, sixth, seventh and eighth order RungeKutta formulas with stepsize control. NASA Technical Report, 1968.

6. S. Gill. A process for the step-by-step integration of differential equations in an automatic computing machine. *Proc. Cambridge Philos. Soc.*, 47:96–108, 1951.

7. K. Heun. Neue Methoden zur approximativen Integration der Differentialgleichungen einer unabhängigen Veränderlichen. *Z. Math. Phys.*, 45:23–38, 1900.

8. R. K. Jain and S. R. K. Iyenger. *Numerical Methods: Advanced Engineering Mathematics.* Narosa Publishing House, New Delhi, 2014.

9. W. Kutta. Beitrag Zurn nöherungsweisen Integration totaler Differentialgleichungen. *Z. Math. Phys.*, 46:435–453, 1901.

10. J. D. Lawson. An order five Runge-Kutta process with extended region of stability. *SIAM J. Numer. Anal.*, 3:593–597, 1966.

11. R. H. Merson. An operational method for the study of integration processes. *Proceedings Symposium on Data Processing. Weapons Research Establishment*, Salisbury, Australia, pp. 110-1–110-25, 1957.

12. E. J. Nystrom. Uber die numerische Integration von Differentialgleichungen. *Acta Soc. Sci. Fennicae*, 50(13):1–55, 1925.

13. C. Runge. On the numerical solution of differential equations. *Math. Ann.*, 46:167–178, 1895.

14. G. Söderlind. *Numerical Methods for Differential Equations.* London University, London, 2008.

10 A Criterion Space Decomposition Method for a Tri-objective Integer Program

Masar Al-Rabeeah and Ali Al-Hasani
RMIT University
Basrah University

Santosh Kumar
University of Melbourne
RMIT University

Andrew Eberhard
RMIT University

CONTENTS

10.1 INTRODUCTION

In many practical situations, one is required to optimize several objective functions simultaneously. Since these objective functions may conflict with each other, finding a solution that can optimize all objective functions at the same time is impossible. In such cases, non-dominated solutions are of interest which represent points where value of one objective cannot be improved without adversely affecting the value of some other objective function. These problems are called multi-objective optimization problems, and when variables are restricted to integer values, they are called Multi-Objective Integer Programs (MOIP) (Antunes et al., 2016; Ehrgott, 2006b; Greco et al., 2016). Similarly, when variables are continuous, the models are known as Multi-objective Linear Programming (MOLP) problems and they have been discussed in Benson (1998) and Benson and Sun (2000, 2002). In Multi-Objective Mixed-Integer Programming (MOMIP) problems, some variables are continuous and the remaining variables are restricted to integer values (Al-Hasani et al., 2018; Stidsen et al., 2014).

A solution for a MOIP can be identified either in the decision space or in the criterion space. Algorithms that search in the criterion space have advantages compared to the algorithms that search in the decision space (Boland et al., 2017). Basically, the problem in the criterion space requires less computational effort than that in the decision space. Therefore, we are focusing search for the non-dominated points in the criterion space. In this chapter, we consider problems that involve three objective functions with integer restricted variables, which are called Tri-objective Integer Programming (TOIP) problems.

In this study, we develop a criterion-based decomposition method to solve a TOIP problem. The algorithm by Boland et al. (2017) motivated us; when their idea is combined with Ozlen and Azizoglu (2009) and Ozlen et al. (2014), it has further scope of improvement. This modified algorithm minimizes the number of constraints and the binary variables that are added to the problem to decompose the criterion space, which resulted in a decrease in the number of integer problems (IPs) solved and hence a decrease in the central processing unit (CPU) time.

In Section 10.2, literature is reviewed. Some important concepts for TOIP are presented in Section 10.3. The proposed algorithm is discussed in Section 10.4. Computational experiments are presented in Section 10.5 and finally, the chapter is concluded in Section 10.6.

10.2 LITERATURE REVIEW

Existing literature on Bi-objective Integer Programming is abundant; but when the number of objectives is more than two, the situation changes. Only a few methods have discussed the TOIP model. Here is a brief review of some of those methods that have analyzed TOIP problems.

TOIP had been addressed by Sylva and Crema (2004), Ehrgott (2006a), Chinchuluun and Pardalos (2007), and Chankong and Haimes (2008). They have used different strategies to find the whole set of non-dominated points. Elimination of the dominated region method has been used by Sylva and Crema (2004), where instead of decomposing the criterion space, the authors eliminated the dominated part from further consideration. In each iteration, new constraints and binary variables are added to the problem, which makes the problem larger and larger afterwards. That means more time is required to get the whole set of solution. Further modification of Sylva and Crema (2004) has been discussed by Lokman and Koksalan (2013) where the authors provided two new algorithms to find the non-dominated set for a MOIP. The number of constraints and binary variables required in (Sylva and Crema, 2004) has been decreased in the first algorithm, and the complexity is further reduced in the second algorithm by considering bounds for the objective functions. Thus, their second algorithm outperformed the first one according to their computational experiments.

Some methods have used decomposition in the criterion space to create smaller sub-problems by adding constraints and binary variables depending on the recent non-dominated point (Dachert and Klamroth, 2015; Boland et al., 2016). Dachert and Klamroth (2015) developed a new algorithm by splitting the area of criterion space into smaller sub-problems by considering neighborhood characteristic to avoid generating redundant sub-problems. Boland et al. (2016) developed an efficient algorithm for solving TOIP which is called the L-shape method (LSM). The authors used the criterion space to find the non-dominated points. Combining ideas from Sylva and Crema (2004) and Dachert and Klamroth (2015) led to the LSM algorithm. There is a disadvantage in this approach; that is, it requires solving more IPs than what is required by the algorithm developed by Dachert and Klamroth (2015).

Another strategy to solve TOIP is using scalarization technique, that is, the ε-constraint method (Ehrgott, 2006a; Ozlen and Azizoglu, 2009; Mavrotas and Florios, 2013; Ozlen et al., 2014) where the multi-objective problem transforms to a single objective problem. By combining the ε-constraint method and the weighted-sum method (Aneja and Nair, 1979), Ehrgott (2006a) proposed a new algorithm with flexible constraints, which is called the method of elastic constraints. In Ozlen and Azizoglu (2009) and Ozlen et al. (2014), more modifications were proposed using the objective bounds to identify the non-dominated solutions. They generated all non-dominated points by shrinking those bounds.

10.3 PRELIMINARIES

10.3.1 BASIC CONCEPT

For completeness, some basic ideas from Ehrgott (2006b), Antunes et al. (2016), Al-Hasani et al. (2019), and Al-Rabeeah et al. (2019) are reproduced here.

TOIP can be expressed as follows:

$$\max z_1(x), z_2(x), z_3(x) \tag{10.1}$$

subject to $x \in X$ where $z_i(x)$ represents the objective functions i, $i = 1, 2, 3$. X represents the feasible set in the decision space, and Y represents the feasible set in the criterion space such that $x_j \in \mathbb{Z}$ for $j = 1, 2, \ldots, n$.

Each objective function: $z_i(x) = \sum_{j=1}^{n} a_{ij} x_j$, where $a_{ij} \in \mathbb{Z}$ such that $i = 1, 2, 3$.

Definition 10.1 A feasible solution $x^\circ \in X$ in the decision space is called efficient solution if there is no $x \in X$ such that $z_i(x) \geq z_i(x^\circ)$ for each i and $z_i(x) > z_i(x^\circ)$ for at least one i. The image $z(x^\circ)$ of efficient point x° in the criterion space is called non-dominated point. Let X_E, Y_N denote the set of all efficient solutions and non-dominated solutions, respectively.

Definition 10.2 Let $x^\circ \in X_E$ be an efficient solution. If there is some positive value λ such that $x^\circ \in X_E$ is an optimal solution of $\min_{x \in X} \lambda^T z(x^\circ)$, then x° is termed as supported efficient solution; otherwise, it is called non-supported efficient solutions, $z(x^\circ)$ is termed as supported non-dominated solutions.

Definition 10.3 An efficient solution $x^\circ \in X$ in the decision space is called strictly efficient if there is no $x \in X$ such that $z_i(x) \geq z_i(x^\circ)$ for each i, image of the strictly efficient solution in the criterion space is called strictly non-dominated solution.

Definition 10.4 A feasible point in the criterion space is called weakly non-dominated point if and only if it is not strictly dominated by any other non-dominated point.

Overall, our interest is to compute the whole set of non-dominated points for a TOIP model that includes non-dominated supported and non-dominated non-supported points.

10.3.2 REVIEW OF SOME RECENT APPROACHES

10.3.2.1 The ε-Constraint Method

The ε-constraint method finds the non-dominated points for a general MOIP. It solves a MOIP by transforming it into a single objective problem by considering one objective and converting other objectives into constraints with restricted value. Thus, the problem (10.1) is reformulated as

$$\max z_3(x)$$
$$\text{s.t.} \quad x \in X, \quad z_i(x) \geq \varepsilon_i \tag{10.2}$$

where ε_i can be set as lower bound for each objective function $i = 1, 2$. $i \neq 3$. The algorithm starts with this initial ε_i value, and then, in each iteration, this value is updated according to the last non-dominated point detected. This process continues until the value of ε_i touches the upper bound, and then, the algorithm terminates. It is well known that in general, the ε-constraint method can only generate weakly non-dominated points (Antunes et al., 2016); therefore, Ozlen and Azizoglu (2009) and Antunes et al. (2016) improved it by combining it with the weighted-sum method (Aneja and Nair, 1979) as follows:

$$\max z_3(x) + w \sum_{i=1}^{2} z_i(x)$$
$$\text{s.t.} \quad x \in X, \quad z_i(x) \geq \varepsilon_i \tag{10.3}$$

for all $i = 1, 2$.

From Equation (10.3), it is clear that the weight specified for the objective function 3 is 1 and small positive weights w are assigned to the rest of objective functions. It may be noted that there is no need to add constraint to the objective 3 which has been considered later in the proposed algorithm.

10.3.2.2 Boland, Charkhgard, and Savelsbergh Method (2017)

Based on Sylva and Crema (2004), Boland et al. 2017 developed a method to find the non-dominated solution for TOIP. This algorithm not only eliminates the dominated part from the search space as was done in Sylva and Crema (2004) but also decomposes the rest of the criterion space into smaller sub-spaces (regions) defined by a set of constraints to avoid solving unnecessary IPs. These small regions are updated by changing the constraints after each step according to the recent non-dominated point determined. A new region is created (p^U, p, p^L), which is denoted by three points that is upper bound of the region, search index point, and lower bound of the region, respectively. The algorithm maintains a queue of regions, and at each iteration, the algorithm selects one region and searches inside that region for a non-dominated point. The search inside the region (p^U, p, p^L) is carried out by

$$\max\left\{ \sum_{i=1}^{3} z_i^a \right\},$$

$$\text{s.t.} \quad x \in X, \quad z_i^a = z_i(x) \quad \forall i = \{1,2,3\}$$

$$z_i^a \geq \left(p_i^j - \varepsilon - m_i \right) b_{ij} + m_i \quad \forall i = \{1,2,3\}, \quad j = \{1,\dots,t\}, i \in \theta(p^j)$$

$$\sum_{i\in\theta(p)} b_{ij} = 1, \quad \forall i = \{1,\dots,t\}, \quad b_{ij} \in \{0,1\} \qquad (10.4)$$

$$z_i^a \geq (p_i - \varepsilon - m_i) b_i' + m_i \quad \forall i = \{1,2,3\}, \quad j = \{1,\dots,t\}, i \in \theta(p^j)$$

$$\sum_{i\in\theta(p)} b_i' = 1, \quad \forall j = \{1,\dots,t\} \quad b_i' \in \{0,1\}$$

where a refers to an item in the list of the non-dominated points in this specific region, m is a large constant, and ε is a small positive value; the two sets $\theta(p^j)$ and $\theta(p)$ used by Boland et al. (2017) were just for computational purpose. It may be noted that for each objective function, there are at least three or more constraints associated with it as shown in the numerical example discussed by Boland et al. (2017). Our goal, in this chapter, is to utilize the advantages of method used by Boland et al. (2017) along with other current algorithms and develop a new algorithm that can outperform all of them.

10.4   THE PROPOSED ALGORITHM

In the proposed approach, the non-dominated points for the problem (10.1) are determined by using an augmented weighted-sum objective function similar to problem (10.3) along with some constraints from the formulation Equation (10.4). The algorithm deals with a queue of regions based on the latest non-dominated point detected. At each iteration, the algorithm searches inside one of these regions. Suppose that we start our search with the initial search region $\left(p^U, p, p^L\right)$ where $pU = (p_1^U, p_2^U)$, $p = (p_1, p_2)$, and $p^L = (p_1^L, p_2^L)$, the modified problem for each iteration becomes

$$\max\{w_1 z_1 + w_2 z_2 + z_3\}$$
$$\text{s.t. } x \in X, \quad z_1 = z_1(x), \ z_2 = z_2(x) \text{ and } z_3 = z_3(x)$$

$$z_1 \geq p_1^L, \ z_2 \geq p_2^L \tag{10.5}$$

$$z_1 \geq (p_1 - \varepsilon - m_1) b_1$$

$$z_2 \geq (p_2 - \varepsilon - m_2) b_2$$

$$b_1 + b_2 = 1, b_i \in \{0,1\}$$

where $w_1 = \left(1/\left(\left(Z_1^{UB} - Z_1^{LB} + 1\right)\left(Z_2^{UB} - Z_2^{LB} + 1\right)\right)\right)$ and $w_2 = \left(1/\left(Z_2^{UB} - Z_2^{LB} + 1\right)\right)$, m is a small constant, and ε is a small positive value. From Equation (10.5), it is clear that we have saved some constraints with respect to the third objective function in Equation (10.4). In addition, the proposed formulation simplifies by reducing the number of constraints and binary variables.

For the initial step, the algorithm starts by setting $p^L = p = (m_1, m_2) = (Z_1^{UB} - 1, Z_2^{UB} - 1)$. The values of p^L will be updated according to the new non-dominated points identified. The variables $b_1, b_2, \ldots$ are binary variables. The algorithm starts by finding the global lower and upper bound for the three objective functions $\left(Z_1^{LB}, Z_1^{UB}\right), \left(Z_2^{LB}, Z_2^{UB}\right)$ and $\left(Z_3^{LB}, Z_3^{UB}\right)$. First search region $\left(p^U, p, p^L\right)$ will be set accordingly such that $\left(p^U, p, p^L\right) = \left(\left(Z_1^{UB}, Z_2^{UB}\right), M, M\right)$, $M = (m_1, m_2)$, $m_1 = \left(Z_1^{LB} - 1\right)$, $m_2 = \left(Z_2^{LB} - 1\right)$. Through our explanation, we call the set of search region R and the set of non-dominated points as Nd. First search region $\left(p^U, p, p^L\right)$ is represented by $pU = (p_1^U, p_2^U)$, $p = (p_1, p_2)$, and $p^L = (p_1^L, p_2^L)$.

If the algorithm finds a new non-dominated point inside this search region, then it is added to R; otherwise, it sets temporary variable $p^{\wedge U}$ to find another search region, and it will check some steps to avoid creating repeated search regions.

The steps of the proposed algorithm are given in Algorithm 10.1.

Theorem 10.1 For $x° \in X$, if $y° = z(x°)$ is a non-dominated solution for the proposed weighted-sum objective Equation (10.5), then it is also a non-dominated solution for the TOIP in Equation (10.1). For proof, see Chankong and Haimes (2008).

Algorithm 10.1 Pseudo-code for the Proposed Algorithm

Step 1. Set R and Nd as a null set.

Step 2. Create an initial search region (p^U, p, p^L) and add it to R.

Step 3. While R is not empty, pick a first search region and solve Equation (10.5), If it is feasible then add the new non-dominated point z^{new} to Nd, otherwise go to Step 4. If it is empty go to Step 6.

Step 4. Set temporary variables $p^{\wedge U}, p^{\wedge L}$ as follow:
$p^{\wedge U} = (p_1^{\wedge U}, p_2^{\wedge U}) = (\min(\max(p_1, z_1^{new}), p_1^U), \min(\max(p_2, z_2^{new}), p_2^U))$.
$p^{\wedge L} = (p_1^{\wedge L}, p_2^{\wedge L}) = ((\min(p_1, z_1^{new}), \min(p_2, z_2^{new}))$ If $(z_1^{new} < p_1)$ or $(z_2^{new} < p_2)$.
Then add $(p^{\wedge U}, p, p^{\wedge L})$ to R, otherwise go to Step 6.

Step 5. Check

5.1: If $p_1^{\wedge U} < p_1^U$ and $p_2^{\wedge U} < p_2^U$ then add $(p^U, p^{\wedge U}, p^L)$ to R.

5.2: Else if $p_1^{\wedge U} \geq p_1^U$ and $p_2^{\wedge U} < p_2^U$ then add $(p^U, M, (p_1^L, p_2^{\wedge U}))$ to R.

5.3: Else if $p_1^{\wedge U} < p_1^U$ and $p_2^{\wedge U} \geq p_2^U$ then add $(p^U, M, (p_1^{\wedge U}, p_2^L))$ to R.

Then go Step 3.

Step 6. Stop.

10.4.1 NUMERICAL ILLUSTRATION

In this section, we explain the steps of the algorithm by solving a knapsack problem with ten variables:

$$\max z_i(x) = P_i X^T, \ i = 1, 2, 3$$

$$\text{s.t. } AX^T = W_{r,j} X^T \leq 295 \tag{10.6}$$

where
$P_1 = [21 \ 69 \ 26 \ 92 \ 77 \ 30 \ 96 \ 80 \ 60 \ 61]$,
$P_i = [52 \ 92 \ 19 \ 10 \ 63 \ 34 \ 100 \ 60 \ 11 \ 12]$,
$P_i = [37 \ 100 \ 74 \ 17 \ 60 \ 69 \ 49 \ 69 \ 49 \ 59]$,
$W_{r,j} = [84 \ 49 \ 68 \ 20 \ 97 \ 74 \ 60 \ 30 \ 13 \ 95]$,
$X = [x_1, x_2, x_3, x_4, x_5, x_6, x_7, x_8, x_9, x_{10}]$.

Solution process:

1. Find the global upper and lower bounds for the first and second objective function. $(Z_1^{UB}, Z_2^{UB}) = (474, 336), (Z_1^{LB}, Z_2^{LB}) = (0, 0) \Longrightarrow m_1 = Z_1^{LB} - 1 = -1$ similarly $m_2 = Z_2^{LB} - 1 = -1$

2. For the initial search part $(p^U, p, p^L) = ((474, 336), (-1, -1), (-1, -1))$, these values are initialized by setting $p^L = p = (m_1, m_2) = (Z_1^{LB} - 1, Z_2^{LB} - 1)$.

3. Find a non-dominated point inside the search region in the second step using Equation (10.5). This results in $z^{new} = (361, 316, 410)$.

4. Set $p^{\wedge U} = \min(\max(-1, 361), 474), \min(\max(-1, 316), 336) \Rightarrow p^{\wedge U} = (361, 316)$. Now check if $(z_1^{new} < p_1)$ or $(z_2^{new} < p_2)$ and the answer is NO, then go to Step 5.

5. Check

- If $p_1^{\wedge U} < p_1^U$ and $p_2^{\wedge U} < p_2^U$, and the answer is YES, add $((474,336)$, $(361,316), (-1,-1))$ to R.
- Else if $p_1^{\wedge U} \geq p_1^U$ and $p_2^{\wedge U} < p_2^U$, and the answer is NO, continue checking.
- Else if $p_1^{\wedge U} < p_1^U$ and $p_2^{\wedge U} \geq p_2^U$, the answer is NO.

So, from Step 5, one new search region is created, that is, $((474,336), (361,316)$, $(-1,-1))$; hence, find a non-dominated point in this search region and get $z = (404,255,369)$. Repeat Steps 4 and 5 until infeasible region has reached, and then, stop. The outcome of all other iterations is in Table 10.1.

10.5 COMPUTATIONAL EXPERIMENTS

The proposed algorithm was implemented using Cplex Callable Library. CPLEX 12.5 was used as a solver for integer programming problems in C environment. All experiments were carried out on a Dell Inc. OptiPlex 9020 with processor Intel (R) Core (TM) i7-4770 CPU@ 3.40 GH and RAM 4.00 GB. Lubuntu operating system 16.04.01 was used for these experiments.

In order to investigate the performance of the proposed algorithm, some recently developed methods for TOIP were selected; they are: (Boland et al., 2017), IRM (Ozlen et al., 2014) and EIRM (Al-Rabeeah et al., 2020). For fair comparison with Boland et al. (2017), some of their instances as in http://hdl.handle.net/1959.13/1062187 were used (see Tables 10.2 and 10.6). For comparison with IRM and EIRM, few instances are same that were used in those papers and the rest instances are different (see Tables 10.3 and 10.4).

Instances that are same as in all earlier three papers, viz., Ozlen et al. (2014), Boland et al. (2017), and Al-Rabeeah et al. (2020), are denoted by an asterisk mark. These instances are assignment problems (ASP) and knapsack problems (KP). The proposed algorithm discussed in this chapter has been referred to as TODM in the computational experiments.

10.5.1 BEFORE RELAXATION

The computational experiment in Tables 10.2, 10.3, 10.4, 10.7, and 10.8 shows encouraging results. The proposed algorithm performed better with respect to the CPU time when compared with (Ozlen et al., 2014; Boland et al., 2017; Al-Rabeeah et al., 2020) (see Figures 10.1–10.3). It may also be noted that the number of IPs solved is less compared with IRM and the EIRM in all instances (see Tables 10.3, 10.4, and 10.8). However, for smaller instances, the proposed algorithm required slightly more IPs than what were required by algorithm proposed by Boland et al. (2017), whereas for larger instances, it required less number of IPs (see Tables 10.2 and 10.8).

Table 10.1
All Iteration for Solving the Numerical Example above before Relaxation

Steps	Nsr	p_1^U	p_2^U	p_1	p_2	p_1^l	p_2^l	z_1^{new}	z_2^{new}	z_3^{new}
				p List				Non-Dominated Point		
1	1	474	336	−1	−1	−1	−1	361	316	410
2	2	474	336	361	316	−1	−1	404	255	369
3	3	404	316	−1	−1	361	255	408	270	364
	4	474	336	404	316	−1	−1			
4	5	474	336	404	316	−1	−1	408	270	364
	6	404	316	−1	−1	361	270			
5	7	404	316	−1	−1	361	270	423	292	358
	8	408	316	−1	−1	404	270			
	9	474	336	408	316	−1	−1			
6	10	408	316	−1	−1	404	270	423	292	358
	11	474	336	408	316	−1	−1			
	12	404	316	−1	−1	361	292			
7	13	474	336	408	316	−1	−1	423	292	358
	14	404	316	−1	−1	361	292			
	15	408	316	−1	−1	404	292			
8	16	404	316	−1	−1	361	292	427	307	353
	17	408	316	−1	−1	404	292			
	18	423	316	−1	−1	408	292			
	19	474	336	423	316	−1	−1			
9	20	408	316	−1	−1	404	292	427	307	353
	21	423	316	−1	−1	408	292			
	22	474	336	423	316	−1	−1			
	23	404	316	−1	−1	361	307			
10	24	423	316	−1	−1	408	292	427	307	353
	25	474	336	423	316	−1	−1			
	26	404	316	−1	−1	361	307			
	27	408	316	−1	−1	404	307			
11	28	474	336	423	316	−1	−1	427	307	353
	29	404	316	−1	−1	361	307			
	30	408	316	−1	−1	404	307			
	31	423	316	−1	−1	408	307			
12	32	404	316	−1	−1	361	307	474	336	344
	33	408	316	−1	−1	404	307			
	34	423	316	−1	−1	408	307			
	35	427	316	−1	−1	423	307			
	36	474	336	427	316	−1	−1			
13	37	408	316	−1	−1	404	307	474	336	344
	38	423	316	−1	−1	408	307			
	39	427	316	−1	−1	423	307			
	40	474	336	427	316	−1	−1			
14	41	423	316	−1	−1	408	307	474	336	344
	42	427	316	−1	−1	423	307			
	43	474	336	427	316	−1	−1			
15	44	427	316	−1	−1	423	307	474	336	344

Table 10.2

Comparison between the Proposed Method and Boland et al. Algorithm (2017) with Respect to CPU Time (Seconds) and IPs

Problems	Solutions	Boland et al. Algorithm		TODM-BR	
		CPU Time	IP	CPU Time	IP
ASP100*	221	20.481	462	13.095	524
ASP225	2198	200.345	4149	150.262	4361
ASP400*	1942	679.179	3853	243.501	3905
ASP625	6928	1066.386	12987	847.401	13095
ASP900*	5195	4061.29	10253	1110.8	10654
ASP1225	9142	3984.567	15335	2537.775	15552
ASP1600*	14733	6963.232	27737	4830.799	23355
AP2025	22714	11300.347	43653	9314.681	41563
ASP2500*	29193	14885.142	54420	12857.901	50430
DKP10	9	0.047	18	0.021	19
DKP20	61	5.768	129	3.263	134
DKP30	195	75.439	429	55.917	450
DKP40	389	320.221	872	198.472	723
DKP50	1048	1085.411	2424	910.502	2109
DKP100	6500	44.651	14752	39.730	13909

Table 10.3

Comparison between the Proposed Method and the IRM with Respect to CPU Time (seconds) and IPs

Problems	Solutions	IRM		TODM-BR	
		CPU Time	IP	CPU Time	IP
ASP100*	221	13.2190	1158	13.095	524
ASP225	2198	164.912	9994	150.262	4361
ASP400*	1942	330.007	9055	243.501	3905
ASP625	6928	1026.464	26349	847.401	13095
ASP900*	5195	1378.103	22410	1110.8	10654
ASP1225	9142	2957.021	37133	2537.775	15552
ASP1600*	14733	4972.023	55935	4830.799	23355
AP2025	22714	10029.568	84853	9314.681	41563
ASP2500*	29193	15225.142	109142	12857.901	50430
KP10	6	0.055	34	0.047	16
KP20	20	0.351	119	0.299	46
KP30	35	1.095	208	0.911	84
KP40	117	5.402	639	4.332	283
KP60	578	46.814	3068	40.416	1566
KP80	1082	200.561	5430	181.557	2752

Table 10.4

Comparison between the Proposed Method and the EIRM with Respect to CPU Time (Seconds) and IPs

Problems	Solutions	EIRM		TODM-BR	
		CPU Time	**IP**	**CPU Time**	**IP**
ASP100*	221	13.481	802	13.095	524
ASP225	2198	158.318	6727	150.262	4361
ASP400*	1942	310.041	6105	243.501	3905
ASP625	6928	881.162	17668	847.401	13095
ASP900*	5195	1239.154	15045	1110.8	10654
ASP1225	9142	2603.277	24878	2537.775	17589
ASP1600*	14733	4866.612	27348	4830.799	23355
AP2025	22714	9532.994	56726	9314.681	41563
ASP2500*	29193	13235.731	72842	12857.901	50430
KP10	6	0.043	26	0.047	16
KP20	20	0.341	87	0.299	46
KP30	35	1.071	150	0.911	84
KP40	117	4.981	452	4.332	283
KP60	578	41.859	2115	40.416	1566
KP80	1082	183.379	3711	181.557	2752

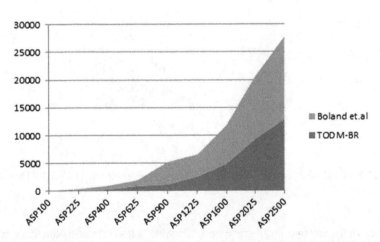

Figure 10.1 CPU time comparison between the proposed method (TODM-BR) and Boland et al. algorithm (Boland et al., 2017).

10.5.2 AFTER RELAXATION

The performance of the proposed approach can further be improved by using the idea of relaxation that was used in Ozlen et al. (2014). Basically, one is utilizing the history of already solved sub-problems to avoid repeat calculations. This means

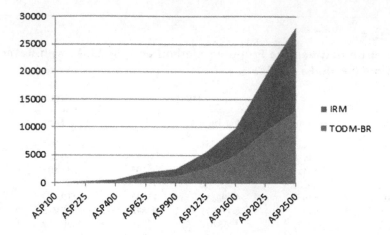

Figure 10.2 CPU time comparison between the proposed method (TODM-BR) and IRM.

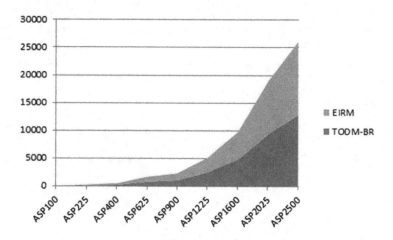

Figure 10.3 CPU time comparison between the proposed method (TODM-BR) and EIRM.

that when there are two problems, one is feasible with respect to other, there is no need to solve both but instead solve only one. In Table 10.1, one can see that the algorithm solves in some cases four or five search regions in maximum which have been avoided using relaxation. Table 10.5 shows the improvement after implementing the relaxation idea when comparison has been performed between the proposed algorithm before and after relaxation (see Figure 10.4). In Tables 10.6, 10.7, and 10.8, all algorithms have been compared in terms of the CPU time (see Figure 10.5) and the number of IPs.

Table 10.5

Comparison between the Proposed Method before and after Relaxation with Respect to CPU Time (Seconds) and IPs

Problems	Solutions	TODM-BR		TODM-AR	
		CPU Time	CPU Time	CPU Time	IP
ASP100*	221	13.095	524	11.465	461
ASP225	2198	150.262	4361	125.992	3745
ASP400*	1942	243.501	3905	226.682	3459
ASP625	6928	847.401	13095	776.936	11399
ASP900*	5195	1110.8	10654	969.249	9303
ASP1225	9142	2537.775	17589	2325.063	15552
ASP1600*	14733	4830.799	23355	4743.477	17415
AP2025	22714	9314.681	50430	8424.913	45548
ASP2500*	29193	12857.901	41563	12731.716	37575
KP10	6	0.047	16	0.025	10
KP20	20	0.299	46	0.272	41
KP30	35	0.911	84	0.861	74
KP40	117	4.332	283	4.188	242
KP60	578	40.416	1566	35.196	1298
KP80	1082	181.557	2752	160.862	2270

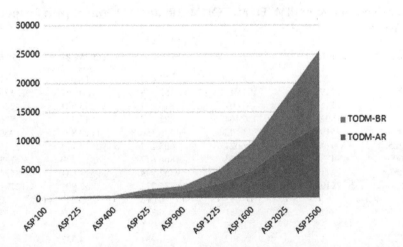

Figure 10.4 CPU time comparison between the proposed method before and after relaxation (TODM-BR and TODM-AR).

Table 10.6

Comparison between the Proposed Method before and after Relaxation and Boland et al Algorithm (2017) with Respect to CPU Time (Seconds)

Problems	Solutions	Boland et al. Algorithm	TODM-BR	TODM-AR
ASP100*	221	20.481	13.095	11.465
ASP225	2198	200.345	150.262	125.992
ASP400*	1942	679.179	243.501	226.682
ASP625	6928	1066.386	847.401	776.936
ASP900*	5195	4061.29	1110.8	969.249
ASP1225	9142	3984.567	2537.775	2325.063
ASP1600*	14733	6963.232	4830.799	4743.477
AP2025	22714	11300.347	9314.681	8424.913
ASP2500*	29193	14885.142	12857.901	12731.716
KP10	6	0.047	0.047	0.025
KP20	20	5.768	0.299	0.272
KP30	35	75.439	0.911	0.861
KP40	117	320.221	4.332	4.188
KP60	578	1085.411	40.416	35.196
KP80	1082	44.651	181.557	160.862

Table 10.7

Comparison between IRM, EIRM, TODM-BR, and TODM-AR with Respect to CPU Time

Problems	Solutions	IRM	EIRM	TODM-BR	TODM-AR
ASP100*	221	13.2190	13.481	13.095	11.465
ASP225	2198	164.912	158.318	150.262	125.992
ASP400*	1942	330.007	310.041	243.501	226.682
ASP625	6928	1026.464	881.162	847.401	776.936
ASP900*	5195	1378.103	1239.154	1110.8	969.249
ASP1225	9142	2957.021	2603.277	2537.775	2325.063
ASP1600*	14733	4972.023	4866.612	4830.799	4743.477
AP2025	22714	10029.568	9532.994	9314.681	8424.913
ASP2500*	29193	15225.142	13235.731	12857.901	12731.716
KP10	6	0.055	0.043	0.047	0.025
KP20	20	0.351	0.341	0.299	0.272
KP30	35	1.095	1.071	0.911	0.861
KP40	117	5.402	4.981	4.332	4.188
KP60	578	46.814	41.859	40.416	35.196
KP80	1082	200.561	183.379	181.557	160.862

Table 10.8

Comparison between IRM, EIRM, TODM-BR, and TODM-AR with Respect to Number of IPs

Problems	Solutions	IRM	EIRM	TODM-BR	TODM-AR
ASP100*	221	1158	802	524	461
ASP225	2198	9994	6727	4361	3745
ASP400*	1942	9055	6105	3905	3459
ASP625	6928	26349	17668	13095	11399
ASP900*	5195	22410	15045	10654	9303
ASP1225	9142	37133	24878	17589	15552
ASP1600*	14733	55935	27348	23355	17415
AP2025	29193	109142	72842	50430	45548
ASP2500*	22714	84853	56726	41563	37575
KP10	6	34	26	16	10
KP20	20	119	87	46	41
KP30	35	208	150	84	74
KP40	117	639	452	283	242
KP60	578	3068	2115	1566	1298
KP80	1082	5430	3711	2752	2270

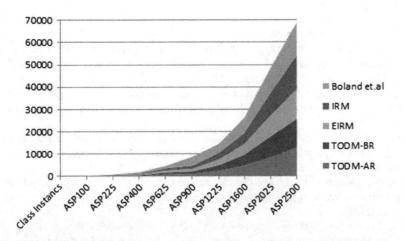

Figure 10.5 CPU time comparison between all previously mentioned algorithms.

10.6 CONCLUSION

In this chapter, a new algorithm has been developed where efficiency to find all set of non-dominated points for a TOIP model has improved. The proposed algorithm combines ideas from some recent algorithms where the requirements on additional number of constraints and binary variables are reduced. The improvement is reflected in reduction on the number of IPs and the CPU time. The performance of the proposed algorithm has been further improved by implanting the idea of relaxation. Computational experiments support the proposed claim.

REFERENCES

Al-Hasani, A., Al-Rabeeah, M., Kumar, S., and Eberhard, A. (2018) An improved CPU time in triangle splitting method for solving a bi-objective mixed integer program, *International Journal of Mathematical, Engineering and Management Sciences* 3(4), pp. 351–364.

Al-Hasani, A., Al-Rabeeah, M., Kumar, S., and Eberhard, A. (2019) Finding all non-dominated points for a bi-objective generalized assignment problem, *International Conference on Mathematics: Pure, Applied and Computation, Journal of Physics: Conference Series 1218 (2019) 012004 IOP Publishing*. doi: 10.1088/1742-6596/1218/1/012004.

Al-Rabeeah, M., Kumar, S., Al-Hasani, A., Munapo, E., and Eberhard, A. (2019) Bi-objective integer programming analysis based on the characteristic equation, *International Journal of Systems Assurance, Engineering and Management*. doi: 10.1007/s13198-019-00824-7.

Al-Rabeeah, M., Al-Hasani, A., Kumar, S., and Eberhard, A. (2020) Enhancement of improved recursive method for multi-objective integer programming problem, *Paper Presented at the International Conference of Mathematics: Pure, Applied and Computation*, 2019 and subsequently the paper has been submitted to IOP proceedings on 10 November 2019.

Aneja, Y.P. and Nair, K.PK. (1979) Bicriteria transportation problem, *Management Science* 25(1), pp. 73–78.

Antunes, C.H., Alves, M.J., and Climaco, J. (2016) *Multi-Objective Linear and Integer Programming*. Springer, Cham.

Benson, H.P. (1998) An outer approximation algorithm for generating all efficient extreme points in the outcome set of a multiple objective linear programming problem, *Journal of Global Optimization* 13(1), pp. 1–24.

Benson, H.P. and Sun, E. (2000) Outcome space partition of the weight set in multi-objective linear programming, *Journal of Optimization Theory and Applications* 105(1), pp. 17–36.

Benson, H.P. and Sun, E. (2002) A weight set decomposition algorithm for finding all efficient extreme points in the outcome set of a multiple objective linear program, *European Journal of Operational Research* 139(1), pp. 26–41.

Boland, N., Charkhgard, H., and Savelsbergh, M. (2016) The L-shape search method for tri-objective integer programming, *Mathematical Programming Computation* 8(2), pp. 217–251.

Boland, N., Charkhgard, H., and Savelsbergh, M. (2017) A new method for optimizing a linear function over the efficient set of a multi-objective integer program, *European Journal of Operational Research* 260(3), pp. 904–919.

Chankong, V. and Haimes, Y.Y. (2008) *Multi-Objective Decision Making: Theory and Methodology*. Courier Dover Publications, Mineola, NY.

Chinchuluun, A. and Pardalos, P.M. (2007) A survey of recent developments in multi-objective optimization, *Annals of Operations Research* 154(1), pp. 29–50.

Dachert, K. and Klamroth, K. (2015) A linear bound on the number of scalarizations needed to solve discrete tricriteria optimization problems, *Journal of Global Optimization* 61(4), pp. 643–676.

Ehrgott, M. (2006a) A discussion of scalarization techniques for multiple objective integer programming, *Annals of Operations Research* 147(1), pp. 343–360.

Ehrgott, M. (2006b) *Multicriteria Optimization.* Springer Science & Business Media, Berlin, Germany.

Greco, S., Figueira, J, and Ehrgott, M. (2016) *Multiple Criteria Decision Analysis.* Springer, Berlin, Germany.

Lokman, B. and Koksalan, M. (2013) Finding all non-dominated points of multi-objective integer programs, *Journal of Global Optimization* 57(2), pp. 347–365.

Mavrotas, G. and Florios, K. (2013) An improved version of the augmented ε-constraint method (AUGMECON2) for finding the exact pareto set in multi-objective integer programming problems, *Applied Mathematics and Computation* 219(18), pp. 9652–9669.

Ozlen, M. and Azizoglu, M. (2009) Multi-objective integer programming: A general approach for generating all non-dominated solutions, *European Journal of Operational Research* 199(1), pp. 25–35.

Ozlen, M., Burton, B.A., and MacRae, C.A.G. (2014) Multi-objective integer programming: An improved recursive algorithm, *Journal of Optimization Theory and Applications* 160(2), pp. 470–482.

Stidsen, T., Andersen, K.A., and Dammann, B. (2014) A branch and bound algorithm for a class of bi-objective mixed integer programs, *Management Science* 60(4), pp. 1009–1032.

Sylva, J. and Crema, A. (2004) A method for finding the set of non-dominated vectors for multiple objective integer linear programs, *European Journal of Operational Research* 158, pp. 46–55.

11 Link-Weight Modification for Network Optimization: Is it a Tool, Philosophy, or an Art?

Santosh Kumar
RMIT University
University of Melbourne

Elias Munapo
North West University

*'Maseka Lesaoana, Philimon Nyamugure,
and Phillemon Dikgale*
University of Limpopo, Turf loop Campus

CONTENTS

11.1 INTRODUCTION

Network optimization problems arise in many real-life situations that can be represented by a network diagram $G(n,m)$, where "n" represents the set of nodes and "m" the set of links; each link joins a pair of the nodes in that network. These nodes represent some kind of physical entities, and links represent the relationship among those entities. The relationship between the two entities is measured by assigning some weight to that link. In many different fields of science, engineering, technology, management, business, and commerce, we often come across situations that can be represented by a network and, therefore, that situation can be analyzed using an appropriate network methodology. A large volume of literature exists in Operations Research (OR), where many real-life situations have been conceptually represented by a network or analyzed as a network, and the results have been interpreted back with regard to the original situation (see Hastings, 1973). Therefore, networks and their associated methods have become a tool for analysis and optimization. The links in the networks are assigned weights to represent the object of the physical situation. These network models can be directed or undirected, and the objective can be to maximize or minimize a given objective function under some given conditions.

This chapter is concerned only with those situations where the associated link-weights were modified in certain ways to get to the required optimum solution for the given objective function. The aim of the chapter is to briefly review some of the known instances where link-weights were modified to get to the optimal solution. For each case, we briefly discuss the modification process and analyze how that modification helped to reach the optimal solution. We then leave it to the reader to advance this philosophy of "link-weight modification" further on other situations as a problem-solving tool. The purpose is to emphasize that the "link-weight modification" is a concept that can be used as a problem-solving tool. Some available applications have been examined closely with regard to its modification process and analysis to reach the optimal solution.

The concept of link-weight modification has been in existence but has not been explored as a potential optimization tool. We first review a few known cases and identify the associated philosophy behind that modification in the context of that specific problem. These applications include

A. An assignment model solved by the well-known Hungarian method of assignment.
B. A re-visitation to the transportation model.
C. Determination of the shortest route in a directed network.
D. The shortest route in a non-directed network.

    E. The minimum spanning tree under an index restriction.

    F. The traveling salesmen tour.

    G. The shortest path passing through "k" number of specified nodes.

The chapter has been organized in five sections. Section 11.2 deals with the broad classification of link-weight modification approaches. In Sections 11.3 and 11.4, we briefly review the existing applications. Some numerical illustrations are discussed in Section 11.5, and the chapter is concluded in Section 11.6, where we leave it to the reader as a challenge: "to advance the link-weight modification philosophy as a problem solving tool to solve other network optimization problems."

11.2 BROAD CLASSIFICATION

The link-weight modification has been successfully applied for solving many network routing instances as mentioned above. We broadly classify them into two categories.

Category 1: Some link-weights are altered to zero value by maintaining relative merits of the links. The purpose of altering weight to zero is to avoid test for the optimality when total cost is zero, which becomes a natural minimum among non-negative quantities. Avoiding the test of optimality is beneficial as in many instances, the identification of the optimality of that solution is not easy. The applications A to D above are in this category. The Hungarian method of assignment is a well-known approach, but the unification of assignment and transportation (application B) has been achieved recently by extending the Hungarian method of assignment to solve the transportation model. The shortest route in a directed network has been obtained by changing the link-weights and the non-directed network is also analyzed by making use of implied directions. Details are provided under each case separately.

Category 2: The link-weights are increased in a controlled way to create alternative possibilities. Once again, relative merits of links are maintained. This situation arises when one is required to find a minimum spanning tree under index restriction of less than or equal to two index value. This kind of network has applications in the traveling salesman tour (application F) and the path through the k-specified nodes (application G).

11.3 NETWORK OPTIMIZATION BY "LINK-WEIGHT MODIFIED TO ZERO VALUE": A CLOSE LOOK AT SOME PROBLEMS IN CATEGORY 1

11.3.1 A CLASSICAL APPLICATION: THE ASSIGNMENT PROBLEM SOLVED BY THE HUNGARIAN METHOD OF ASSIGNMENT

The Hungarian method is a well-documented method in the OR literature. A mathematical model for an assignment problem can be stated as

$$\text{Minimize } Z = \sum_{i=1}^{n} \sum_{j=1}^{n} c_{ij} x_{ij},$$

$$\text{Subject to } \sum_{i=1}^{n} x_{ij} = 1 \quad \text{for} \quad j = 1, 2, \ldots, n$$

$$\sum_{j=1}^{n} x_{ij} = 1 \quad \text{for} \quad i = 1, 2, \ldots, n$$

$$x_{ij} = 0 \text{ or } 1 \quad \forall ij. \tag{11.1}$$

Since the assignment model is a degenerate linear program and also a degenerate transportation model, the usual linear programming and transportation approaches encounter difficulties in the identification of the optimal solution for the model Equation (11.1). The Hungarian method attempts to solve the problem by the link-weight modification and avoids the test of optimality. The modification process starts by subtracting the minimum in each row from all other elements in that row, and works as follows:

1. "It creates a zero in each row and also maintains a relative merit of each element in that row." The same process is repeated for each row and each column. The zero element will be an independent element, if it is the only zero element either in the row or in the column.
2. If the number of independent zeros is insufficient, more zeros are created without changing the relative merits of the cost elements.
3. The purpose behind all these link-weight modifications is to obtain the total assignment cost to zero, that is, an assignment is made only to cells where the modified cost is zero. The zero total cost is a natural minimum among all positive numbers, thus eliminating the need for any test to establish optimality of that solution.

11.3.2 UNIFICATION OF AN ASSIGNMENT AND THE TRANSPORTATION PROBLEMS

The mathematical model for the transportation problem is given by

$$\text{Minimize } Z = \sum_{i=1}^{m} \sum_{j=1}^{n} C_{ij} x_{ij}$$

$$\text{Subject to } \sum_{j=1}^{n} x_{ij} = a_i \quad \text{for} \quad i = 1, 2, \ldots, m$$

$$\sum_{i=1}^{m} x_{ij} = b_j \quad \text{for} \quad j = 1, 2, \ldots, n$$

$$x_{ij} \geq 0 \text{ and integers } \forall (i, j). \tag{11.2}$$

Once again, this is a well-documented model in the OR literature. Here, we propose to discuss only a recent approach, where the transportation model Equation (11.2) has been reconsidered as an assignment model with a_i number of duplicated rows, for row $i, i = 1, 2, \ldots, m$ and b_j number of duplicated columns, for $j, j = 1, 2, \ldots, n$.

Details of this approach are in Munapo et al. (2012), where the link-weight modification and philosophy remain the same as that of the Hungarian method of assignment. The approach does not require any test for optimality of the solution, as the total cost in the reduced cost matrix is zero. This approach does not encounter any difficulty when the transportation model has a degenerate solution.

11.3.3   SHORTEST ROUTE IN A DIRECTED NETWORK

The shortest route in a given network $G(n,m)$ is a classical problem which has many applications. Many different solution procedures have been developed to solve this problem, from time to time. The problem is to find a path between two nodes in the given network such that the sum of the weights of its constituent edges is minimized. Pollack and Wiebenson (1960) reviewed methods for solving the shortest route problem. The directed network discussed here assumes to satisfy the following:

a. Link (i, j) is a directed link that starts at node i and ends at node j, $(i < j)$; in other words, the nodes are numbered in topological order.
b. All nodes, except the source and the destination nodes, have at least one incoming link to it and at least one outgoing link from it.
c. The source has all outgoing links, and the destination has all incoming links. Node "1" is the origin, and "n" is the destination node.

The link-weight modification is carried out as follows:

a. Subtract the minimum weight from all incoming links, and add the same minimum weight to all outgoing links. The total length between the source node and the destination node does not alter by implementing this process. However, if a path can be identified with total cost zero, that path becomes a natural minimum cost path.
b. Each iteration results in at least one zero-modified link-weight. The process can be carried out at nodes 2, 3, ..., etc., until we reach the destination node.
c. A label (m, w) is associated with a node, where m represents its sequential position in the network and w represents the minimum weight from the source node to that node.

Algorithmic steps are as follows:

Step 1: Label the source node. The label will be (1,0), set $k \rightarrow 2$, go to Step 2.
Step 2: Find the min{incoming weight}, w_k associated with node k, i.e., $w_k = \min\{w_{l1k}, w_{l2k}, \ldots, w_{lkk}\}$, where lk is the number of incoming links to node k. Assume that the minimum weight is denoted by w, and the link-weights are modified as follows:
Weights of incoming arc to node k will be modified as $(w_{lik} - w)$, and one of them will be equal to zero.
Weights for the outgoing arc from node k will be $(w_{kj} + w)$ for $j = 1, 2, \ldots$
Step 3: If $k < (n-1)$, set $k = k+1$ and return to Step 2.

Step 4: All nodes have been labeled. Find the label for the destination node. The shortest distance is w at the destination node, and the path can be traced back by passing through the links with modified weights as zero.

The philosophy is to trace a path passing through the links with zero link-weight. That path becomes the required minimum cost path. For details, see Munapo et al. (2008).

11.3.4   SHORTEST ROUTE IN A NON-DIRECTED NETWORK

A non-directed network cannot be analyzed on lines of a directed network, because conversion of a non-directed network to a directed network by replacing each link by two separate links in opposite directions is not feasible as the network will become unwieldy to deal with and nodes will not be able to maintain the topological order that we had in a directed network. Therefore, a new approach is required (Wikipedia, 2019). However, the following properties hold for the non-directed network.

11.3.4.1   Label Associated with a Node

Since the shortest path has to start from the origin node, the shortest distance from origin to origin is a known distance, which is 0. Hence, the origin node can be labeled as was done in the case of a directed network. Since topographical order of nodes is missing in a non-directed network, we need a label that can not only provide the total distance from the origin but also trace the path back from the origin to that node. In other words, we need the history associated with that label. Note that in the case of a directed network, a two-element label was sufficient, but for a non-directed network, a three-element label is required. It is given by (i, j, d), where i indicates the order, j indicates the previous node, and d denotes the distance from the origin node to that node. In this sense, the origin node will have a label $(0, O, 0)$; that is, the label on the origin node will be $O_{(0,O,0)}$.

11.3.4.2   Notations and Definitions

We need some notations that will facilitate the subsequent discussion. Define a few sets of nodes and links as follows:

L_0 = set of labeled nodes, which at the start, has only one element in this set, i.e., = $\{O\}$

L_1 = set of links connected to the labeled nodes. Initially, all links $(O \rightarrow j)$ will belong to this set, as origin node "O" is the only labeled node.

L_2 = set of links that will never participate in the shortest path. Initially, we have no knowledge about these links; hence, it will be a null set = $\{\varnothing\}$.

$L_{3(O,k)}$ = set of links that form the shortest path between the node O and the node k. Initially, the origin node is the only labeled node, hence, when $k = O$, $L_{3(O,k)} = \{(O,O)\}$, i.e., no movement, since it is a path from the node "O" to "O." The path is $(O \rightarrow O)$.

Since the first link on the shortest path will be a link starting from the origin node, one may assume the links from the origin as directed $(O \rightarrow j)$, where $(O \rightarrow j)$ are directly connected links with the origin node. Although the links are non-directed, in theory, one can return to the origin node O from any other node, but the path from O to j will only increase its distance. Thus, returning to a labeled node will definitely increase the total distance. All links, directed or non-directed, between two labeled nodes, will only increase the distance. These links will belong to the set L_2.

11.3.4.3 Link-Weight Modification Using the Implied Direction

Find the minimum weight associated with a link that belongs to the set $\{L_1\}$. Let the corresponding node associated with the minimum weight be denoted by the node j and the minimum weight be denoted by $w_{(Oj)}$. Now label the node j by $j_{(1,O,w_{(oj)})}$. The number 1 in the label indicates that it is the first label after the initial label at node O. The minimum distance from the node O to j is $w_{(Oj)}$, and the shortest path from the origin to node j is formed by the link (O, j).

The link-weights will be modified as follows:

New weight of the link (O, j) will be $= w_{(Oj)} - w_{(Oj)} = 0$

New weights associated with links (j, k) will be given by $w_{(j,k)} + w_{(Oj)}$. The weight of the link (O, j) in the direction (j, O) will be $= w_{(Oj)} + w_{(Oj)} = 2w_{(Oj)}$.

New weights of the remaining links will remain unchanged, and one has to upgrade the sets L_1, L_2, and L_3.

Since nodes O and j are labeled, once again, one can assume directions from the labeled nodes to unlabeled nodes. Find the minimum weight of all links in the set L_2, which will have all unlabeled links joining the nodes O and j. Note that a new node will be labeled each time. The process will terminate in n steps.

Thus, once again, the link-weights have been altered to make link-weight zero by identifying some implied directions as we did with directed links. For details, see Kumar et al. (2013).

11.4 LINK-WEIGHT MODIFICATION APPROACH TO FIND A MINIMUM SPANNING TREE WITH NODE INDEX ≤ 2

In Sections 11.3, link-weights were altered by subtracting the minimum weight to create a link with altered weight equal to zero. However, in the following, instead of subtracting, we add a certain weight to achieve a different goal. We consider the well-known minimum spanning tree (MST) problem under an index restriction (see Munapo et al., 2016). For a connected network $G(n, m)$, where n denotes the number of nodes and m denotes the number of links, the conventional MST deals with a selection of $(n - 1)$ links such that the n node network remains connected and total weight of the selected links is minimum. Such an MST graph can be obtained by any greedy approach (see Anupam, 2015; Garg and Kumar, 1968). In such an MST, the index of a node i denoted by n_i can be an integer value such that $1 \leq n_i \leq (n-1)$. Since the MST under index restriction will have to satisfy the index condition, that is, $1 \leq n_i \leq 2, i = 1, 2, \ldots, n$ and $n_i = 1$ for the origin and the destination node, the MST obtained by the greedy approach will have to be modified to satisfy the index restriction.

A link (i, j) is basic if it belongs to the MST, and it is non-basic if it does not belong to the MST. It means that the index value for the node i denoted by n_i is high if it is >2, i.e., if the number of basic links is more than two. For the high index nodes, some basic links have to be removed and added elsewhere, as the number of links will have to remain equal to $(n-1)$ for an n-node network. It may be noted that in the first place, they were selected since they satisfied the selection criteria by application of the greedy approach. To undo that selection, we add to the link-weight and make an alternative selection possible. In other words, we remove links from a high index node and add links to a low index node.

11.4.1　INDEX BALANCING THEOREMS

Theorem 11.1 Since a conventional MST can easily be obtained by any greedy approach, it can be assumed that index of each node i denoted by n_i is a known number. Therefore, high and low index nodes are known for the given network. The index balancing theorem states that adding the same constant to all arcs emanating from a high index node does not alter their relative merit, but can create an alternative for a link to be replaced to reduce the imbalance. The value of the constant is governed by the outgoing and incoming link-weights.

Theorem 11.2 Total index of a MST in a network $G(n,m)$ is a constant given by $2(n-1)$. Thus, in every attempt when the index of a high index node is reduced by 1, it also balances one low index node. Thus, convergence is guaranteed.

The index-restricted MST has potential applications in finding a path through "K" specified nodes. It means that we need a path joining the origin node to the destination node that passes through all specified nodes with index of these nodes equal to 2 (see details in Kumar et al. (2014)).

The index-restricted MST also has applications in the determination of a traveling salesman tour (see Kumar et al., 2014, 2016, 2018a, 2019). For other innovative approaches, see Kumar et al. (2018b).

11.5　NUMERICAL ILLUSTRATIONS

Since the Hungarian method of assignment is well known, we escape its illustration and move to other models.

11.5.1　UNIFICATION OF THE TRANSPORTATION AND ASSIGNMENT MODELS BY THE HUNGARIAN APPROACH

Consider a transportation model as given in Table 11.1.

Applying the normal row reduction of minimum as carried out in assignment method, one gets

$$
\begin{array}{ccc}
0 & 3 & 4 \\
3 & 0 & 2 \\
0 & 5 & 8
\end{array},
$$

Table 11.1
A 3 × 3 Transportation Model

Source/Demand	1	2	3	Supply
A	4	7	8	55
B	9	6	8	25
C	5	10	13	25
Demand	30	50	25	105

and then column reduction leads to

$$0_{30} \quad 3 \quad 2$$
$$3 \quad 0_{25} \quad 0_0 \; .$$
$$0_0 \quad 5 \quad 6$$

Here, the zero cost cells display the allocation $0_{x_{ij}}$. Thus, the allocation at the zero cost cell (A,1) is equal to 30 in transportation, and we have 30 vertical lines in the Hungarian sense of assignment. Also the current number of zeros is insufficient to assign 105 units; hence, additional number of zeros will have to be created. It is like the Hungarian approach with a different interpretation to horizontal and vertical lines. The new table with allocation will be

$$0_{25} \quad 1 \quad 0_{25}$$
$$5 \quad 0_{25} \quad 0 \; .$$
$$0_5 \quad 3 \quad 4$$

Once again, the number of zeros is insufficient; therefore, one more effective zero in cell (1,2) is created, and the result is the cost matrix given by

$$0_5 \quad 0_{25} \quad 0_{25}$$
$$6 \quad 0_{25} \quad 1 \; ,$$
$$0_{25} \quad 2 \quad 4$$

which is the required optimal solution with a total of 105 allocations, as per the above plan. From the above, optimal values of the required transportation plan can be easily obtained, and the corresponding optimal cost can be easily established.

11.5.2 SHORTEST PATH IN A DIRECTED NETWORK

Consider the graph in Figure 11.1.

Directions of the links are given by (i, j) where $i < j$. Let the link-weights be as given in Table 11.2.

After the initial label at the origin node 1, the next label will be at node 2. It will be the $\min\{3(1,2)\} = 3$, and altered weights will be given by: link $(1,2) = 3 - 3 = 0$, link $(2,3) = 4 + 3 = 7$, link $(2,4) = 5 + 3 = 8$, and link $(2,6) = 7 + 3 = 10$. These values are indicated by a # mark in Table 11.2.

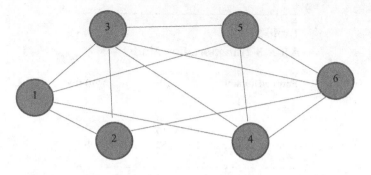

Figure 11.1   Given network with link-weights as shown in Table 11.2.

Table 11.2
Link-Weights and Labelling Process for the Solution

S. No	Directed Link	Link-Weights, Altered Link-Weights	Node and Label	Minimum Cost and Path
1	(1,2)	3, 3 − 3 = 0#	1(0,1)a	0, path 1->1
2	(1,3)	2, 2 − 2 = 0*	2(1,3)	3, path 1->2
3	(1,4)	4, 4 − 4 = 0**	3 (1,2)*	2, path 1->3
4	(1,5)	5, 5 − 5 = 0***	4(1,4)**	4, path 1->4
5	(2,3)	4, 4 + 3 = 7#	4(3,5)***	4, path 1->3->5
6	(2,4)	5, 5 + 3 = 8#	6(3,6) or (4,6)##	6, path 1->3->6 or 1->4->6
7	(2,6)	7, 7 + 3 = 10#		
8	(3,4)	5, 5 + 2 = 7*		
9	(3,5)	2, 2 + 2 = 4*, 4 − 4 = 0***		
10	(3,6)	4, 4 + 2 = 6*		
11	(4,5)	3, 3 + 4 = 7**		
12	(4,6)	2, 2 + 4 = 6**		
13	(5,6)	4, 4 + 5 = 9***		

The next label will be node 3. It will be $\min\{2(1,3), 7(2,3)\} = 2$, and altered weights will be given by: link $(1,3) = 2 - 2 = 0$; and for other links, it will be: link $(3,4) = 5 + 2 = 7$; link $(3,5) = 2 + 2 = 4$; link $(3,6) = 4 + 2 = 6$. These values are shown by an * mark in Table 11.2.

The next label will be on node 4. It will be $\min\{4(1,4), 8(2,4), 7(3,4)\} = 4$. Thus, the altered weights will be: $4 - 4 = 0$ for the link $(1,4)$; and for other links, it will be: link $(4,5) = 3 + 4 = 7$; link $(4,6) = 2 + 4 = 6$. These altered values are shown by ** in Table 11.2.

The next label will be assigned to node 5. It will be given by $\min\{5(1,5), 4(3,5), 7(4,5)\} = 4$. Altered values are shown by *** marks in Table 11.2.

Finally, we label the node 6. It is given by $\min\{10(2,6),6,(3,6),6(4,6),9(5,6)\} = 6$, which is obtained from two alternatives, i.e., from node 3 or node 4. The altered values are shown by ## mark in Table 11.2.

11.5.3 SHORTEST PATH IN THE NON-DIRECTED NETWORK BY LABELLING APPROACH

Reconsider the network in Figure 11.1. Assume that each link is a non-directed network with initial weight as given in Table 11.2 in both directions, i.e., link-weight (i, j) = the link-weight (j, i). This is given in Table 11.3.

Initially, only the origin node, i.e., node 1 is labeled. Therefore, we assign directions to links $(1,2),(1,3),(1,4)$ and $(1,5)$, i.e., from the node 1 to k, where $k = 2,3,4,5$. The minimum of these links from node 1 is $= \{3,2,4,5\} = 2$, which corresponds to the node 3. Hence, this node will be the next to be labeled as shown in Table 11.4. The modified weight for various links will be:

$$\text{Link } (1,3) = 2-2 = 0, \text{ link } (3,1) = 2+2 = 4,$$

Table 11.3

Link-Weights for the Non-Directed Network

From\To	1	2	3	4	5	6	Label (i,j,d)
1	-	3	2	4	5	-	$(0, O, 0)\ O_{(0,0,0)}$
2	3	-	4	5	-	7	
3	2	4	-	5	2	4	
4	4	5	5	-	3	2	
5	5	-	2	3	-	4	
6	-	7	4	2	4	-	

Table 11.4

Link-Weights after Labeling the Node 3 (altered values are shown in bold)

From\To	1	2	3	4	5	6	Label (i,j,d)
1	-	3	**$2-2=0$**	4	5	-	$(0, O, 0)\ O_{(0,0,0)}$
2	3	-	4	5	-	7	
3	**$2+2=4$**	**$4+2=6$**	-	**$5+2=7$**	**$2+2=4$**	**$4+2=6$**	$3_{(1,1,2)}$
4	4	5	5	-	3	2	
5	5	-	2	3	-	4	
6	-	7	4	2	4	-	

Link $(3,2) = 4+2 = 6$, link $(3,4) = 5+2 = 7$, link $(3,5) = 2+2 = 4$ and link $(3,6) = 4+2 = 6$

These modified weights are shown in Table 11.4.

For the next label, we find minimum of links connected to nodes 1 and 3. This minimum is $= \min\{3(1,2),4(1,4),5(1,5),6(3,2),7(3,4),4(3,5),6(3,6)\} = 3(1,2)$. Thus, node 2 will be labeled. Modified weights as well as the label on node 2 will be as shown in Table 11.5.

Note that the link-weights will be altered only for the links (1,2), (2,1), (2,3), (2,4), and (2,6). This is given in Table 11.5.

For the next label, we find $\min\{4(1,4),5(1,5),8(2,4),10(2,6),7(3,4),4(3,5), 6(3,6)\} = 4(1,4)$ or link (3,4). After labeling the node 4, the modified weights are given in Table 11.6.

Checking for the next label, we find minimum of links from node 1, 2, 3, and 4, which is given by $\min\{5(1,5),10(2,6),4(3,5),6(3,6),7(4,5),6(4,6)\} = 4(3,5)$. The modified weight will be $4+4 = 8$, as shown in Table 11.7.

Table 11.5
Modified Link-Weights after Labelling the Node 2

From\To	1	2	3	4	5	6	Label (i,j,d)
1	-	$3-3=0$	$2-2=0$	4	5	-	$(0, O, 0)\ O_{(0,0,0)}$
2	$3+3=6$	-	$4+3=7$	$5+3=8$	-	$7+3=10$	$2_{(2,1,3)}$
3	$2+2=4$	$4+2=6$	-	$5+2=7$	$2+2=4$	$4+2=6$	$3_{(1,1,2)}$
4	4	5	5	-	3	2	
5	5	-	2	3	-	4	
6	-	7	4	2	4	-	

Table 11.6
Modified Weights after Labelling the Node 4

From\To	1	2	3	4	5	6	Label (i,j,d)
1	-	$3-3=0$	$2-2=0$	$4-4=0$	5	-	$(0, O, 0)\ O_{(0,0,0)}$
2	$3+3=6$	-	$4+3=7$	$5+3=8$	-	$7+3=10$	$2_{(2,1,3)}$
3	$2+2=4$	$4+2=6$	-	$5+2=7$	$2+2=4$	$4+2=6$	$3_{(1,1,2)}$
4	$4+4=8$	$5+4=9$	$5+4=9$	-	$3+4=7$	$2+4=6$	$4_{(3,1,4)}$
5	5	-	2	3	-	$4+4=8$	$4_{(4,1,5)}$
6	-	7	4	2	4	-	

Table 11.7
Modified Links Weights after Labeling the Node 5

From\To	1	2	3	4	5	6	Label (i,j,d)
1	-	3 − 3 = 0	2 − 2 = 0	4 − 4 = 0	5	-	(0, O, 0) $O_{(0,0,0)}$
2	3 + 3 = 6	-	4 + 3 = 7	5 + 3 = 8	-	7 + 3 = 10	$2_{(2,1,3)}$
3	2 + 2 = 4	4 + 2 = 6	-	5 + 2 = 7	**4 − 4 = 0**	4 + 2 = 6	$3_{(1,1,2)}$
4	4 + 4 = 8	5 + 4 = 9	5 + 4 = 9	-	3 + 4 = 7	2 + 4 = 6	$4_{(3,1,4)}$
5	**5 + 4 = 9**	-	**2 + 4 = 6**	**3 + 4 = 7**	-	**4 + 4 = 8**	$4_{(4,1,5)}$
6	-	7	4	2	4	-	$6_{(5,3,6)}$ or $6_{(5,4,6)}$

For the next label, we find min$\{10(2,6), 6(3,6), 6(4,6), 8(5,6)\} = 6$ via the nodes 3 or 4. Thus, two alternative paths exist, which are $1 \rightarrow 3 \rightarrow 6$ and $1 \rightarrow 4 \rightarrow 6$, and cost for each path will be 6.

11.5.4 MINIMUM SPANNING TREE

Let us reconsider the network in Figure 11.1. The conventional minimum spanning tree using any greedy approach will be as shown in Figure 11.2.

The MST in Figure 11.2 occurs to be a MST path between node 2 and node 6. However, if we were looking for an MST path between nodes 1 and 6, the node 1 is a high index node and node 2 is a low index node. Thus, we have to remove an arc from the high index node 1 and select an arc joining the low index node 2. By application of the index balancing theorem, if we add two to all links joining the node 1, the link-weight of the link (1,3) will become $2 + 2 = 4$ and therefore, a tie exists between the link (1,3) and the link (2,3); hence, we select the link (2,3), replacing the link (1,3). Thus, an alternative has been developed, and we selected the one which balances the node index. The required MST path joining nodes 1 to 6 is shown in Figure 11.3.

The total weight is 14, whereas in Figure 11.2, the total weight was 12.

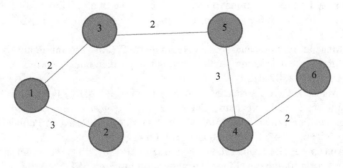

Figure 11.2 MST of the network in Figure 11.1.

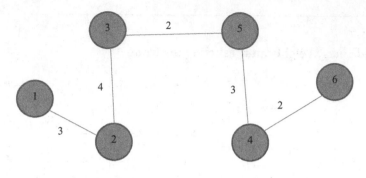

Figure 11.3   The MST path joining nodes 1 and 6.

11.6   CONCLUDING REMARKS

In this chapter, we have cited a few instances where a desired goal was achieved through the link-weight modification, and the actual process of link-weight modification was dependent on that situation. There is no set of rules or steps, the process of link-weight changes from problem to problem. The concept of link-weight modification is an optimization tool, yet its execution is an art which depends on our own ability to exploit the given situation to develop a suitable process. Therefore, we conclude that the link-weight modification is a philosophy, applicable for optimization in networks, but the execution of this tool remains an art.

REFERENCES

Anupam, G. (2015), Deterministic MST, Advanced Algorithms, CMU, Spring 15-859E, pp. 1–6.

Garg, R.C. and Kumar, S. (1968), Shortest connected graph through dynamic programming, *Maths Magazine*, Vol. 41, No. 4, pp. 170–173.

Hastings, N.A.J. (1973), *Dynamic Programming with Management Applications*. Crane, Russak, New York.

Kumar, S., Munapo, E., Ncube, O., Sigauke, C. and Nyamugure, P. (2013), A minimum weight labeling method for determination of a shortest route in a non-directed network, *International Journal of System Assurance Engineering and Management*, Vol. 4, No. 1, pp. 13–18.

Kumar, S., Munapo, E., Lesaoana, M. and Nyamugure, P. (2014), A minimum spanning tree approximation to the routing problem through k specified nodes, *Journal of Economics*, Vol. 5, No. 3, pp. 307–312.

Kumar, S., Munapo, E., Lesaoana, M. and Nyamugure, P. (2016), Is the travelling salesman problem actually NP hard? Chapter 3. In: *Engineering and Technology: Recent Innovations and Research*, (Ed.) A. Matani. International Research Publication House, New Delhi, pp. 37–58. ISBN 978-93-86138-06-4.

Kumar, S., Munapo, E., Lesaoana, M. and Nyamugure, P. (2018a), A minimum spanning tree based heuristic for the travelling salesman tour, *Opsearch*, Vol. 55, No. 1, pp. 150–164. doi: 10.1007/s12597-017-0318-5.

Kumar, S., Munapo, E., Lesaoana, M. and Nyamugure, P. (2018b), *Some Innovations in OR Methodology: Linear Optimization*, LAP Lambert Academic Publishing, ISBN: 978-613-7-38007-9.

Kumar, S., Munapo, E., Sigauke, C. and Al-Rabeeah, M. (2019), The minimum spanning tree with node index ≤ 2 is equivalent to the minimum travelling salesman tour, Chapter 8. In *Mathematics in Engineering Sciences: Novel Theories, Technologies, and Applications*, (Ed.) Mange Ram, CRC Press, pp. 227–244.

Munapo, E., Jones, B.C. and Kumar, S. (2008), A minimum incoming weight label method and its applications in CPM Networks, *ORiON*, Vol. 24, No. 1, pp. 37–48.

Munapo, E., Nyamugure, P., Lesaoana, M. and Kumar, S. (2012), A note on unified approach to solving the transportation and assignment models in Operations Research, *International Journal of Mathematical modelling, Simulation and Applications*, Vol. 5, No. 2, pp. 140–149.

Munapo, E., Kumar, S., Lesaoana, M. and Nyamugure, P. (2016), A Minimum spanning tree with node index ≤ 2, *ASOR Bulletin*, Vol. 34, No. 1, pp. 1–14.

Pollack, M. and Wiebenson, W. (1960). Solution of the shortest-route problem-a review. *Operations Research*, Vol. 8, No. 2, pp. 224–230. doi: 10.1287/opre.8.2.224.

Wikipedia. (2019), Shortest path problem. https://en.wikipedia.org/wiki/Shortest_path_problem. Accessed 21 April 2019.

12 Residual Domain-Rich Models and their Application in Distinguishing Photo-Realistic and Photographic Images

Prakhar Pradhan*, *Vinay Verma**, *Sharad Joshi,*
Mohit Lamba, and Nitin Khanna

Indian Institute of Technology Gandhinagar (IITGN)

CONTENTS

---

*The first two authors contributed equally.

12.1 INTRODUCTION

Rich model is built by assembling various diverse submodels which are formed by the co-occurrences of the residuals [1]. Residuals are higher-order pixel differences, or in other words, the residual domain is obtained when an image is filtered with a denoising filter. The purpose of considering various diverse submodels is that each submodel tries to capture different relationships between neighboring pixels in the residual domain.

12.2 NEED FOR RESIDUAL DOMAIN

In this section, we will discuss the residual domain and why feature extraction from the residual domain is preferred. The residual domain of an image denotes the output image filtered through a high-pass filter. Processing an image using a high-pass filter is equivalent to highlighting the details of an image as well as suppressing the smooth regions such as the blue sky/flat regions lacking any texture. Feature extracted from the residual domain is beneficial as the extracted features tend to be independent of the content for the classification tasks. The decision of a classifier should not be biased by the image content, and hence, features obtained from the residual domain are robust and more generalized. Images captured from the most of the cameras have certain dependencies in the neighboring pixels due to the natural scene complexity as well as various digital signal processing operations such as color filter array demosaicing, gamma correction, and filtering on the irradiance values [2]. These spatial dependencies between the neighboring pixels are graphically shown in [3]. Authors in [3] used 10,700 grayscale images from the BOWS2 dataset [3] to show that the joint probability of occurrences of two adjacent pixels follows a near-linear profile. With this fact, we can deduce that for natural images, the joint probability distribution of adjacent pixels will not vary much, or in other words, the pixels differences of neighboring pixels will be smaller for natural or uncorrupted images. It was further shown in [3] that the shape of joint probability distribution remains unchanged with the pixel value variation using information-theoretic tools like entropy, mutual information, etc. Mathematically, the joint probability $P(I_{i,j}, I_{i,j+1})$ between two adjacent pixels is measured, where $I_{i,j}$ and $I_{i,j+1}$ are the two pixel values adjacent to each other in the horizontal direction. It can be deduced that pixel values $I_{i,j}$ and $I_{i,j+1}$ should be somewhere close to each other for uncorrupted images. Histograms of a double, triple, and large group of neighboring pixels can be used to model the dependencies among pixels in the natural images. But this method of modeling dependencies is less efficient due to the following reasons [3].

- Consider the case of an Eight-bit grayscale image in which pixel values lie between 0 and 255, in the joint histogram of two neighboring pixels, there could be $256^2 = 65,536$ bins.
- There are such color combinations whose probability of occurrence together is very less, for example, a pixel value of 255 and 0 adjacent to each other in an eight-bit grayscale image. The corresponding bins in the histogram will be empty and thus act as a noise in the features.
- The features obtained using the histogram are image content dependent.

Now, we will show how feature dimension can be reduced by using the residual domain. Consider the case of eight-bit grayscale images. This means that each pixel can choose from $2^8 = 256$ unique values. The implication is that, as a pair, there can be $(256)^2 = 65,536$ combinations for $(I_{i,j}, I_{i,j+1})$. Let us now consider the second case, i.e., considering only the differences between the pixel intensities, where intensity pairs $(I_{i,j}, I_{i,j+1})$ such as $(1,2)$, $(2,3)$, and $(250,251)$ all fall in the same bucket. Empirically, the minimum difference possible is $0 - 255 = -255$ and maximum difference possible is $255 - 0 = 255$. Hence, the total number of combinations are $255 - (-255) + 1 = 511$, instead of 65536. One such application of residual domain is presented in [3], in which pixel differences are modeled as Markov chains, and then, sample transition probability matrix is used as features for steganalysis of digital images.

12.3 RICH MODELS OF NOISE RESIDUAL

Formation of rich models [1] aims to capture various kinds of dependencies among the neighboring pixels. The construction of a rich model [1] starts by computing various diverse submodels. Submodels capture different spatial relationships in the neighboring pixels of the noise residuals. The procedure described below gives a complete description of the construction of submodels in the residual domain.

12.3.1 NOISE RESIDUAL

Consider a single-component eight-bit grayscale image **I**, of size $M \times N$, where $0 \leq I(i,j) \leq 255$, for $i \in \{1 \dots M\}$, $j \in \{1 \dots N\}$. For the given image **I**, let $p_{i,j}$ represent a pixel in the spatial domain, $r_{i,j}$ be its residual domain counterpart, and $\mathcal{N}(p_{i,j})$ the neighborhood of pixel $p_{i,j}$, with pixel $p_{i,j}$ excluded. Noise residual $\mathbf{R} \in \mathbb{R}^{M \times N}$ of an image is defined [1] in Equation 12.1.

$$r_{i,j} = \Psi(\mathcal{N}(p_{i,j})) - c \cdot p_{i,j} \tag{12.1}$$

Here, Ψ is a function which accepts $\mathcal{N}(p_{i,j})$ as input and outputs an estimate of $c \cdot p_{i,j}$, where i, j span the rows and columns of the image and c denotes the residual order. The function $\Psi(\mathcal{N}(p_{i,j}))$ is implemented using denoising filters and is aimed to estimate $c \cdot p_{i,j}$; thus, the center pixel $p_{i,j}$ is excluded in calculating the residual. As otherwise the estimator-related function $\Psi(\mathcal{N}(p_{i,j}))$ will converge to the trivial solution of giving weight c to the central pixel and zero to all other neighbors.

Tables 12.1–12.4 show some of the linear and non-linear filters/functions [1] used for calculating noise residuals of a given image. For linear filters, central pixel $p_{i,j}$ is always marked with *. The integer multiplied with the central pixel $p_{i,j}$ is the residual order c.

Two types of residuals have been defined as "spam" residuals and "minmax" residuals. Residuals obtained from the filters/functions shown in Table 12.1 are of type "spam" as they contain exactly one linear operation. Residuals with two or more linear operations combined together using minimum or maximum operation are called "minmax" type of residuals. The final filters/functions corresponding to "minmax" type of residuals are non-linear in nature. Hence, for the "minmax" type residuals, we can have two outputs by differently (using a minimum or maximum operation) combining two sets of linear filters, whereas for "spam" residuals, we will have one output. The advantage of using "minmax" residual is that it introduces non-linearity in the features extracted from the residual domain. For example, residual 2a in Tables 12.1 corresponds to second-order "spam"-type residual where $r_{i,j}$ can be written as in Equation 12.2.

$$r_{i,j} = p_{i,j-1} + p_{i,j+1} - 2p_{i,j} \tag{12.2}$$

Residual 1a (Table 12.1) corresponds to the first-order "spam"-type residual, where $r_{i,j}$ can be written as in Equation 12.3. In first-order residuals, the central pixel is predicted as an adjacent neighbor pixel.

$$r_{i,j} = p_{i,j+1} - p_{i,j} \tag{12.3}$$

Residual 2b (Table 12.3) corresponds to "minmax"-type residual which provides two residuals as described in Equations 12.4 and 12.5.

Table 12.1

Mathematical form of Some of the Linear Filters in the Filter Space $\mathcal{B}$ (*Denotes the Central Pixel)

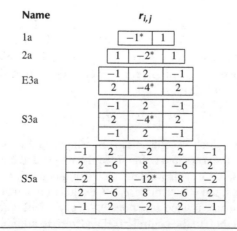

Name	$r_{i,j}$

1a: -1^* | 1

2a: 1 | -2^* | 1

E3a:
| -1 | 2 | -1 |
| 2 | -4^* | 2 |

S3a:
-1	2	-1
2	-4^*	2
-1	2	-1

S5a:
-1	2	-2	2	-1
2	-6	8	-6	2
-2	8	-12^*	8	-2
2	-6	8	-6	2
-1	2	-2	2	-1

Table 12.2

Mathematical form of some of the first order non-linear filters in the filter space $\mathcal{B}$

Name	$r_{i,j}$	
1b	$\max \begin{cases} p_{i,j-1} - p_{i,j} \\ p_{i,j+1} - p_{i,j} \end{cases}$,	$\min \begin{cases} p_{i,j-1} - p_{i,j} \\ p_{i,j+1} - p_{i,j} \end{cases}$
1c	$\max \begin{cases} p_{i-1,j} - p_{i,j} \\ p_{i,j+1} - p_{i,j} \end{cases}$,	$\min \begin{cases} p_{i-1,j} - p_{i,j} \\ p_{i,j+1} - p_{i,j} \end{cases}$
1d	$\max \begin{cases} p_{i-1,j} - p_{i,j} \\ p_{i,j+1} - p_{i,j} \\ p_{i,j-1} - p_{i,j} \end{cases}$,	$\min \begin{cases} p_{i-1,j} - p_{i,j} \\ p_{i,j+1} - p_{i,j} \\ p_{i,j-1} - p_{i,j} \end{cases}$
1e	$\max \begin{cases} p_{i-1,j} - p_{i,j} \\ p_{i,j+1} - p_{i,j} \\ p_{i,j-1} - p_{i,j} \\ p_{i+1,j} - p_{i,j} \end{cases}$,	$\min \begin{cases} p_{i-1,j} - p_{i,j} \\ p_{i,j+1} - p_{i,j} \\ p_{i,j-1} - p_{i,j} \\ p_{i+1,j} - p_{i,j} \end{cases}$
1f	$\max \begin{cases} p_{i-1,j} - p_{i,j} \\ p_{i-1,j+1} - p_{i,j} \\ p_{i,j+1} - p_{i,j} \end{cases}$,	$\min \begin{cases} p_{i-1,j} - p_{i,j} \\ p_{i-1,j+1} - p_{i,j} \\ p_{i,j+1} - p_{i,j} \end{cases}$
1g	$\max \begin{cases} p_{i-1,j-1} - p_{i,j} \\ p_{i-1,j} - p_{i,j} \\ p_{i-1,j+1} - p_{i,j} \\ p_{i,j+1} - p_{i,j} \end{cases}$,	$\min \begin{cases} p_{i-1,j-1} - p_{i,j} \\ p_{i-1,j} - p_{i,j} \\ p_{i-1,j+1} - p_{i,j} \\ p_{i,j+1} - p_{i,j} \end{cases}$
1h	$\max \begin{cases} p_{i-1,j-1} - p_{i,j} \\ p_{i-1,j} - p_{i,j} \\ p_{i-1,j+1} - p_{i,j} \\ p_{i,j+1} - p_{i,j} \\ p_{i+1,j+1} - p_{i,j} \end{cases}$,	$\min \begin{cases} p_{i-1,j-1} - p_{i,j} \\ p_{i-1,j} - p_{i,j} \\ p_{i-1,j+1} - p_{i,j} \\ p_{i,j+1} - p_{i,j} \\ p_{i+1,j+1} - p_{i,j} \end{cases}$

$$\max \begin{cases} p_{i-1,j} + p_{i+1,j} - 2p_{i,j} \\ p_{i,j-1} + p_{i,j+1} - 2p_{i,j} \end{cases} \tag{12.4}$$

$$\min \begin{cases} p_{i-1,j} + p_{i+1,j} - 2p_{i,j} \\ p_{i,j-1} + p_{i,j+1} - 2p_{i,j} \end{cases} \tag{12.5}$$

All the residuals are categorized into six residual classes [1], namely, first-order, second-order, third-order, SQUARE, EDGE$_{3\times3}$, and EDGE$_{5\times5}$. The second- and third-order residuals give better predictions in the region of complex textures, whereas EDGE$_{3\times3}$ gives better estimates of the central pixel at edges, and SQUARE consider more number of neighboring pixels to estimate central pixel.

Each residual is further categorized on the basis of its symmetric nature. The advantage of the symmetric residual is that it reduces the number of submodels and in turn reduces the feature dimension. Moreover, features obtained from symmetric residuals are statistically robust. One category of residuals is called directional residuals, and the other is called non-directional residuals. If the rotation of given image by 90° does not change the resulting residuals (with respect to the given image), then the corresponding residual is categorized as non-directional residuals; else, it is

Table 12.3

Mathematical form of Some of the Second Order Non-Linear Filters in the Filter Space $\mathcal{B}$

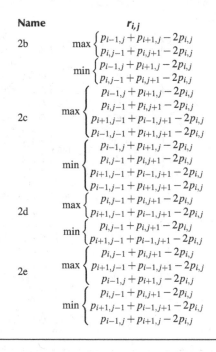

Name	$r_{i,j}$
2b	$\max \begin{cases} p_{i-1,j} + p_{i+1,j} - 2p_{i,j} \\ p_{i,j-1} + p_{i,j+1} - 2p_{i,j} \end{cases}$
	$\min \begin{cases} p_{i-1,j} + p_{i+1,j} - 2p_{i,j} \\ p_{i,j-1} + p_{i,j+1} - 2p_{i,j} \end{cases}$
2c	$\max \begin{cases} p_{i-1,j} + p_{i+1,j} - 2p_{i,j} \\ p_{i,j-1} + p_{i,j+1} - 2p_{i,j} \\ p_{i+1,j-1} + p_{i-1,j+1} - 2p_{i,j} \\ p_{i-1,j-1} + p_{i+1,j+1} - 2p_{i,j} \end{cases}$
	$\min \begin{cases} p_{i-1,j} + p_{i+1,j} - 2p_{i,j} \\ p_{i,j-1} + p_{i,j+1} - 2p_{i,j} \\ p_{i+1,j-1} + p_{i-1,j+1} - 2p_{i,j} \\ p_{i-1,j-1} + p_{i+1,j+1} - 2p_{i,j} \end{cases}$
2d	$\max \begin{cases} p_{i,j-1} + p_{i,j+1} - 2p_{i,j} \\ p_{i+1,j-1} + p_{i-1,j+1} - 2p_{i,j} \end{cases}$
	$\min \begin{cases} p_{i,j-1} + p_{i,j+1} - 2p_{i,j} \\ p_{i+1,j-1} + p_{i-1,j+1} - 2p_{i,j} \end{cases}$
2e	$\max \begin{cases} p_{i,j-1} + p_{i,j+1} - 2p_{i,j} \\ p_{i+1,j-1} + p_{i-1,j+1} - 2p_{i,j} \\ p_{i-1,j} + p_{i+1,j} - 2p_{i,j} \end{cases}$
	$\min \begin{cases} p_{i,j-1} + p_{i,j+1} - 2p_{i,j} \\ p_{i+1,j-1} + p_{i-1,j+1} - 2p_{i,j} \\ p_{i-1,j} + p_{i+1,j} - 2p_{i,j} \end{cases}$

directional residuals. For example 1a, 1b, 2a, 2e, and E3c are directional residuals in the Tables 12.1–12.4, whereas 1e, 2b, 2c, S3a, and E3d residuals correspond to non-directional residuals.

Further each of the residuals described in Tables 12.1–12.4 have a symmetric index [1], σ associated with it. This parameter tells about the number of different residuals that can be obtained by rotating an image before calculating residual. For example, the residuals 1c and 1g in Table 12.2 have symmetric index 4 and 8, respectively.

12.3.2 TRUNCATION AND QUANTIZATION

The residual space can have a high dynamic range, forming a multi-dimensional co-occurrence which can lead to a sparsely populated array. This problem is solved with the help of truncation, which can limit the dynamic range of residuals. If there were no truncation, there would be a whole bunch of values outside $[-T, T]$ and consequently less number of pairs satisfying the condition of co-occurrence, leading to underpopulated bins. Further, the residuals obtained $r_{i,j}$ from higher-order pixel differences will have higher dynamic range; hence, the features obtained from the

Table 12.4

Mathematical form of Some of the Fourth-Order Non-Linear Filters in the Filter Space $\mathcal{B}$

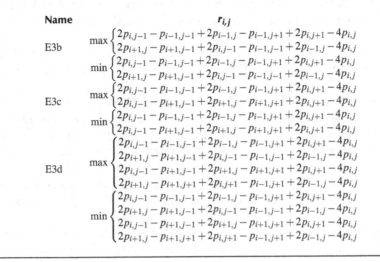

Name	$r_{i,j}$
E3b	$\max \begin{cases} 2p_{i,j-1} - p_{i-1,j-1} + 2p_{i-1,j} - p_{i-1,j+1} + 2p_{i,j+1} - 4p_{i,j} \\ 2p_{i+1,j} - p_{i+1,j-1} + 2p_{i,j-1} - p_{i-1,j-1} + 2p_{i-1,j} - 4p_{i,j} \end{cases}$
	$\min \begin{cases} 2p_{i,j-1} - p_{i-1,j-1} + 2p_{i-1,j} - p_{i-1,j+1} + 2p_{i,j+1} - 4p_{i,j} \\ 2p_{i+1,j} - p_{i+1,j-1} + 2p_{i,j-1} - p_{i-1,j-1} + 2p_{i-1,j} - 4p_{i,j} \end{cases}$
E3c	$\max \begin{cases} 2p_{i,j-1} - p_{i-1,j-1} + 2p_{i-1,j} - p_{i-1,j+1} + 2p_{i,j+1} - 4p_{i,j} \\ 2p_{i,j-1} - p_{i+1,j-1} + 2p_{i+1,j} - p_{i+1,j+1} + 2p_{i,j+1} - 4p_{i,j} \end{cases}$
	$\min \begin{cases} 2p_{i,j-1} - p_{i-1,j-1} + 2p_{i-1,j} - p_{i-1,j+1} + 2p_{i,j+1} - 4p_{i,j} \\ 2p_{i,j-1} - p_{i+1,j-1} + 2p_{i+1,j} - p_{i+1,j+1} + 2p_{i,j+1} - 4p_{i,j} \end{cases}$
E3d	$\max \begin{cases} 2p_{i,j-1} - p_{i-1,j-1} + 2p_{i-1,j} - p_{i-1,j+1} + 2p_{i,j+1} - 4p_{i,j} \\ 2p_{i+1,j} - p_{i+1,j-1} + 2p_{i,j-1} - p_{i-1,j-1} + 2p_{i-1,j} - 4p_{i,j} \\ 2p_{i,j-1} - p_{i+1,j-1} + 2p_{i+1,j} - p_{i+1,j+1} + 2p_{i,j+1} - 4p_{i,j} \\ 2p_{i+1,j} - p_{i+1,j+1} + 2p_{i,j+1} - p_{i-1,j+1} + 2p_{i-1,j} - 4p_{i,j} \end{cases}$
	$\min \begin{cases} 2p_{i,j-1} - p_{i-1,j-1} + 2p_{i-1,j} - p_{i-1,j+1} + 2p_{i,j+1} - 4p_{i,j} \\ 2p_{i+1,j} - p_{i+1,j-1} + 2p_{i,j-1} - p_{i-1,j-1} + 2p_{i-1,j} - 4p_{i,j} \\ 2p_{i,j-1} - p_{i+1,j-1} + 2p_{i+1,j} - p_{i+1,j+1} + 2p_{i,j+1} - 4p_{i,j} \\ 2p_{i+1,j} - p_{i+1,j+1} + 2p_{i,j+1} - p_{i-1,j+1} + 2p_{i-1,j} - 4p_{i,j} \end{cases}$

residuals will have high dimensions. To avoid the growth of feature dimension, the obtained residuals are quantized with quantization step, q as shown in Equation 12.6. Here, [.] denotes the rounding operation.

$$r_{i,j} \leftarrow \Upsilon\left(\left[\frac{r_{i,j}}{q}\right]; T\right) \tag{12.6}$$

$$\Upsilon(x;T) = \begin{cases} T & \text{if } x \geq T \\ x & \text{if } |x| < T \\ -T & \text{if } x \leq -T \end{cases} \tag{12.7}$$

Hence, each element of $\mathbf{R}$, $r_{i,j}$ can be any integer value between $-T$ and $+T$, for example if $T = 2$, then $r_{i,j} \in \{-2, -1, 0, +1, +2\}$

12.3.3 CO-OCCURRENCES

Co-occurrences matrices from the quantized and truncated noise residual $\mathbf{R}$ are calculated to form the models. Co-occurrence matrix is defined in Equation 12.8.

$$\mathbf{C_u^h} = \sum_i \sum_j \mathbb{1}_{[r_{i,j}=u_1, r_{i,j+1}=u_2, \cdots > r_{i,j+n}=u_n]},$$

$$u_1, u_2, \cdots, u_n \in [-T, T] \tag{12.8}$$

In the Equation 12.8, co-occurrences are calculated in the horizontal direction, denoted as $\mathbf{C^h}$, and similarly, one can obtain co-occurrence matrix in the vertical ($\mathbf{C^v}$), diagonal ($\mathbf{C^d}$), and minor diagonal direction ($\mathbf{C^m}$). In the Equation 12.8, the defined co-occurrence matrix, $\mathbf{C}$, is an n-dimensional array indexed with $\mathbf{u}$, $u = (u_1, u_2 ..... u_n)$ and $\mathbf{C_u^h}$ represents u^{th} element of $\mathbf{C^h}$.

In the Equation 12.8, $\mathbb{1}$ is the indicator function which is 1 when the condition inside [ ] is satisfied; otherwise, it is 0. Therefore, from each residual, four co-occurrence matrix can be obtained, in the horizontal, vertical, diagonal, and in minor diagonal direction. Since each element of $\mathbf{R}$, $r_{i,j} \in [-T, +T]$ and hence there would be $(2T+1)^n$ elements in a co-occurrence matrix, $\mathbf{C}$, where T is the threshold used for truncating residuals, $r_{i,j}$, and n is the order of co-occurrence matrix.

As mentioned in [1], correlation between the neighboring pixels reduces gradually in the diagonal directions, and hence, out of the four directions mentioned above, only horizontal and vertical directions are considered for building rich models. Therefore for each residual $\mathbf{R} \in \{\mathbf{R^h}, \mathbf{R^v}\}$, there exist two co-occurrence matrices one in horizontal direction ($\mathbf{C^h}$) and one in vertical direction ($\mathbf{C^v}$).

As mentioned in Section 12.3.1, there exist directional and non-directional residuals, going one-step forward, for any residual $\mathbf{R}$, if its horizontal and vertical co-occurrence matrices can be added to form a single matrix, such residuals are known as hv-symmetrical residuals [1]. This implies that all the non-directional residuals are hv-symmetrical residuals because, for non-directional residuals, the residual value does not change when the image is rotated. On the other hand, those residuals whose horizontal and vertical co-occurrence matrices cannot be added to form a single matrix are known as hv-nonsymmetrical residuals. Therefore, hv-symmetrical residual produces a single co-occurrence matrix (sum of horizontal and vertical co-occurrence matrices), whereas hv-nonsymmetrical residual produces two co-occurrence matrices.

Another important point to discuss is the selection of threshold, T, and co-occurrence order, n. Each co-occurrence matrix contains $(2T+1)^n$ elements. If we choose a larger value of T, then there can be a case in which the co-occurrence matrix is sparsely populated and thus results in a poor selection of submodels.

Since it has already been discussed in the definition of rich models that rich models consist of various diverse submodels, each submodel tries to capture a different kind of relationship between the neighboring pixels of an image. The use of different types of residuals (six classes of residuals) captures different types of relationship between neighboring pixels and then forming one or more populated co-occurrence matrix (submodels) results in diverse submodels. The point to concern is to choose the correct value of T and n, which does not lead to a sparsely populated co-occurrences matrix.

Rich models can be widely applicable in the image forensic problems. Consider the case of pixel-based image forgeries where the statistical correlation between the pixels of an image changes. The advantage of applying rich models to such problems is that rich models consider the different types of relationship between the neighboring pixels and hence features obtained using rich models would be effective in distinguishing forged images.

In this work, first, the use of rich models in steganalysis of digital images will be explained, which already exist in the literature, and later, the use of rich models in distinguishing natural images from computer-generated (CG) images will be explained.

12.4   RICH MODELS FOR STEGANALYSIS OF DIGITAL IMAGES

Steganography is the art of hiding information in the information. Various file formats can be used to hide the information. Those file formats which have a high degree of redundancy are best suitable for steganography, and hence, digital images are very well suitable for steganography as images contain various types of redundancies. In images, redundant bits are those bits which do not affect the visual perception of human beings. Therefore even if redundant bits are altered by hiding information, it is hardly noticeable.

Image steganography techniques are divided into two categories

1. **Spatial Domain**
   In this category, information is hidden in the pixels of an image. For example, the least significant bit embedding.
2. **Transform Domain**
   In this, an image is first transformed into another domain such as the frequency domain, and then, information is embedded in the transformed domain. The steganography done in the frequency domain is more robust.

Steganalysis is the art of detecting the messages hidden using steganography techniques. Since hiding the information in the images may alter the statistical properties of the images, such as the correlation between the neighboring. Therefore, these can act as signatures to detect the hidden messages.

In recent times, the content-adaptive steganography is an emerging area in steganography research. Adaptive steganography distorts the pixels of an image very less so that the detection of hidden messages becomes difficult. Adaptive steganography is mainly divided into two categories: adaptive steganography in the spatial domain and adaptive steganography in the frequency domain. HUGO (Highly Undetectable steGO) is an example of content-adaptive steganography in the spatial domain and is one of the recent advanced steganography algorithms. The embedding algorithm HUGO considers a complex model consisting of various submodels, each capturing different embedding artifacts [1]. Also, HUGO preserves the joint distribution of first-order differences of neighboring pixels, and hence, considering only first-order residuals for detecting, HUGO will not give better results. Therefore, forming submodels using higher-order residuals (considering more neighboring pixels) will be a better choice in identifying HUGO.

Moreover as mentioned above, there are two types of residuals, "spam" and "minmax." The use of "minmax" residuals is very effective in detecting adaptive steganography because the residuals computed from the pixels in the direction perpendicular to the edge will result in a large value and residuals computed from the pixels in the direction parallel to the edge will result in a small value, and hence, "minmax" residual plays a major role in detecting HUGO. The role of quantizing residual also plays a major in steganalysis. Other than reducing feature dimension, quantization also makes the features more sensitive to embedding changes at discontinuities in the image [1].

12.5   RICH MODELS FOR DISTINGUISHING PHOTO-REALISTIC AND PHOTOGRAPHIC IMAGES

In the 1980s, pictures were formed on the photosensitive films, doctoring which was not easy, and any attempt at manipulation was easily noticeable [4]. With the outburst of digital technology, images are reduced to mere numbers organized as matrices, which makes them easier to manipulate. One issue concerning digital forensics is the rampant use of CG images disguised as photographic images (PIM)/natural or real images and vice versa. Such type of CG images is referred to as photo-realistic computer-generated (PRCG) images.

Historically, the problem of automatic differentiation of PRCG and PIM has not received much attention as other forensic problems such as detection of copy-move forgery in a given image. It is mainly because differentiating between CG image and PIM was a task of less difficulty due to limitations of the systems used for generating CG images. Nevertheless, the artificial intelligence (AI) and computer vision community have made great strides to simulate PRCG images visually close to the natural images (Figure 12.1).

Besides producing fake photos, PRCG images are also of an utmost requirement in many situations. One possible situation can be a scene which has enough physical description available, in such cases, generating a CG image is much easier where viewing angles are difficult to capture using conventional photography techniques. Observing the trend, PRCG images are bound to become ubiquitous. Therefore, there is a need to develop methods to distinguish between PRCG and PIM.

The pipeline of the generation of natural images and CG images is very different from each other. In the case of natural images, light from a real scene falls on the image sensor after passing through lenses, optical filter, and a thin layer of color filter array (CFA) pattern. The purpose of this thin layer is to extract specific color components from the light; for example, in the case of RGB images, CFA pattern will extract red, green, and blue color wavelengths from the light. In digital cameras, an image signal is first generated at the image sensor, and either red, green, or blue color is present at each location. The other two color components at each location are interpolated from the neighboring pixels. In this way, each location in an RGB image has three values present at each pixel location. Later image is passed through various in-camera processing operations such as white balancing and contrast saturation. Due to CFA interpolation and various in-camera processing, there exists a correlation between neighboring pixels in a natural image, whereas for CG images, this is not the case as there is no interpolation in their generation. Motivated from this fact, intuitively we can say that features obtained from rich models might be able to

Figure 12.1 Visual comparison of photo-realism with a PIM. The image (a) is a genuine PIM from Google dataset [26,27]. The image (b) is CG image from PRCG dataset [26,27].

distinguish CG and real images because the rich model captures the various types of complex dependencies among the neighboring pixels, which exists in the natural images and not with the CG images.

12.5.1 EXISTING WORKS

One of the earliest works addressing automated detection of CG images devised simple metrics such as texture, saturation level, edge sharpness, and histogram proximity [5]. This was the time of the late 1990s, a time when photorealism was just picking up. Consequently, much more sophisticated measures were introduced in [4,6]. In [4], the natural-scene quality (NSQ) modeled using natural-imaging quality (NIQ) was used to distinguish between PRCG and PIM. The proposed NIQ captures information of imaging process such as white balancing and demosaicing, and

NSQ is typical of the scene like an illumination in a real scenario. Of the two, NSQ is more characteristic of PIM as duplicating complex natural scene illumination is difficult in CG images because it needs precise and intricate mathematical modeling of light and other materials.

Use of statistical features was further extended in [6] by utilizing higher-order statistics of photographic and photorealistic images such as skewness and kurtosis. This method used the wavelet transform instead of Fourier transform as the latter is not localized in spatial and frequency domain simultaneously. It was shown that wavelet sub-band coefficients of natural images follow generalized Laplacian distribution, characterized by a sharp peak and a long tail, justifying the use of higher-order statistics in [6]. Along with these statistics, a linear predictive model was also used to exploit the fact that natural images bear higher spatial correlation. This established the observation that irrespective of the degree of perceptual similarity between PIM and PRCG images, the underlying statistics of the two are quite different owing to the multitude of differences in the image forming procedure.

An entirely different approach was used in [7]. Instead of finding statistical differences between PRCG and PIM, the authors concentrated on one fundamental difference between the two, the presence of demosaicing artifacts. Almost all cameras today use CFAs for the demosaicing purpose to save hardware space and bypass the sensor registration problem [8]. For a PIM capturing a natural scene, abrupt changes in pixel values are most unlikely. It can be construed in this way; given the information about the neighboring pixels in a natural image, one is in a good position to predict the value of the pixel whose neighbors are being analyzed. As a result, for any of the three color channels–red, green, or blue, digital cameras do not record color values at every pixel location. Pixel values are recorded at only a few locations, and the rest is obtained via CFA demosaicing. This small concept leads to a significant simplification in camera manufacturing. This simplification, however, is not a restriction nor a necessity while generating computer graphics. Hence, CFAs are not used altogether for CG images. Detecting CFA traces in images turned out to be a beneficial technique with accuracies reaching more than 98%. A reasonably new approach for classification was proposed in [9] where features were extracted from the noisy signal. First, the image under consideration was convolved with a denoising filter. The denoised image was then subtracted from the original image to obtain the noise signal. A 15-dimensional feature based upon the normalized correlation values was then obtained from the noise signal and used for classification.

Local binary patterns (LBP) have emerged as a simple yet an effective texture descriptor [10] with application ranging from image-based face-recognition [11] to video-based motion-detection [12]. However, texture-based PIM versus PRCG classification using the LBP was first proposed in [13]. The technique relied on uniform LBP features which can take over 58 different combinations and accounted for the 58 bins of the 59-dimensional histogram feature. The last bin accounted for non-uniform LBP features. A fresh approach to classification, which proved more effective than [13], using homomorphic filtering was introduced in [14] exploiting the illumination reflectance model. According to the illumination reflectance model, the images captured by digital cameras are the product of illumination falling on it, and the light reflected. Illumination dropping in a scene does not change suddenly; it is only the reflectance which may vary depending upon the surface properties of the objects in the scene. Accordingly, for PIM, illumination is captured in the low frequency, and reflectance constitutes the high frequency. Consequently, it was shown that after homomorphic filtering, the details in PIM were boosted, while in PRCG image, they were blurred since PRCG images do not adhere to such illumination reflectance model. Contour transform was then applied to homomorphic images. Properties of contour transform are multiresolution, localization, critical sampling, directionality, and anisotropy. A detailed description of contour transform and its advantages over Fourier and wavelet transforms can be found in [15]. Once the contour transform sub-bands of the homomorphic filtered images were obtained, a co-occurrence matrix was formed. Instead of using the entire co-occurrence matrix as features, the co-occurrence matrix was characterized by energy, contrast, homogeneity, maximum column mean, and texture similarity. In [16], it was proposed that a CG image would have features similar to that of a "completely tampered image." It used the tampering localization features proposed by [17] for classification giving an average accuracy of 96%. A convolutional neural networks(CNN)-based method [18] for distinguishing PIM versus PRCG images divides the input image into patches of size 100×100. A special pooling layer, instead of the max-pooling layer, is used to compute the statistical values from the filtered images. The final predicted class of an input image is decided by the classification probability of the 100×100 patches. A weighted voting strategy is used to predict the class of an input image. Authors in [19] had also presented a more exhaustive CNN-based approach to distinguish natural images and CG images. The input to the CNN is an RGB image, and hence, 3D convolutional filters are used at the input layer. A local-to-global strategy is used in which the CNN is trained on patches and the final class of an image is predicted by the majority voting of the decisions obtained on the patches. Patches from the images are cropped randomly using maximal Poisson-disk sampling (MPS) [20] technique. The CNN architecture consists of three convolutional layers: two fully connected layers and one softmax layer. Before feeding the input image into CNN, each image is resized such that the shorter edge of an image has 512 pixels, and 200 patches each of size 240×240 are extracted from each image for training and 30 patches of the same size are used for testing and the final decision is made by majority voting of the decisions of the 29 patches, thus discarding the last patch.

In the present work, a successful attempt has been made to distinguish PIM and PRCG using rich models in the residual domain. In [1], rich models were used to build steganalysis detectors. Motivated from the application of rich models, an attempt has been made to apply the concepts of rich models in the differentiation of natural images and CG images. As discussed in Section 12.3, building a rich model requires various residuals, a threshold T, and co-occurrence order n. The value of the threshold, T, and co-occurrence order, n, is chosen to be 2 and 3, respectively. Hence, as mentioned in Section 12.3.3, the number of bins will be $(2 \times 2 + 1)^3 = 125$.

Residuals are truncated to curb the dynamic range of residuals and quantized to reduce the feature dimension. But the open question is, what should be the value of the quantization step $q > 0$ which gives better results? Due to truncation of residuals, there is an undesirable loss of information, and hence to compensate for this loss, a possible way is to consider different values of q for various submodels obtained from one residual. In [1], the value of quantization step, q, is obtained experimentally and best performance is achieved according to the following equations:

$$q \in \begin{cases} \{c, 1.5c, 2c\} & c > 1 \\ \{1, 2\} & c = 1 \end{cases}$$

where c is the residual order or the number associated with the central pixel $p_{i,j}$ in Tables 12.2–12.4. In Table 12.1, central pixel $p_{i,j}$ is denoted by *. In the proposed system for distinguishing PIM from PRCG, similar rules for the selection of q are followed.

12.5.2 FEATURE EXTRACTION

The final feature dimension for each image with co-occurrence order $n = 4$ is 34,671 (please refer to [1] for detailed calculation). There are primarily three ways to reduce this high feature dimensionality. First is to cut short the number of filters in filter space $\mathcal{B}$. However, this reduces the kind of features which can be learned by the classifier. It can be interpreted that each filter in $\mathcal{B}$ tries to capture a different aspect of spatial correlation among the neighboring pixels of the input image. This way, one does not fine-tune the detection to some particular traits of the input image. It is due to this reason only that removing some filters from $\mathcal{B}$ was not considered a viable way for dimensionality reduction. Second, a lower truncation value, T, can be used. This, however, leads to heavy loss of information for images having high contrast and large dynamic range. Finally, the order of the co-occurrence matrix could be reduced. Increasing the order n of the co-occurrence matrix is susceptible to having underpopulated bins resulting in less descriptive features. Accordingly, the order n of the co-occurrence matrix is reduced from four to three. This reduced the feature dimensionality from 34,671 to 8,001.

12.5.3 CLASSIFICATION

For classifying very high dimensional features, the proposed system uses an ensemble classifier as proposed in [21]. This was necessary since kernel support vector machine (SVM) scales poorly with increasing dimensionality, making it difficult to use for a feature size of 8001. A good comparison of ensemble classifier and SVM can be found in [21]. It was shown that the ensemble classifier reaches SVM's classification accuracy rate and sometimes outperforms it with the only fraction of computation time required by kernel SVM.

The ensemble classifier is composed of several base learners. Each base learner is a weak and unstable [22,23] binary classifier and acts on a random (but uniformly distributed) subspace of the original feature space. During the testing phase, each base learner gives its prediction of whether the input image is a CG or PIM. The final decision is taken by getting a consensus of each base learner. In this way collectively, all the base learners behave as a single entity and act as a single binary classifier.

The particular choice considered for binary classifier was Fisher's linear discriminant (FLD) [24,25]. FLD is a binary classification technique which tries to find a mapping w such that when features of two classes are mapped to a one-dimensional subspace, their respective mean is separated by maximum with the least within-class variance. Each FLD works on a small subset of original feature space. This makes the ensemble classifier easily scalable with changing feature dimension. This is in contrast to kernel SVM where computation time increases with feature dimension. As the feature dimension increases, more base learners can be added to the ensemble classifier. Since each FLD acts only on a small subset of the bootstrap feature space, FLDs are made to run in parallel and thereby not increase the computation time as high as that of the kernel SVM.

12.5.4 SYSTEM DESCRIPTION

The exact scheme is now presented algorithmically. The aim here will be to amalgamate all the ideas, concepts, and motivation listed so far as a single unit.

- Let **I** denote a single-component input image of size $M \times N$.
- Set up the filter bank $\mathcal{B}$, and fix the filters to be used for calculating residuals. Let N_{filters} denote the number of filters.
- Fix the parameters, threshold, $T = 2$ and order of co-occurrence matrix, $n = 3$.
- For each filter, f_l in $\mathcal{B}$, where $l \in \{1, 2, ..., N_{\text{filters}}\}$, find the filtered image.

$$I_{\text{filtered},l} = I * f_l \tag{12.9}$$

  here "$*$" denotes the convolution operation. This will be l number of filtered images. The obtained filtered images are the noise residuals.
- Quantize and truncate the obtained residual images using Equation 12.6.
- Calculate the third-order co-occurrence matrix using Equation 12.8

12.6 EXPERIMENTS AND DISCUSSION

In this section, results on various datasets described below are shown, and the proposed method is compared with the state-of-the-art methods. In all the experiments, the input image to the proposed system is re-sized to 512×512 to speed up the execution time. All the experiments in this work are performed on MATLAB R2018a.

12.6.1 DATASETS

To show the application of the proposed system in distinguishing natural images from CG images, experiments are performed on three datasets.

1. **Columbia Image Database**

    Columbia image dataset [26,27] contains two types of natural images, "Google" and "Personal" and one type of CG images, say "PRCG Columbia."

    a. **Google** : The images in this dataset are natural images.

    b. **Personal** : Consists of 800 natural images which are captured by the author using the professional single-lens-reflex cameras. The content of the images in this dataset is diverse in nature.

    c. **PRCG Columbia** : Consists of 800 CG images, which are collected from 40 3D graphics websites as mentioned in [26,27]. Maya, 3-ds MAX, Softimage-xsi, and so on are used as rendering software. Other than the abovementioned rendering software, geometry modeling and high-end rendering techniques are also used.

2. **RAISE**

    The RAISE dataset [28] consists of $8,156$ high-resolution uncompressed images. The images of this dataset are collected by four photographers. The images are captured by three different cameras. All the images in this dataset are of high resolution $3,008 \times 2,000, 4,288 \times 2,848$ and $4,928 \times 3,264$ pixels. Each image in the dataset belongs to either of the seven categories, as mentioned in [28]. But for this work, experiments have been performed on the images shared by the author [29]. The images of the RAISE dataset serve the purpose of natural images in experiments.

3. **Level-Design Reference Database**

    The images in this dataset [30] are CG images. The dataset consists of $63,368$ images of resolution, $1,920 \times 1,080$ pixels. These images are the screenshots of video games and are stored in JPEG format.

12.6.2 ANALYSIS FOR DIFFERENT COLOR CHANNELS

As input to the proposed algorithm is a single-channel image. Therefore, a possible solution is to convert an RGB image into a grayscale image. However, the use of grayscale image inhibits the possibility of carrying out experimentation on all the three channels independently. Now, the choice is to select either red channel, blue channel, or green channel of an RGB image. However, the green channel should be chosen because it is visually most consistent as compared to the other two color channels, but the results described later show that considering the effect of all the three channels gives better results. Therefore, red, green, and blue channels of an RGB image are independently processed and classified, and the final predicted class of the color image is decided by the majority voting. The advantage of using all the three channels is that it can capture the effect of photorealism in all channels and gives a better chance to distinguish PIM and PRCG.

Input to the proposed system is an RGB (three-component) image. As input to the algorithm (Section 12.5.4) is a single-channel image; hence, feature extraction process and classification should be done independently for the red, green, and blue channel. So, the final predicted class of an image will be decided by taking a majority vote of the individual predictions obtained using the three channels. Figures 12.2 and 12.3 show the block diagram of the training and testing procedure, respectively, of the proposed system.

For the experiments performed on the Columbia image database, 75% images of the all the three sub-databases of Columbia image database are randomly selected for training, and the remaining images are used for testing. As mentioned above, three sub-databases are "Google" and "Personal" which have PIM images and CG images, "PRCG Columbia." Table 12.5 shows the result of distinguishing images from the "Google" versus "PRCG Columbia" as well as "Personal" versus "PRCG Columbia." Table 12.5 reports the median and mean accuracies of seven iterations on the 25% testing images.

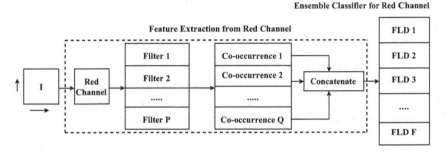

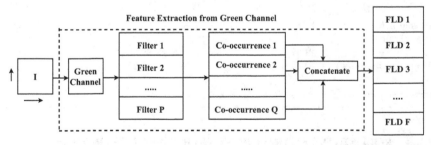

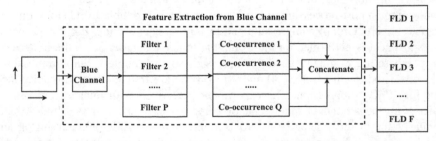

Figure 12.2 Block diagram corresponding to training phase of the proposed system.

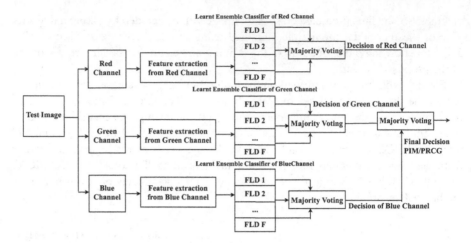

Figure 12.3   Block diagram corresponding to testing phase of the proposed system.

Table 12.5

Median and Mean Accuracies of Seven Iterations for Different Color Channels (Red, Green, Blue) and for the Combined Three-Channel Image

	Google versus PRCG Columbia		Personal versus PRCG Columbia	
	Median	**Mean**	**Median**	**Mean**
Red channel	82.83	82.72	98.75	98.89
Green channel	83.08	82.94	99	98.71
Blue channel	81.57	82.11	99	98.96
RGB (majority voting)	83.84	83.73	99.25	99.04

Results are reported on the Columbia Image Database

12.6.3   ROBUSTNESS AGAINST POST-PROCESSING OPERATIONS

To check the robustness of the proposed system, we test the proposed system against some common post-processing image operations such as double JPEG (Joint Photographic Experts Group) compression and histogram equalization. For this purpose, an experiment is performed on the Columbia image database. Since the images of the database are originally single compressed, so for double-JPEG compression, we considered quality factors 80 and 90 as these are frequently used quality factors for JPEG compression. The experiment performed with images compressed twice with the secondary quality factor being 80 is referred to as JPEG-80 and similarly for quality factor being 90 is referred to as JPEG-90. For histogram equalization, all the images of the Columbia database are histogram equalized. For all the three experiments, 75% images are randomly selected for training,

Table 12.6

Robustness Against JPEG Compression and Histogram Equalized Images

	JPEG-80		JPEG-90		Histogram Equalization	
	Median	Mean	Median	Mean	Median	Mean
Google versus PRCG Columbia	80.81	80.59	82.58	82.40	81.57	81.39
Personal versus PRCG Columbia	96.50	96.21	98	98	97.25	97.29

Median and mean accuracies of seven iterations are shown.

and the remaining images are used for testing. In Table 12.6, median and mean accuracies of seven iterations are reported. As can be seen from Table 12.6, the proposed system is robust against double-JPEG compression and histogram equalization. From the experimental results shown above, it can be concluded that rich models can be used to distinguish real images and CG images, although they are originally used for steganalysis of digital images. The results obtained from rich models are comparable to CNN-based methods and features obtained from rich models are very less computationally expensive as compared to the two CNN-based methods.

12.6.4   COMPARISON WITH STATE OF THE ART

Performance of the proposed system is compared with two methods based on CNN, proposed by Rahmouni et al. [18] and Quan et al. [19]. For different experiments related to comparison, all the methods are provided with the same training and testing datasets. For reporting the comparative performance with [18], and [19] all the results are directly taken from [19]. A patch size of 100×100 is used in Rahmouni et al. [18]. While the method proposed by Quan et al. [19] obtains the best accuracy on using patches of size 240×240. The accuracies on the full-size image are reported based on the majority voting scheme used by the respective methods.

12.6.4.1   Comparison on Columbia Database

The experiments on Columbia database are compared with [18] and [19], and median classification accuracies of seven iterations are reported in the Table 12.7. The dataset is randomly split into 75% of training images and 25% of testing images. The performance of the proposed system on the Personal dataset (Personal versus PRCG Columbia) is better than [18], and [19]. On the Google dataset (Google versus PRCG Columbia) proposed system performs better than [18], but method in [19] outperforms both the methods. The reason behind the lack of performance in Google dataset is that it is an uncontrolled dataset; that is, it has a diverse image content

Table 12.7

Comparison on Columbia Image Database

	Rahmouni et al. [18]	Quan et al. [19]	Proposed
Google versus PRCG Columbia	75.31	93.2	83.84
Personal versus PRCG Columbia	75.75	98.5	99.25

Median accuracies of seven iterations are shown.

Table 12.8

Comparison on RAISE versus Level-Design Reference Dataset on Full Size Images

Rahmouni et al. [18]	Quan et al. [19]	Proposed
99.3	99.58	99.86

Median accuracies of seven iterations are shown.

and involves more types of cameras, photographer styles, and lighting conditions, whereas the Personal dataset has limited diversity in camera and photographer style factors. Moreover, the ground truth of the Google dataset might not be reliable [30], whereas the images in Personal dataset have reliable sources.

12.6.4.2 Comparison on RAISE versus Level-Design

Experiments are performed on the list of images provided with the code [29] shared by the authors [18]. The experiments on this database are compared with [18], and [19]. For this comparison, method by Quan et al. [19] uses the patch size of 60×60 instead of their best patch size of 240×240.

As shown in the Table 12.8, on the full-size test images, the proposed system based on rich models performs slightly better than [18], and [19]. To, summarize, results presented in the Tables 12.5–12.8 show that the rich models are capable to distinguish between PIM and PRCG images.

12.7 CONCLUSION

Given the current eruption of photorealistic CG images, it has become quintessential to verify the credibility of images. To this end, a new method for PIM versus PRCG images classification has been proposed. Good classification accuracies, close to 99%, have been obtained on standard publicly available datasets, for PRCG versus natural image classification. A useful extension would be to locate the patches in a given image which have been computer-generated.

ACKNOWLEDGMENT

This material is based upon work partially supported by the Department of Science and Technology (DST), Government of India under the Award Number ECR/2015/000583. Any opinions, findings, and conclusions or recommendations expressed in this material are those of the author(s) and do not necessarily reflect the views of the funding agencies. Address all correspondence to Nitin Khanna at nitinkhanna@iitgn.ac.in.

REFERENCES

1. J. Fridrich and J. Kodovsky. Rich models for steganalysis of digital images. *IEEE Transactions on Information Forensics and Security*, 7(3):868–882, 2012.
2. R. Szeliski. *Computer Vision: Algorithms and Applications.* Springer Science & Business Media, Berlin, Heidelberg, 2010.
3. T. Pevny, P. Bas, and J. Fridrich. Steganalysis by subtractive pixel adjacency matrix. *IEEE Transactions on Information Forensics and Security*, 5(2):215–224, 2010.
4. T.-T. Ng and S.-F. Chang. Classifying photographic and photorealistic computer graphic images using natural image statistics. ADVENT Technical Report# 220–2006–6, Columbia University, New York, 2004.
5. V. Athitsos, M. J. Swain, and C. Frankel. Distinguishing photographs and graphics on the World Wide Web. *In Proceedings IEEE Workshop on Content-Based Access of Image and Video Libraries*, San Juan, Puerto Rico, pp. 1017, June 1997.
6. S. Lyu and H. Farid. How realistic is photorealistic? *IEEE Transactions on Signal Processing*, 53(2):845–850, 2005.
7. A. C. Gallagher and T. Chen. Image authentication by detecting traces of demosaicing. *In 2008 IEEE Computer Society Conference on Computer Vision and Pattern Recognition Workshops*, Anchorage, AK, pp. 1–8, June 2008.
8. B. K. Gunturk, J. Glotzbach, Y. Altunbasak, R. W. Schafer, and R. M. Mersereau. Demosaicking: Color filter array interpolation. *IEEE Signal Processing Magazine*, 22(1): 44–54, 2005.
9. N. Khanna, G. T. C. Chiu, J. P. Allebach, and E. J. Delp. Forensic techniques for classifying scanner, computer generated and digital camera images. *In 2008 IEEE International Conference on Acoustics, Speech and Signal Processing*, Las Vegas, NV, pp. 1653–1656, March 2008.
10. Z. Guo, L. Zhang, and D. Zhang. A completed modeling of local binary pattern operator for texture classification. *IEEE Transactions on Image Processing*, 19(6):1657–1663, 2010.
11. T. Ahonen, A. Hadid, and M. Pietikainen. Face description with local binary patterns: Application to face recognition. *IEEE Transactions on Pattern Analysis and Machine Intelligence*, 28(12):2037–2041, 2006.
12. M. Heikkila and M. Pietikainen. A texture-based method for modeling the background and detecting moving objects. *IEEE Transactions on Pattern Analysis and Machine Intelligence*, 28(4):657–662, 2006.

13. Z. Li, J. Ye, and Y. Q. Shi. Distinguishing computer graphics from photographic images using local binary patterns. *In the International Workshop on Digital Forensics and Watermarking 2012*, Springer, Berlin, Germany, pp. 228–241, 2013.

14. X. Wang, Y. Liu, B. Xu, L. Li, and J. Xue. A statistical feature based approach to distinguish PRCG from photographs. *Computer Vision and Image Understanding*, 128:84–93, 2014.

15. M. N. Do and M. Vetterli. The contourlet transform: An efficient directional multiresolution image representation. *IEEE Transactions on Image Processing*, 14(12):2091–2106, 2005.

16. R. S. Ayyalasomayajula and V. Pankajakshan. Differentiating photographic and PRCG images using tampering localization features. *In Proceedings of International Conference on Computer Vision and Image Processing*, Springer, Singapore, vol. 2, pp. 429–438, 2017.

17. J. Zuo, S. Pan, B. Liu, and X. Liao. Tampering detection for composite images based on re-sampling and JPEG compression. *In the First Asian Conference on Pattern Recognition*, Beijing, China, pp. 169–173, 2011.

18. N. Rahmouni, V. Nozick, J. Yamagishi, and I. Echizen. Distinguishing computer graphics from natural images using convolution neural networks. *In IEEE Workshop on Information Forensics and Security, WIFS 2017*, Rennes, France, 2017.

19. W. Quan, K. Wang, D.-M. Yan, and X. Zhang. Distinguishing between natural and computer-generated images using convolutional neural networks. *IEEE Transactions on Information Forensics and Security*, 13(11):2772–2787, 2018.

20. W. Quan, D.-M. Yan, J. Guo, W. Meng, and X. Zhang. Maximal Poisson-disk sampling via sampling radius optimization. *In SIGGRAPH ASIA 2016 Posters*, ACM, Macao, South Coast of China, pp. 22, 2016.

21. J. Kodovsky, J. Fridrich, and V. Holub. Ensemble classifiers for steganalysis of digital media. *IEEE Transactions on Information Forensics and Security*, 7(2):432–444, 2012.

22. L. Breiman. Heuristics of instability and stabilization in model selection. *The Annals of Statistics*, 24(6):2350–2383, 1996.

23. S. Shalev-Shwartz and S. Ben-David. *Understanding Machine Learning: From Theory to Algorithms*. Cambridge University Press, Cambridge, 2014.

24. C. M. Bishop. *Pattern Recognition and Machine Learning (Information Science and Statistics)*. Springer-Verlag, New York, 2006.

25. R. O Duda and P. E. Hart. *Pattern Classification and Scene Analysis*, vol. 3. Wiley, New York, 1973.

26. T.-T Ng, S.-F. Chang, J. Hsu, and M. Pepeljugoski. Columbia photographic images and photorealistic computer graphics dataset. Technical Report 205-2004-5, ADVENT, Columbia University, New York, 2004.

27. T.-T. Ng, S.-F. Chang, J. Hsu, L. Xie, and M.-P. Tsui. Physics-Motivated Features for Distinguishing Photographic Images and Computer Graphics. *In ACM Multimedia*, Singapore, pp. 239248, 2005.

28. D.-T. Dang-Nguyen, C. Pasquini, V. Conotter, and G Boato. RAISE: A raw images dataset for digital image forensics. *In Proceedings of the 6th ACM Multimedia Systems Conference*, ACM, Portland, OR, pp. 219224, 2015.

29. N. Rahmouni. Computer graphics vs real photographic images: A deep-learning approach. https://github.com/NicoRahm/CGvsPhoto/, October 2018.

30. M. Piaskiewicz. Level-design reference database. http://level-design.org/referencedb/, 2017.

13 Swirling Subsonic Annular Circular Jets

Gopinath Shanmugaraj and Sahil Garg
Lovely Professional University

Mohit Pant
National Institute of Technology-Hamirpur

CONTENTS

13.1 INTRODUCTION

Annular jets with and without swirl have been used in many combustion-related applications for many years. Generally, annular jets are issued from a straight or convergent duct with an axisymmetric central body or bluff body of a diameter which is smaller than the diameter of the straight duct. A recirculation zone is formed behind this central body which plays a crucial role in mixing and flame stabilization.

Finite element method (FEM)-based analysis is one of the most extensively used computational methods for numerical simulation of a variety of engineering problems [1–3]. Del Taglia et al. [4] studied the flow emanating from an annular duct with a central blockage and observed that the flow field inside the recirculation region is asymmetric about the jet axis. Memar et al. [5] studied the effect of swirling on the convective heat transfer of the coaxial jets. The inner jet swirl number was varied from 0 to 1.2, and outer jet swirl number was approximately equal to 1.3. Local heat transfer had shown a strong dependency on the inner jet swirl number, outer jet swirl number, and Reynolds number ratio. For high annular flow rates, increasing the inner jet swirl number appeared to have less impact on heat transfer enhancement. Nikjooy et al. [6] applied standard k-ε and algebraic stress turbulence model to predict the mean and turbulence characteristics of non-swirling axisymmetric coaxial jets. The mean, turbulence flow properties, and shear stress predicted using algebraic stress model agreed well with the experimental results than the results predicted using k-ε turbulence model. To some extent, the effectiveness of the turbulence model can be obscured by the flow boundary conditions. Buresti et al. [7] characterized the mean and fluctuating flow field of coaxial jets using laser-Doppler anemometry (LDA) and hotwire anemometry.

Nejad and Ahmed [8] investigated the effects of different swirl type (constant angle, free vortex, and forced vortex) with constant swirl number on an isothermal suddenly expanded dump combustor flow properties. Imposing swirl to the flow significantly altered the flow field properties of the dump combustor and resulted in a marked reduction of corner recirculation length. Due to the large-scale motion of the vortex core, turbulence present in the constant angle swirl flow was the greatest.

A complex mathematical model was developed by Aleksandar et al. [9] for predicting combustion dynamics of the pulverized coal combustion in an axisymmetric combustion chamber with swirl burner. The study reveals that a simple k-ε model can easily be used to predict the multiphase turbulent reacting flows without compromising the accuracy of the solution. Harris [10] computed the decay of swirl flow in a pipe using the order of magnitude analysis for solving an approximation of the Navier–Stokes equations. A simple expression has been derived to predict the swirl decay rate which proved to be in good agreement with the experimental results. It was found that the swirl decay rate is proportional to the pipe friction factor. Pipes with smooth surface have extremely persistent swirl flows when the flow Reynolds number is very high. It was predicted that at a sufficient distance from the swirler, the swirl flow rotates like a solid body. Park and Shin [11] studied the entrainment characteristics of free swirling jets using Schlieren flow visualization and developed a new technique for measuring entrainment velocity, which was found to be reasonably accurate. Visualization studies revealed that mass entrainment rate was proportional to the swirl strength. Processing vortex core induced a large-scale periodic motion which was in synchronization with fluctuation velocity signal. This model successfully predicts the sheet trajectory, velocity, and thickness of the flow. The important observation was the non-swirling annular sheet starts to converge as it moves downstream of the issuing nozzle. But, the swirling annular jet tends to diverge from the flow axis. For swirling annular sheets, the thickness reduces as the flow moves downstream of the nozzle. Percin et al. [12] used experimental particle image velocimetry (PIV) technique to investigate the vortex structures of the swirling annular flow. This study reveals that the vortex core is formed inside the swirling jet. Inside this swirling vortices, there is a point at which the vortex breakdown occurs and the pressure fluctuations are maximum before this vortex breakdown point. Sheen et al. [13] presented an experimental study which investigates the flowfields of confined swirling jets and unconfined swirling jets. Reynolds number and swirl numbers were used as control parameters. Smoke streak visualization technique was used to identify the flow patterns, and velocity measurements were taken by using laser anemometry. The flowfield was classified into seven different zones.

Each zone has distinctive variation based on the flow condition. The length of the recirculation zone is correlated with the swirl number. Garcia-Vallalba and Jochen Frohlich [14] demonstrated a large Eddy simulation (LES) of unconfined swirling annular jets. The swirl number variation is the main control parameter between different flow conditions. It is varied from 0 to 1.2. The mean flow characteristics and the vortex structure were studied. They introduced a pilot co-annular jet at the jet center axis. This introduction causes an additional swirl at the jet centerline which observed to be eliminating the coherent structures of the jet. Koen et al. [15] studied the combustion characteristics of the compact combustion chamber in which they used different variations of confined swirl mixer. They identified five different flame states, and each of these flame states has distinct properties. This variation arises only because of the change in swirl numbers. Internal recirculation zone was identified, and its role in flame stabilities was found. Marsik et al. [16] studied the stability of the swirling annular flow. A new approach is proposed presented to stabilize the swirling flow. The flow stability can be attained by the attenuation of the kinetic energy of the disturbances. To demonstrate this concept, they used an isothermal coaxial swirl flow with constant viscosity and density was used. Parra et al. [17] used Reynolds averaged Navier–Stokes equation (RANS) simulation to study the mixing and combustion characteristics of turbulent coaxial swirling flows. The interaction between two reacting swirling flows was studied, and its complex flow patterns were revealed. Stronger swirling flows generate recirculation zones and thinner flame front, whereas weaker swirl flows create thicker flame fronts. Ibrahim and Mckinney [18] developed a mathematical model to study annular non-swirling and swirling liquid sheets.

From the previous studies, it is evident that the studies related to confined swirling annular free jet have not extensively analyzed. The future scope of work will entrap the potential of meshfree methods [19] for analysis of more complex domains and problems.

The work presented is based on previous studies from which it is evident that the swirling annular free jet has not extensively studied. The studies discussed above are either experimental or numerical LES studies except the study reported by Parra et al. [15]. LES is a very time-consuming technique when compared to simpler RANS simulations. Since it is very important for the engineering community to understand the flow physics swirling flow to improve the performance of future combustor applications, this study is mainly focused to investigate the flowfield of the swirling annular jet issuing from a straight annular circular duct with the help of RANS simulation.

13.2 COMPUTATIONAL MODEL

Although more advanced computational techniques such as LES and DNS are available to study complex flow phenomenon like swirling flows, in this study a simple RANS simulation was used. The main aim of using RANS simulation is to demonstrate its ability to accurately predict the mean flow properties of the swirling flows. By doing this, one can able to easily characterize the swirler design without much need of computing power and enormous amount of time. ANSYS CFX Software is used to perform numerical analysis. For solving the system of linearized governing equations, it uses a multigrid accelerated incomplete lower-upper factorization technique. It is an iterative solver which means the exact solution of the equations is approached during the course of several iterations. For higher accuracy, it heavily relies on pressure-based solution technique for broad applicability. To discretize the domain and to integrate the fluid flow equations in its solution strategies, it uses FEM.

13.2.1 GEOMETRY

A circular solid cylinder of smaller diameter is fixed inside the hollow circular duct of larger diameter which forms an annular duct as shown in Figure 13.1. A solid cylinder is used as a bluff body which creates a recirculation zone downstream of the duct exit. A stationary swirler is placed inside the annular region to induce the incoming fluid flow to swirl around its axis. Stationary swirler consists of ten angular vanes fixed circumferentially to the inner circular wall of the swirler. Three different vane angles (swirl angles φ) of $0°$, $25°$, and $50°$ are considered for this study. All the swirlers are designed such that they have the same vane length, thickness, and wetted surface area. Swirler is positioned 25 mm upstream of the duct exit in order to establish a fully developed swirling flow at the annular duct. The annular duct has the inner diameter, outer diameter, and annular area of $D_i = 10$ mm, $D_o = 15$ mm, and $A = 98$ mm^2, respectively. It has an equivalent diameter (D_{eq}) of 11.18 mm. For all the swirl configurations, inlet total pressure has been maintained constant. Detailed dimensions of the annular duct and swirler are shown in Figure 13.1.

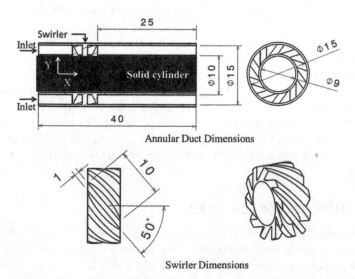

Figure 13.1   Dimensions of circular annular duct and swirler.

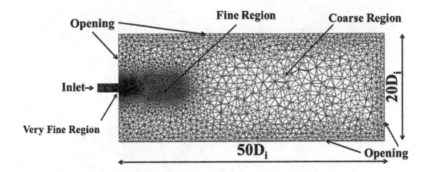

Figure 13.2 Domain dimensions, mesh distribution, and boundary conditions.

13.2.2 MESHING

Domain dimensions, mesh distribution, and boundary conditions are presented in Figure 13.2. An unstructured three-dimensional grid constructed by tetrahedral elements has been used to simulate the flowfield. To accurately capture the swirl properties, very fine grid elements with the size of 0.2 mm are used at the immediate downstream location of the annular duct exit. This region is followed by a fine region with an element size of 0.8 mm, and the remaining part of the mesh is constructed with coarse elements. Extreme care has been taken while setting mesh parameters at the swirler geometry in order to achieve error-free mesh at the regions between the angular vanes. Approximately four million elements are used in mesh construction. The size of the inflation layers near the wall region is adjusted appropriately to achieve the Yplus value of <1.

13.2.3 PHYSICS DEFINITION

To simulate the flowfield CFX solver is used. For solving the RANS, CFX solver uses an iterative method to obtain an exact solution during the course of the iteration sequence. It is a pressure-based solver which suits well for our problem. At the inlet, total pressure and static temperature have been specified. At the opening boundary, opening pressure and static temperature have been specified. Shear Stress Transport (SST) turbulence model is used to simulate the turbulence present in the flow field. It provides an optimal switching between the k-ε and k-ω turbulence model. Standard sea level pressure (101,325 Pa) has been considered as the reference pressure for the entire analysis.

13.3 RESULTS AND DISCUSSIONS

Figure 13.3 shows the fully developed swirl flow issuing from the annular duct with 50° swirler. Swirling flow is characterized by a dimensionless parameter called swirl number (S). It is defined as the ratio of the axial flux of angular momentum to the axial flux of axial momentum.

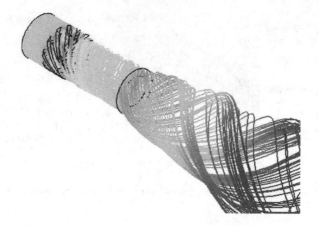

Figure 13.3   Swirl flow emanates from the annular duct with 50° swirler.

$$S = \frac{\int_{R_i}^{R_o} \rho u w_{\tan} r^2 dr}{\int_{R_i}^{R_o} \rho u^2 r^2 dr} \tag{13.1}$$

S – Swirl number
R_i and R_o – Inner and outer radius of the annular duct
ρ – Fluid density
u – Axial velocity at the exit plane
$W_{\tan}$ – Tangential velocity at the exit plane
r – Radial distance from the centerline

Swirl number calculated for different swirl angles by using Equation 13.1 is plotted in Figure 13.4. From the plot, it is evident that the swirl number shows a linear

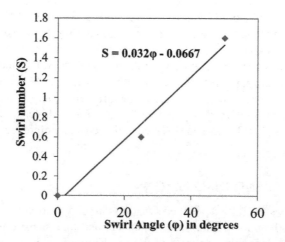

Figure 13.4   Variation of swirl number with swirl angle.

relation with the swirl angle (φ). A simple linear equation is used to fit the data obtained by the computation which has the slope ($dS/d\varphi$) of 0.032.

Figure 13.5 shows static pressure distribution along the jet centerline (X-axis). Static pressure value obtained for all the jet configurations has been normalized by the standard atmospheric pressure ($P_{atm} = 101,325$ Pa) at sea level. To validate the simulated results, the experimental data for the swirling annular flow are taken from the study carried out by Gopinath [20] and compared with the computational results, which shows that they are in good agreement with each other. The experiments are conducted on the scaled-up version of the swirling annular duct of the same design.

For 0° swirl flow, the static pressure initially reduces and reaches a normalized minimum pressure of $P_{min}/P_{atm} = 0.96$ at (location of minimum pressure) $X_{pmin}/D_{eq} = 0.27$. Beyond the minimum pressure location, the static pressure starts increasing till it finally reaches the atmospheric pressure at $X/D_{eq} = 2.147$. Swirl flow results from a 25° swirler have the normalized minimum pressure value of $P_{min}/P_{atm} = 0.8$ at $X_{mp}/D_{eq} = 0.626$. Pressure distribution of 50° swirl flow also exhibits the same tendency as 0° and 25° swirl flows. It has the minimum pressure value of $P_{min}/P_{atm} = 0.96$ at $X/D_{eq} = 0.626$. Among the three configurations considered, 25° swirl flow has the lowest minimum pressure value, which is 16.6% less than the value of 0° and 50° swirl flow.

Velocity distribution along jet centerline is shown in Figure 13.6. From the plot, it is evident that the velocity distribution has significant dependency on the swirl conditions. Among the three jet configurations, 0° swirl flow has the maximum centerline velocity of $U_{max} = 150.78$ m/s at $X/D_{eq} = 3.22$. Velocities obtained for 25° and 50° swirl flows were also normalized by using the U_{max} value obtained for 0° swirl flow (it is the maximum velocity reached by the flow emanates from the 0° swirler). Maximum reverse flow velocity (U_{rev}) and location of maximum reverse flow velocity (XU_{rev}) vary with swirl angle which is presented in Table 13.1. The other two

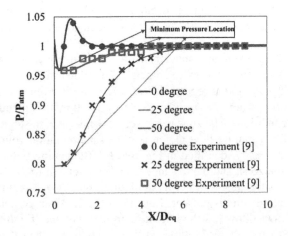

Figure 13.5 Pressure distributions along jet centerline.

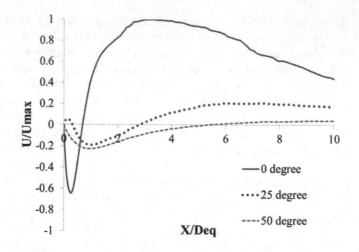

Figure 13.6 Velocity (X component) distributions along jet centerline.

Table 13.1

Comparison between the Flow Properties of Different Swirl Flows

φ	S	P_{min}/P_{atm}	U_{rev}/U_{max}	$X_{p\ min}/D_{eq}$	$X_{U\ rev}/D_{eq}$	X_s/D_{eq}
0°	0	0.96	−0.64	0.27	0.27	0.67
25°	0.6	0.8	−0.19	0.63	0.98	2.8
50°	1.6	0.96	−0.22	0.63	1.07	5.25

components of the velocity (Y and Z) oscillate very closer to zero and the oscillation range (0.4 m/s > v & w > −0.4 m/s) is very less when compared to the X component velocity.

The flow exiting from the annular duct starts to converge and joints at a downstream location from the exit geometry. This downstream location is generally a stagnation point and is represented as a Red dot in Figure 13.7. Location of this stagnation point where the velocity is zero ($U = 0$ m/s) is significantly altered by the swirling conditions. The zone in between the nozzle exit, stagnation point and the boundary of the annular flow region is called recirculation zone. This recirculation zone can be easily identified in Figure 13.7. Stagnation point marks the end of the recirculation zone, and the distance from the duct exit to the stagnation point may be denoted as the length of the recirculation zone (X_s). Streamline pattern inside the recirculation zone is presented in Figure 13.7. When compared with other swirl flows, 50° swirl flow has the largest recirculation zone (Figure 13.7c), which is almost twice that of the 25° swirl and five times larger than the 0° swirl flow. Recirculation pattern is almost symmetrical about the jet axis for 0° and 50° swirl

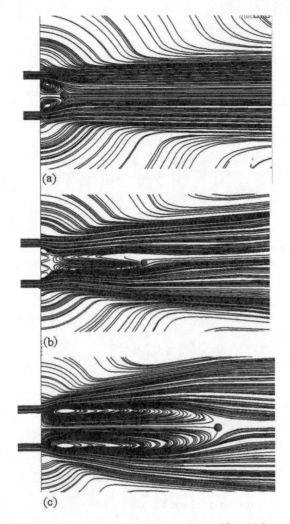

Figure 13.7 Streamline pattern in the recirculation zone. (a) 0° swirl flow, (b) 25° swirl flow, and (c) 50° swirl flow.

flow. But, the recirculation pattern of 25° swirl flow is asymmetrical about the jet axis (Figure 13.8).

$$V = \sqrt{u^2 + v^2 + w^2} \qquad (13.2)$$

Figure 13.9 presents lateral velocity distribution at $X/D_{eq} = 0$. In this figure, Y-axis represents the nondimensionalized form of resultant velocity (V/V_{max}). The equation that used to calculate V is shown in Equation 13.2 in which u, v, and w represent the local velocity component at X, Y, and Z directions. V_{max} is equal to 150.78 m/s. All jets have a double peak of positive maximum velocity. Although the jet flow is swirling about its axis, it is observed that the velocity distribution is symmetric

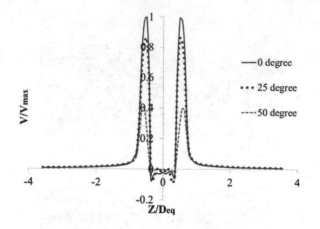

Figure 13.8   Lateral velocity distribution (Z axis) at $X/D_{eq} = 0$.

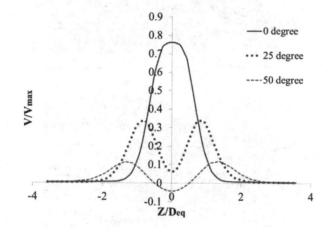

Figure 13.9   Lateral velocity distribution $X/D_{eq} = 5$.

about the jet axis. As observed in centerline velocity distribution, in lateral velocity distribution plot also 0° swirl flow has the maximum velocity among the three configurations. All jet configurations almost have zero exit velocity at the center. Comparing 0°, 25°, and 50° swirl flows, the 50° swirl has less maximum velocity than the other two. The maximum velocity of 50° swirl flow is 61% < 0° swirl flow and 55% < 25° swirl flow. Figure 13.10 shows the velocity distribution at $X/D_{eq} = 5$. The swirling flow of 0° vane angle has a single peak maximum velocity at $Z/D_{eq} = 0$. Swirling flow of 25° and 50° vane angle has double peak maximum velocity at $Z/D_{eq} = 1$ and $Z/D_{eq} = 1.5$, respectively.

Mass entrainments of different swirling annular jet configurations are plotted in Figure 13.10. Mass flow at different X/D_{eq} is obtained and normalized by the inlet mass flow rate (m_i). Mass flow rate at different locations for different swirl conditions has been fitted by using a simple linear expression (13.3).

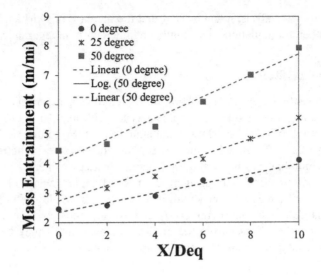

Figure 13.10   Mass entrainment.

$$\frac{m}{m_i} = K_m \left(\frac{X}{D_{eq}} \right) + C_m \tag{13.3}$$

In Equation 13.3, K_m denotes streamwise mass entrainment rate. Obtained values of K_m are plotted against swirl angle in Figure 13.11. It is observed that K_m increases linearly with increase in swirl angle. Slope of the linear curve in Figure 13.11 $dK_m/d\varphi = 0.039$. It is interesting to note that $dK_m/d\Phi$ (K_m vs φ) is approximately equal to $dS/d\Phi$ (S vs φ).

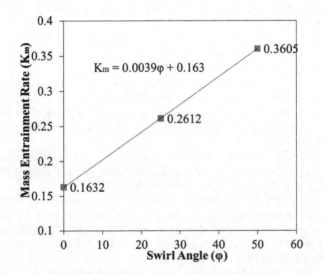

Figure 13.11   Variation of streamwise mass entrainment rate with swirl angle.

From the above analysis, it can be said that the increase in swirl angle increases the swirl number, recirculation zone length, and mass entrainment rate.

13.4 CONCLUSIONS

This work presents the effect of swirling on the circular annular jets issuing from a straight annular circular duct. Stationary swirler with angular vanes of $0°$, $25°$, and $50°$ was used to induce swirling in the annular duct. From the results, it is observed that the $25°$ swirl flow has the lowest value of pressure in the recirculation region. Swirl number and mass entrainment rate increases linearly at the rate of ($dK_m/d\Phi$ and $dS/d\Phi$) 0.03 with an increase in swirl angle. Length of the recirculation region also increases linearly with increase in swirl angle. Fifty-degree swirl flow has the largest recirculation region zone, maximum mass entrainment rate, and highest swirl number. Hence from the preliminary analysis, it may be stated that the $50°$ swirl flow is more suitable for the applications where enhanced mixing is required.

REFERENCES

1. V.M. Puri, R.C. Anantheswaranb, The finite-element method in food processing: A review, *J. Food Eng.* 19 (1993) 247–274.
2. T.Y. Chao, W.K. Chow, H. Kong, A review on the applications of finite element method to heat transfer and fluid flow, *Int. J. Archit. Sci.* 3 (2002) 1–19.
3. S.L. Ho, W.N. Fu, Review and future application of finite element methods in induction motors, *Electr. Mach. Power Syst.* 26 (2007) 111–125. doi:10.1080/07313569808955811.
4. C. Del Taglia, L. Blum, J. Gass, Y. Ventikos, D. Poulikakos, Numerical and experimental investigation of an annular jet flow with large blockage, *J. Fluids Eng.* 126 (2004) 375. doi:10.1115/1.1760533.
5. H. Memar, J.P. Holman, P.A. Dellenback, The effect of a swirled annular jet on convective heat transfer in confined coaxial jet mixing, *Int. J. Heat Mass Transf.* 36 (1993) 3921–3930. doi:10.1016/0017-9310(93)90142-S.
6. M. Nikjooy, K.C. Karki, H.C. Mongia, V.G. McDonell, G.S. Samueisen, A numerical and experimental study of coaxial jets, *Int. J. Heat Fluid Flow.* 10 (1989) 253–261.
7. G. Buresti, A. Talamelli, P. Petagna, Experimental characterization of the velocity field of a coaxial jet configuration, *Exp. Therm. Fluid Sci.* 9 (1994) 135–146. doi:10.1016/0894-1777(94)90106-6.
8. A.S. Nejad, S.A. Ahmed, Flow field characteristics of an axisymmetric sudden-expansion pipe flow with different initial swirl distribution, *Int. J. Heat Fluid Flow.* 13 (1992) 314–321. doi:10.1016/0142-727X(92)90001-P.
9. A.R. Milićević, S.V. Belošević, I.D. Tomanović, N.D. Crnomarković, D.R. Tucaković, Development of mathematical model for co-firing pulverized coal and biomass in experimental furnace, *Therm. Sci.* 22 (2018) 709–719. doi:10.2298/TSCI170525206M.
10. M.J.R. Harris, The decay of swirl in a pipe, *Int. J. Heat Fluid Flow.* 15 (1994) 212–217. doi:10.1016/0142-727X(94)90040-X.
11. S.H. Park, H.D. Shin, Measurements of entrainment characteristics of swirling jets, *Inr. J. Hear Mass Transf.* 36 (1993) 4009–4018.

12. M. Percin, M. Vanierschot, B.W. van Oudheusden, Analysis of the pressure fields in a swirling annular jet flow, *Exp. Fluids.* 58 (2017) 1–13. doi:10.1007/s00348-017-2446-3.

13. H.J. Sheen, W.J. Chen, S.Y. Jeng, Recirculation zones of unconfined and confined annular swirling jets, *AIAA J.* 34 (1996) 572–579. doi: 10.2514/3.13106.

14. M. García-Villalba, J. Fröhlich, LES of a free annular swirling jet: Dependence of coherent structures on a pilot jet and the level of swirl, *Int. J. Heat Fluid Flow.* 27 (2006) 911–923. doi:10.1016/j.ijheatfluidflow.2006.03.015.

15. K.P. Vanoverberghe, E.V. Van Den Bulck, M.J. Tummers, Confined annular swirling jet combustion, *Combust. Sci. Technol.* 175 (2003) 545–578. doi:10.1080/00102200302388.

16. F. Marsik, Z. Travnicek, P. Novotny, E. Werner, Stability of a swirling annular flow, *J. Flow Vis. Image Process.* 17 (2011) 267–279. doi:10.1615/jflowvisimageproc.v17.i3.70.

17. T. Parra, R. Perez, M.A. Rodriguez, A. Gutkowski, R. Szasz, F. Castro, Mixing and combustion of turbulent coaxial jets: An application of computational fluid dynamics to swirling flows (2014) 545–550. doi:10.5220/0005009005450550.

18. E.A. Ibrahim, T.R. McKinney, Injection characteristics of non-swirling and swirling annular liquid sheets. *Proc. Inst. Mech. Eng. C: J. Mech. Eng. Sci.* 220(2) (2006) 203–214. https://doi.org/10.1243/09544062C02505.

19. S. Garg, M. Pant, Meshfree methods: A comprehensive review of applications, *Int. J. Comput. Methods.* 15 (2018) 1830001. doi:10s.1142/S0219876218300015.

20. S. Gopinath, Effects of swirling co-flow on non-circular supersonic jets, Anna University Chennai, 2013.

14 Computations of Linear Programming Problems in Integers

Sunil Kumar
Graphic Era Hill University

Seema Saini
Graphic Era (Deemed to be) University

CONTENTS

14.1 INTRODUCTION

Linear programming (LP) is mathematical modeling of a real-world problem related to business industries. This method is used to find optimum solutions of the problems in which objective function and constraints appear as linear functions of decision variables. Linear programming formulations of real-life problems are essentially required, if more than two variables appear in the problem. A large number of variables in a LP problem make its solution quite complex. Therefore, a more efficient method and a solving tool of finding optimal solutions are required. In 1947, Dantzig [3,4], a mathematical scientist and member of the U.S. Air Force, formulated some general LP problems and developed simplex method for solving them. This method has become very significant to bring LP into broader use. Nadar [18] highlighted that simplex method has played a vital role during these many years in many real-world problems and still improving in order to get the optimum solution. Taha [22] checked the convergence of the process. Simplex method is considered as a significant development for making optimal decisions in complex situations.

There are so many applications of LPs. The approach is being used in the petroleum refineries since a long time. Generally, a refinery has a choice to purchase crude oil from different sources which differs in compositions and prices. Refinery can manufacture different products, such as petrol, diesel, and different types of lubricants in varying quantities. The constraints may be due to the restrictions on the quantity of the crude oil available from a particular source, the capacity of the refinery to produce a particular product, and so on. A mix of the purchased crude oil and manufactured products is sought that gives the maximum profit [18]. Optimal solutions of such problems can be obtained in fractions by using simplex, two-phase or dual simplex method.

Another one important application is the handling of a cutting stock problem that derives by the different constraints of a paper production mill. System accepts orders from customers and plan to produce accomplished within due time. The aim is to find an optimal solution for cutting master reels of standard size into ancillary reels of different sizes in a manner to satisfy a conglomeration of customers insistences with minimum possible waste. Such problems can be handled easily by LP approach [11]. The approach is also associated with the number of cutting blades to cut master reel into auxiliary reels of required widths simultaneously. Some constraints must be framed during the determination of the cutting patterns. This approach determines feasible combinations of widths with a minimal waste production. Such problems can be solved by various methods, but the solutions are required in integers. For this purpose, some approaches such as Gomory's cutting-plane algorithm or branch-and-bound techniques are generally used. However, manual implementations of above-mentioned methods for solving a LP system having large number of variables and constraints are quite difficult for an individual or a team.

After inclusion of digital computers, some clever algorithms are introduced by researchers for solving complex LP problems. Some defense departments have shown their intense interest in LP application [5]. The U.S. National Bureau of Standards with Pentagon funding did first experiment on techniques for solving linear programs primarily by Hoffman et al. [10]. The progress is continuing till date for increasing the efficiency of algorithms to solve more and more complicated problems.

Nowadays, many commercial tools such as Linear Interactive and Discrete Optimization (LINDO), General Interactive Optimizer (GINO), TORA, AMPL, and EXCEL SOLVER, packages, as well as many other commercial and academic packages are available to find solutions of LP problems, but no efficient tool is available to find integer solutions of different types of LP problems. Neuman [19] used MATLAB to solve such problems, but approach could not give satisfactory results on some problems, especially, for the problems of minimization in integers. At this stage, a reliable tool/code is essential for significant uses of LP problems of real world, specially, if integer solutions are required. In this chapter, review of methods and codes developed in C language are given to solve any type of LP problem in fractions as well as in integers.

14.2 REVIEW ON TERMS RELATED TO LINEAR PROGRAMMING PROBLEMS AND SIMPLEX METHOD

Let a LP problem be of the type:

$$\text{Maximize} \quad C^T \cdot x \tag{14.1}$$

subject to

$$Ax \le b, x_i \ge 0, \tag{14.2}$$

where $x = (x_1, x_2, ..., x_n)$ is a set of variables and $C = (C_1, C_2, ..., C_n)$ be the set of involved constants. A is a $p \times n$ matrix and $B = (B_1, B_2, ..., B_p)$ with $B_i \ge 0$.

The objective of the problem is to find a feasible region to achieve an optimal solution. In geometric terms, the feasible region defined by all values of x such that $Ax \le B, x_i \ge 0$ is a (possibly unbounded) convex polytope. There is a simple characterization of the extreme points or vertices of this polytrope; that is, $x = (x_1, x_2, ..., x_n)$ is an extreme point if and only if the subset of column vectors A_i corresponding to the non-zero entries of $x(x_i \ne 0)$ are linearly independent. In this context, such a point is known as a basic feasible solution (BFS). If at least one constraint is violated, solution is known as infeasible solution. The collection of all feasible solutions is called feasible region. A feasible solution is called optimum solution if it is most favorable.

If there are m equality constraints having $m + n$ variables ($m \le n$), a start for an optimum solution is made by putting n unknowns out of $m + n$, equal to zero and then solving for m equations in remaining m unknowns, provided that the solution exists and is unique. The n zero variables are called non-basic variables and the rest m are known as basic variables which form a basic solution. If the solution yields all non-negative basic variables, it is called BFS; otherwise, it is infeasible. This reduces the number of alternatives for the optimal solution from infinite to a finite number, whose maximum limit can be $\binom{m+n}{m} = \frac{(m+n)!}{m!n!}$.

In a LP problem, if all basic variables are (> 0), solution is called non-degenerate solution, and if some of them are zero, solution is called degenerate.

Simplex method is an ancient and reliable algorithm for solving linear programing problems. It also provides the basis for performing various parts of post-optimality analysis very efficiently. Klee and Minty [13] have given a useful geometrical interpretation and concluded that it is an algebraic procedure and its worst-case complexity is exponential. At each iteration, it moves from the current BFS to a better, adjacent feasible solution by choosing both an entering basic variable and a leaving basic variable and using Gaussian elimination method to solve a system of linear equations. When the current solution has no adjacent BFS, it is better the current solution is optimal and algorithm stops.

Maros [17] has given the summary of simplex computational techniques with many experimental observations. The gap between practical effectiveness and theoretical complexity was bridged by Borgwardt [1] and Spielman and Teng [21].

14.2.1   CONVERGENCE OF SIMPLEX METHOD

Let there exist a constant $\theta > 0$ such that for every iteration r, the values of all basic variables x_j^r satisfy $x_j^r \geq \theta > 0$ for all j_i.

In the starting of t_{th}, by eliminating the basic variables from the objective equation, we obtain

$$z_{t-1} - z = \sum (-C_j^{-t}) x_j, \qquad (14.3)$$

where $C_{j_i}^{-t} = 0$ for all basic $j = j_i$.

If $(-C_s^{-t}) = \max(-C_j^{-t}) \leq 0$, the iterative process stops with the current BFS optimal. Otherwise, we increase non-basic x_s, to $x_s = \theta_t \geq \theta$ and adjust basic variables to obtain the BFS to start iteration $t + 1$.

14.2.2   PROGRAMMING FOR SIMPLEX METHOD

If a LP problem has n variables and m constraints and each constraint has"$\leq$" sign, the problem will be solved by simplex method. However, if some constraints have another inequalities such as "$=$" and/or "$\geq$," problem will be solved by another approach, which are discussed in subsequent sections. In this section, it is necessary to introduce main function of the program for understanding the meanings of variables and arrays mentioned. This is common part to call problem-related subroutine; therefore, it is necessary to define variables related to all inequalities "$\leq$," "$=$," and "$\geq$," in main function.

```
int main()
{
int n, m, l, g, e, i, j, x[50], y[50], k, s, retur;
float c[50], b[50], a[50][50], zc[50], d[50], fact, w;
printf("Enter the number of variables in objective function......:");
scanf("%d", &n);
printf("Enter the number of constraints in LPP...................:");
scanf("%i",&m);
printf("Enter the number of inequalities having sign <= ..........:");
scanf("%i",&l);
printf("Enter the number of inequalities having sign >= ..........:");
scanf("%i",&g);
printf("Enter the number of inequalities having sign  = ..........:");
scanf("%i",&e);
if(m!=l+g+e)
```

```
      {
      printf("Incorrect input! Try again....");
      getch();
      return(0);
      }
  if(l!=0 && g==0 && e==0)
      {
      printf("\nThis problem is solving by Simplex Method.");
      printf("\nWelcome by Simplex.");
      }
  if(g!=0 || e!=0)
      {
      printf("\nThis problem is solving by Two-Phase Simplex Method.");
      printf("\nWelcome by Simplex.");
      }
      printf("\nPress 1 for maximization.\nPress 2 for minimization.\n");
      scanf("%d", &k);
      if(k==1) fact=1; else if(k==2) fact=-1;
      else {puts("Invalid choice !\n"); getch(); return(0);}
  for(j=1;j<=n+l+g+g+e;j++)
  x[j]=j;
  for(i=n+1;i<=n+l;i++)
      {
      y[i-n]=i;
      b[i-n]=0;
      }
  printf("Enter the coefficients of the variables in objective
      function.:\n");
  for(j=1;j<=n;j++)
      {
      printf("C[%d]=",j);
      scanf("%f", &c[j+1]);
      }
      printf("Constant term:");
      scanf("%f", &c[1]); printf("\n");
  for(j=n+2;j<=n+l+1;j++)
      c[j+1]=0;
  for(j=1;j<=n+l+1;j++)
      c[j]=c[j]*fact;
  if(l!=0)
      {
      printf("Enter the coefficients of the constraints having
          '<=' sign.....:\n");
      for(i=1;i<=l;i++)
          {
          for(j=1;j<=n;j++)
              {
              printf("Coefficient %i :",j);
```

```
          scanf("%f",&a[i][j+1]);
            }
      printf("Constant term:");
      scanf("%f",&a[i][1]); printf("\n");
      }
    printf("\n");
    }
if(g!=0)
   {
   printf("Enter the coefficients of the constraints having
        '>=' sign.....:\n");
   for(i=l+1;i<=l+g;i++)
      {
        for(j=1;j<=n;j++)
           {
           printf("Coefficient %i :",j);
           scanf("%f",&a[i][j+1]);
           }
        printf("Constant term:");
        scanf("%f",&a[i][1]); printf("\n");
      }
    printf("\n");
    }
if(e!=0)
   {
   printf("Enter the coefficients of the constraints having
        '=' sign......:\n");
   for(i=l+g+1;i<=l+g+e;i++)
      {
        for(j=1;j<=n;j++)
           {
           printf("Coefficient %i :",j);
           scanf("%f",&a[i][j+1]);
           }
        printf("Constant term:");
        scanf("%f",&a[i][1]); printf("\n");
      }
    printf("\n");
    }
for(i=1;i<=m;i++)
      {
      if(a[i][1]<0.0)
        {
        printf("Bad input: Constants bi must be non-negative.\n");
        return(0);
        }
      }
      if(l!=0 && g==0 && e==0)
```

```
        retur=simplex(l, m, n, c, x, y, b, a, zc);
        else
            {
            for(j=1;j<=n+1;j++)
                d[j]=c[j];
            for(j=1;j<=n+l+g+1;j++)
                    c[j]=0;
            for(j=n+l+g+2;j<=n+l+g+g+e+1;j++)
                    c[j]=-1;
            w=ibfs(m, l, g, e, a);
            retur=t_phase(l, g, e, m, n, c, x, y, b, a, zc, d, w);
            }
        if(k==1)
          {
            printf("\nMaximization of the problem is as:\n");
            printf("Maximum of objective function is=%f\n",zc[1]);
          }
        if(k==2)
            {
            printf("\nMinimization of the problem is as:\n");
            printf("Minimum of objective function is=%f\n",-zc[1]);
            }
        for(j=1;j<=n;j++)
            for(i=1;i<=m;i++)
                if(j==y[i])
                    printf("X%d=%f\n",y[i],a[i][1]);
        printf("\nDo you want integer solution of the problem?");
        printf("\nPress 1 for yes!\nPress 2 for no!\n");
        scanf("%d", &k);
        if(k==1 && retur==-20)
          gomor(l, m, n, c, x, y, b, a, zc);
        if(k==1 && retur!=-20)
          {
        printf("\nProblem is not fit for obtaining an integer solution!");
          return(0);
          }
        else if(k==2)
                return(0);
        else if (k!=1 || k!=2)
                {puts("Invalid choice !\n"); getch(); return(0);}
    }
```

This part of the program will perform simplex procedure.

```
int simplex(int l, int m, int n, float c[], int x[],
    int y[], float b[], float a[][], float zc[])
{
int i, j;
for(i=1;i<=m;i++)
   for(j=n+2;j<=n+l+1;j++)
```

```
       {
         if(i==j-n-1)
            a[i][j]=1;
         else
             a[i][j]=0;
       }
simp_comp(l, m, n, c, x, y, b, a,zc);
return(-20);
}


int simp_comp(int l, int m, int n, float c[], int x[],
    int y[], float b[50], float a[50][50], float zc[50])
{
 int i, j, s, r; float dv, temp;
 for(j=1;j<=n+l+1;j++)
    zc[j]=z_c(j, m, c, b, a);
 s=m_neg(n, l, zc);
 if(s==-1)
   {
    printf("Pivot element could not be found\n");
    table(s,l, m, n, c, x, y, b, a, zc);
   }
 else
 {
  r=m_rat(s, m, a);
  printf("Position of Pivot element is (%d, %d)\n", r, s);
  table(s,l, m, n, c, x, y, b, a, zc);
 }
 do
  {
  temp=-1;
  for(i=1;i<=m;i++)
  if(a[i][s+1]>0)
    temp=a[i][s+1];
  if(temp>0)
    {
      dv=a[r][s+1];
      for(j=1;j<=n+l+1;j++)
          a[r][j]=a[r][j]/dv;
      for(i=1;i<=m;i++)
        {
           dv=a[i][s+1];
           if(i!=r)
              for(j=1;j<=n+l+1;j++)
              a[i][j]=a[i][j]-a[r][j]*dv;
        }
      y[r]=x[s]; b[r]=c[s+1];
      for(j=1;j<=n+l+1;j++)
```

```
                zc[j]=z_c(j, m, c, b, a);
         dv=zc[2];
         for(j=2;j<=n+l+1;j++)
             if(dv>zc[j])
                dv=zc[j];
             if(dv>=0)
                {
                    table(s,l, m, n, c, x, y, b, a, zc);
                    printf("\nOptimum Solution Achieved.");
                  return(0);
                }
         s=m_neg(n, l, zc);
         if(s==-1)
           {
             printf("Pivot element could not be found.\n");
             table(s,l, m, n, c, x, y, b, a, zc);
             return(0);
           }
         else
              {
                 r=m_rat(s, m, a);
                 printf("Position of Pivot element is (%d, %d).\n", r, s);
                 table(s,l, m, n, c, x, y, b, a, zc);
              }
         temp=zc[2];
         for(j=2;j<=n+l+1;j++)
             if(temp>=zc[j])
                temp=zc[j];
     }
     else
          {
            if(s!=-1)
            printf("\nThe problem has unbounded solution!");
            return (0);
          }
     }
     while(temp<0);
          dv=zc[n+2];
     for(j=n+2;j<=n+l+1;j++)
         if(dv>=zc[j])
            dv=zc[j];
     if(dv==0)
{
 printf("\nOptimum Solution Achieved");
 printf("\nHowever, alternative optimum solutions will
            exist for this problem!\n");
}
}
```

14.2.3 CODE TO PERFORM OPTIMALITY TEST

Optimality test checks whether the current feasible solution can be improved or not. This is done by computing $Z_j - C_j$, where $Z_j = \sum C_B A_{ij}$. If $Z_j - C_j$ is negative under any column in simplex table, the current feasible solution is not optimal and at least one better solution is possible.

```
float z_c(int j, int m, float c[50], float b[50], float a[50][50])
    {
      int i; float zc[50]; zc[j]=0.0;
      for(i=1;i<=m;i++)
          zc[j]=zc[j]+b[i]*a[i][j];
      zc[j]=zc[j]-c[j];
      return(zc[j]);
    }
```

To determine the selection of entering variable, most negative $Z_j - C_j$ from the columns of simplex table is to be marked. Similarly, leaving variable is decided by marking minimum non-negative ratio of the elements of quantity column, with the elements of the column in which most negative $Z_j - C_j$ exists. Following two codes perform these tasks:

Code to calculate most negative $Z_j - C_j$

```
int m_neg(int n, int l, float zc[50])
    {
      int j, k; float temp; temp=0;
      for (j=2;j<=n+l+1;j++)
      if(temp>zc[j])
        {
        temp=zc[j];
        k=j-1;
        }
        if(temp<0)
           return(k);
        else
         return(-1);
    }
```

Code to calculate minimum ratio

```
int m_rat(int s, int m, float a[50][50])
    {
      int i, k; float mr[50], temp;
      for(i=1;i<=m;i++)
      if(a[i][s+1]>0)
      mr[i]=a[i][1]/a[i][s+1];
      else
      mr[i]=10000;
      temp=mr[1];
```

```
    for(i=1;i<=m;i++)
        if(temp>=mr[i]&& mr[i]>=0)
            {
             temp=mr[i]; k=i;
            }
    return(k);
    }
```

14.2.4   CODE TO PRINT SIMPLEX TABLE FOR EACH ITERATION

```
int table(int s, int l, int m, int n, float c[], int x[],
          int y[], float b[50], float a[50][50], float zc[50])
{
int i,j; float mr[50], dv;
for(j=1;j<=n+l;j++)
printf("----------------");
printf("\n        Cj");
for(j=1;j<=n+l+1;j++)
printf("  %f ",c[j]);
printf("\n");
for(j=1;j<=n+l;j++)
printf("----------------");
printf("\nB.V.   Cb   Xb\t\t");
for(j=1;j<=n+l;j++)
printf("X%d          ",x[j]);
printf("M.R\n");
for(j=1;j<=n+l;j++)
printf("----------------");
for(i=1;i<=m;i++)
{
dv=a[i][s+1];
if(dv!=0)
mr[i]=a[i][1]/dv;
else
mr[i]=10000;
}
for(i=1;i<=m;i++)
    {
    printf("\n");  printf("X%d ",y[i]); printf(" %3.2f",b[i]);
    for(j=1;j<=n+l+1;j++)
    printf("  %f ", a[i][j]);
    if(mr[i]!=10000)
    printf("        %4.1f",mr[i]);
    else
    printf("        ---");
    }
    printf("\n");
for(j=1;j<=n+l;j++)
```

```
printf("----------------"); printf("\n       Z=");
for(j=1;j<=n+1+1;j++)
   printf(" %f  ", zc[j]);
printf("\n");
return(0);
}
```

14.3   TWO-PHASE METHOD

Simplex algorithm is not applicable if all constraints of a LP problem do not have "≤" sign. In this situation, some special variables (artificial variables) are to be added in the given constraints and solution can be found in two phases. Phase I is an attempt to find initial feasible solution by possible removal of these artificial variables under some specific conditions.

Consider a LP problem of the type:

$$\text{Minimize } C^T \cdot x, \tag{14.4}$$

$$\text{subject to } Ax = B, x_i \geq 0. \tag{14.5}$$

By multiplying some rows with -1 if necessary, we can achieve that the right-hand side B satisfies $B \geq 0$. From this, we construct a linear program from which an initial basic solution is readily available:

$$\text{Minimize } z_1 + \cdot + z_m, \tag{14.6}$$

$$\text{subject to } Ax + z = B, x_i, z_i \geq 0. \tag{14.7}$$

The set $B = \{n+1, ..., n+m\}$ is a feasible basis with BFS (x^*, z^*) defined by $x^* = 0$ and $z^* = b$.

At the end of Phase I, artificial variables will be removed only when they are non-basic. If one or more artificial variables are basic, following steps must be performed to remove them before the implementation of Phase II:

Step 1- Select a zero artificial variable (leaving variable) and mark its row as the pivot row.

Step 2- Select an entering variable with a non-zero coefficient from the pivot row, and then, apply simplex iteration.

Step 3- Remove the column of just-leaving artificial variable. If all zero artificial variables are removed, go to Phase II. Otherwise, return to Step 1.

To obtain initial BFS of a LP problem in obtained in Phase I, an objective function w is created to find the sum of all artificial variables.

New objective function is then minimized, under the constraints of the given problem, using simplex method. At the end of Phase I, three cases arise:

1. If minimum value of w is greater than zero and at least one artificial variable appears in the bases at a positive level, given problem has no feasible solution and the procedure terminates.

2. If minimum value of w is zero and all artificial variables have been removed from the bases, a BFS of the given problem is obtained. The artificial variable column(s) is/are deleted before the Phase II computation.

3. If minimum value of w is zero and one or more artificial variables appear in the bases at zero level, a BFS of the problem is obtained. However, we must take care of this artificial variable and see that it never becomes positive during Phase II computations. Zero cost coefficient is assigned to this artificial variable, and it is retained in the initial table of Phase II. If this variable remains in the bases at zero level in all Phase II computation, there is no problem; however, the problem arises if it becomes positive in some iteration. In such a case, a slightly different approach is adopted in the selection of outgoing variable. As artificial variable is selected, simplex method can then be applied as usual to obtain optimal BFS.

However, after adding artificial variables, M-method (also called the Big M-method) can also be used, but a difficulty arises when the problem is to be solved on a digital computer, because M must be assigned some numerical value which is must larger than the values $C_1, C_2, C_3, ... C_n$ in the objective function. However, a computer has fixed number of digits. Therefore, two-phase method is chosen for solving such type of LP problems.

14.3.1 CODE TO CHECK FEASIBILITY OF THE SOLUTION

Following code will check the value of w for starting two-phase method according to abovementioned rules.

```
int ibfs(int m, int l, int g, int e, float a[50][50])
{
  float **A, *tem, app, sum, mult, w;
  int i, j, k, N, P;
  N=l+g+e;
  A = (float**)malloc(N*sizeof(float*));
  for(i=0; i<N; i++)
      A[i] = (float*)malloc(N*sizeof(float));
  tem = (float*)malloc(N*sizeof(float));
  for(i=0; i<N; i++)
      for(j=0; j<N; j++)
          if(i==j)
            A[i][j]=1;
          else
              A[i][j]=0;
  for(i=0; i<N; i++)
      A[i][N]=a[i+1][1];
  for(i=0; i<N; i++)
      {
        app = A[i][i];
        P = i;
        for(k = i+1; k < N; k++)
```

```
            if(fabs(app) < fabs(A[k][i]))
               {
                 app = A[k][i] ; P = k;
               }
        for(j = 0; j <= N; j++)
            {
              tem[j] = A[P][j];
              A[P][j] = A[i][j];
              A[i][j] = tem[j];
            }
        for(j=i+1; j<N; j++)
            {
              mult = A[j][i]/A[i][i];
              for(k=0; k<=N; k++)
                  A[j][k] -= mult*A[i][k];
            }
        }
    for(i=N-1; i>=0; i--)
        {
          sum = 0;
          for(j=i+1; j<N; j++)
              sum += A[i][j]*tem[j];
          tem[i] = (A[i][N]-sum)/A[i][i];
        }
    w=0;
    for(i=1;i<N;i++)
        w = w+tem[i];
    for(i = 0; i < N; i++)
        free(A[i]);
    free(A);
    free(tem);
return w;
}
```

14.3.2   CODE FOR TWO-PHASE METHOD

```
int t_phase(int l, int g, int e, int m, int n, float c[], int x[],
    int y[], float b[50], float a[50][50], float zc[50], float d[50],
       float w)
    {
      int i, j, s, r, check;
      float dv, temp, a1[50][50];
      for(i=1;i<=m;i++)
          for(j=n+2;j<=n+l+1;j++)
              if(i+1==j-n)
                a[i][j]=1;
              else
                   a[i][j]=0;
      for(i=1;i<=m;i++)
```

```
             for(j=n+1+2;j<=n+1+g+1;j++)
                 if(j-i==n+1)
                   a[i][j]=-1;
                 else
                     a[i][j]=0;
    for(i=1;i<=m;i++)
        for(j=n+1+g+2;j<=n+1+g+g+1;j++)
            if(j-i==n+g+1)
               a[i][j]=1;
            else
                a[i][j]=0;
    for(i=1;i<=m;i++)
        for(j=n+1+g+g+2;j<=n+1+g+g+e+1;j++)
            if(j-i==n+g+1)
               a[i][j]=1;
            else
                a[i][j]=0;
    for(i=1;i<=m;i++)
         for(j=n+1+1+g+g+e;j>=n+2;j--)
             if(a[i][j]==1)
             {
             y[i]=j-1;
             if(j>n+1+1+g)
                 b[i]=-1;
             else
                     b[i]=0;
             }
    for(i=1;i<=m;i++)
        for(j=1;j<=n+1+1+g+g+e;j++)
            a1[i][j]=a[i][j];
    for(j=1;j<=n+1+g+g+e+1;j++)
        zc[j]=z_c(j, m, c, b, a);
    s=m_neg(n, 1+g+g+e, zc);
    if(s==-1)
      {
       printf("Pivot element could not be found\n");
       t_phase_t(s, 1, g, e, m, n, c, x, y, b, a, zc);
      }
    else
        {
        r=m_rat(s, m, a);
        printf("Position of Pivot element is (%d, %d)\n", r, s);
        t_phase_t(s, 1, g, e, m, n, c, x, y, b, a, zc);
        }
do
 {
   temp=-1;
   for(i=1;i<=m;i++)
```

```
        if(a[i][s+1]>0)
            temp=a[i][s+1];
    if(temp>0)
      {
      dv=a[r][s+1];
      for(j=1;j<=n+1+g+g+e+1;j++)
            a[r][j]=a[r][j]/dv;
      for(i=1;i<=m;i++)
            {
            dv=a[i][s+1];
            if(i!=r)
              for(j=1;j<=n+1+g+g+e+1;j++)
                    a[i][j]=a[i][j]-a[r][j]*dv;
            }
      y[r]=x[s]; b[r]=c[s+1];
      for(j=1;j<=n+1+g+g+e+1;j++)
            zc[j]=z_c(j, m, c, b, a);
      s=m_neg(n, 1+g+g+e, zc);
      if(s==-1)
        {
         printf("Pivot element could not be found.\n");
         t_phase_t(s, 1, g, e, m, n, c, x, y, b, a, zc);
        }
      else
          {
            r=m_rat(s, m, a);
            printf("Position of Pivot element is (%d, %d).\n", r, s);
            t_phase_t(s, 1, g, e, m, n, c, x, y, b, a, zc);
          }
      temp=zc[2];
      for(j=2;j<=n+1+g+g+e+1;j++)
            if(temp>=zc[j])
                temp=zc[j];
  }
  if(temp>=0)
    {
    printf("Optimization achieved.\n");
    }
 }
while(temp<0);
check=y[1];
for(i=1;i<=m;i++)
    if(check<=y[i])
      check=y[i];
for(j=1;j<=n+1;j++)
c[j]=d[j];
for(j=n+2;j<=n+1+g+g+e+1;j++)
c[j]=0;
```

```
for(i=1;i<=m;i++)
b[i]=c[y[i]+1];

if(check>n+1+g && w>0)
  {
    printf("\nPhase-I of Simplex method ends here.");
    printf("\nAll artificial variables could not be removed.");
    printf("\nThis Problem does not posses a feasible solution.");
    printf("\nNo need to enter in phase-II.\nComputation
          terminated......\n");
    return(-10);
  }
else if(check<=n+1+g && w==0)
      {
        printf("\nPhase-I of Simplex method ends here.");
        printf("\nAll artificial variables have been removed.");
        printf("\n\nWelcome in Phase-II of the Simplex method:-\n");
        l=l+g;
        simp_comp(l, m, n, c, x, y, b, a, zc);
        return(-20);
      }
else if(check<=n+1+g && w>0)
      {
        printf("\nPhase-I of Simplex method ends here.");
        printf("\nAll artificial variables have been removed.");
        printf("\n\nWelcome in Phase-II of the Simplex method:-\n");
        l=l+g;
        simp_comp(l, m, n, c, x, y, b, a, zc);
        return(-20);
      }
else if(check>n+1+g && w==0)
      {
        printf("\nPhase-I of Simplex method ends here.");
        printf("\nAll artificial variables could not be removed.");
        printf("\nHowever, Problem posses a feasible solution.");
        printf("\n\nWelcome in Phase-II of the Simplex method:-\n");
        for(i=1;i<=m;i++)
              for(j=1;j<=n+1+1+g+g+e;j++)
                  a[i][j]=a1[i][j];
        for(i=1;i<=m;i++)
            {
            b[i]=0; y[i]=n+i;
            }
            l=l+g+g+e;
            simp_comp(l, m, n, c, x, y, b, a, zc);
            return(-20);
        }
}
```

14.3.3   CODE FOR TWO PHASE METHOD

```c
int t_phase_t(int s, int l, int g, int e, int m, int n, float c[],
    int x[], int y[], float b[50], float a[50][50], float zc[50])
  {
    int i,j; float mr[50], dv;
    for(j=1;j<=n+l+g+g+e;j++)
        printf("----------------");
    printf("\n        Cj");
    for(j=1;j<=n+l+g+g+e+1;j++)
        printf("  %f ",c[j]);
    printf("\n");
    for(j=1;j<=n+l+g+g+e;j++)
        printf("----------------");
    printf("\nB.V.   Cb   Xb\t\t");
    for(j=1;j<=n+l+g+g+e;j++)
        printf("X%d          ",x[j]);
    printf("M.R\n");
    for(j=1;j<=n+l+g+g+e;j++)
        printf("----------------");
    for(i=1;i<=m;i++)
      {
        dv=a[i][s+1];
        if(dv!=0)
           mr[i]=a[i][1]/dv;
        else
           mr[i]=1000;
      }
    for(i=1;i<=m;i++)
      {
        printf("\n");  printf("X%d ",y[i]); printf(" %3.2f",b[i]);
        for(j=1;j<=n+l+g+g+e+1;j++)
            printf("  %f ", a[i][j]);
        if(mr[i]!=1000)
              printf("      %4.1f",mr[i]);
    if(mr[i]==1000)
            printf("----");
      }
    printf("\n");
    for(i=1;i<=n+l+g+g+e;i++)
        printf("----------------");
        printf("\n      Z=");
    for(j=1;j<=n+l+g+g+e+1;j++)
        printf(" %f ", zc[j]);
    printf("\n");
    return(0);
}
```

14.4 INTEGER CUTTING PLANES

Cutting-plane procedure for convex problems was initially proposed by Kelley [12], and Cheney and Goldstein [2] in 1959–1960 for obtaining an efficient solutions of general problems. Dempster and Merkovsky [6] have given an overview and geometrical interpretation of convergence of the procedure.

In recent years, cutting-plane approaches proved remarkably effective for certain special stochastic programming problems with Integrated Chance Constraints [14], for minimizing Conditional Value at Risk [15] as well as problems involving the second-order stochastic dominance [7,16,20]. These problems can be written as LP problems having a large number of constraints.

Gomorys cutting-plane algorithm is an iterative process which is solved by modifying linear-programming problems until integer solution is not obtained. This process does not partition the feasible region into subdivisions, but instead works with a single linear program, which it refines by adding new constraints. The presence of these new constraints reduces the feasible region until an integer optimal solution is obtained.

14.4.1 BASIC TERMINOLOGY FOR INTEGER PROGRAMMING

Most of common problems are introduced as mixed integer programming problems and can be specified as

$$\min x_0 = C^T x,$$

subject to $Ax = B, x_j \geq 0, j = 1, 2, ..., n$, x_j integer for $j \in N$, where N is some subsets of the set $N_0 = \{0, 1, 2, ..., n\}$. When $N = N_0$, problems are reduced to a pure integer programming problem and quantities C_j, A_{ij}, B_i are reduced to integers.

14.4.2 CUTTING-PLANE ALGORITHM FOR PURE INTEGER PROGRAMMING

Gilmore and Gomory [8,9] introduced a finite algorithm to obtain an integer solution. This approach represents constructions of cutting planes.

$$\text{Let} \quad A_1 X_1 + A_2 X_2 + \cdot + A_n X_n = B \tag{14.8}$$

be a relation formed by some non-negative integers $X_1, X_2, ..., Xn$ and S the set of its solutions.

Let $\lfloor \xi \rfloor$ be the largest integer such that $\lfloor \xi \rfloor \leq \xi$. i.e. $\xi = \lfloor \xi \rfloor + \varepsilon$, $0 \leq \varepsilon \leq 1$, where ξ and ε are real quantities.

On using $A_j = \lfloor A_j \rfloor + f_j$ and $B = \lfloor B \rfloor + f$ in (4.1), we have

$$\sum_{j=1}^{n} (\lfloor A_j \rfloor + f_j) X_j = \lfloor B \rfloor + f \tag{14.9}$$

and hence

$$\sum_{j=1}^{n} f_j X_j - f = \lfloor B \rfloor + f - \sum_{j=1}^{n} \lfloor A_j \rfloor X_j. \qquad (14.10)$$

For any $X \in S$, both sides of Equation (4.3) are integers. Here, $X \geq 0$ and also have $\xi = \sum f_j X_j - f$ and $\xi \geq -f < -1$ are integers and hence deduce that

$$\sum_{j=1}^{n} f_j X_j \geq f \qquad \forall X \in S. \qquad (14.11)$$

However, if solution is not an integer at this stage, consider a basic variable X_i such that

$$X_i + \sum_{j \notin I} B_{ij} X_j = B_{i0}, \qquad B_{i0} \notin I. \qquad (14.12)$$

Putting $f_j = B_{ij} - \lfloor B_{ij} \rfloor$ and $f = B_{i0} - \lfloor B_{i0} \rfloor$ and deduce that

$$\sum_{j \notin I}^{n} f_j X_j \geq f. \qquad (14.13)$$

for all integer solutions to our problem.

Now, $f > 0$ since b_{i0} is not integer, and so Equation (4.6) is not agreed by the current solution since $X_j = 0$ for $j \notin I$ and so Equation (4.6) is a cut.

Gomory cuts $\sum f_j X_j > f$ are deduced to the form $\sum W_j X_j \leq W$ if these are expressed in terms of the original non-basic variables, where W_j and W are integers. The value of $\sum W_j X_j$ after solving a problem is $W + \varepsilon$, where $0 < \varepsilon < 1$ assuming the current solution non-integer. Thus, the cut is obtained by moving a hyper-plane parallel to itself to an extent which cannot exclude an integer solution. It is worth nothing that the plane can usually be moved further without excluding integer points, thus generating deeper cuts. For further discussion and information in this context, specialist books on integer programming can be referred.

14.4.3   CODE FOR GOMORY'S CUTS

This is part of the program to find required Gomory's cuts.

```
int gomor(int l, int m, int n, float c[50], int x[50],
          int y[50], float b[50], float a[50][50], float zc[50])
{
 int i, j, k, r, s;
 float check, g[50][50], temp[50], dv;
 for(i=1;i<=m;i++)
    {
     g[i][1]=a[i][1];
     while (g[i][1]>=1)
```

```
    g[i][1]=g[i][1]-1;
    }
check=0.0;
for(i=1;i<=m;i++)
    if(y[i]<=n && check<=g[i][1])
      check=g[i][1];
if(check==0||(1-check>=0 && 1-check <=0.0001))
  {
  printf("An optimum basic solution is attended.\n");
  for(j=1;j<=n;j++)
      for(i=1;i<=m;i++)
            if(j==y[i])
              printf("X%d=%f\n",y[i],a[i][1]);
    printf("The optimum value of function is:%f\n",zc[1]);
    return(0);
  }
check=g[1][1]; r=1;
for(i=2;i<=m;i++)
    if(check<g[i][1])
        {
        check=g[i][1];
        r=i;
        }
for(j=2;j<=n+1+1;j++)
    g[r][j]=a[r][j];
for(j=1;j<=n+1+1;j++)
    if(g[r][j]<0)
        {
        for(k=1;g[r][j]<=-1;k++)
        g[r][j]=g[r][j]+1;
        }
    else if(g[r][j]>1)
        {
        for(k=1;g[r][j]>1;k++)
      g[r][j]=g[r][j]-1;
        }
    else if(g[r][j]>0 && g[r][j]<1)
                  g[r][j]=g[r][j];
    else
              g[r][j]=0;
  for(j=1;j<=n+1+1;j++)
      if(g[r][j]>0)
        g[r][j]=-g[r][j];
  for(j=1;j<=n+1+1;j++)
      a[m+1][j]=g[r][j];
  c[n+1+2]=0;
  b[m+1]=0;
  for(i=1;i<=m;i++)
```

```
    a[i][n+1+2]=0;
a[m+1][n+1+2]=1;
for(j=1;j<=n+1+2;j++)
zc[j]=z_c(j, m, c, b, a);
printf("\nGomory table is:\n");
gom_t(1, m, n, c, x, y, b, a, zc);
for(j=2;j<=n+1+2;j++)
    if(a[m+1][j]!=0)
        temp[j]=zc[j]/a[m+1][j];
    else
        temp[j]=-10000;
check=temp[2];s=2;
for(j=3;j<=n+1+2;j++)
    if(check<temp[j] && temp[j]<0)
{
  check=temp[j];
  s=j;
}
printf("Position of pivot element is (%d, %d)\n",m+1,s-1);
dv=a[m+1][s];
for(j=1;j<=n+1+2;j++)
    a[m+1][j]=a[m+1][j]/dv;
for(i=1;i<=m;i++)
{
 dv=a[i][s];
 for(j=1;j<=n+1+2;j++)
    a[i][j]=a[i][j]-a[m+1][j]*dv;
}
for(i=1;i<=m+1;i++)
    for(j=1;j<=n+1+2;j++)
      {
        if(a[i][j]==-0.0)
        a[i][j]=0.0;
      }
 x[n+1+1]=n+1+1;
      y[m+1]=x[s]-1; b[m+1]=c[s];
      m++;
for(j=1;j<=n+1+2;j++)
zc[j]=z_c(j, m, c, b, a);
g_table(1, m, n, c, x, y, b, a, zc);
for(i=1;i<=m+1;i++)
    temp[i]=a[i][1];
for(i=1;i<=m+1;i++)
    {
            while(temp[i]>=1)
            temp[i]=temp[i]-1;
    }
dv=0.0;
```

```
if(y[m+1]<=m)
  {
    for(i=1;i<=m+1;i++)
       if(temp[i]!=0.0)
         dv=temp[i];
       }
else
     {
       for(i=1;i<=m;i++)
          if(temp[i]!=0.0)
              dv=temp[i];
     }
if(y[m+1]<=m)
  {
    for(i=1;i<=m+1;i++)
    if(dv==0 || (1-dv>=0 && 1-dv<=0.0001))
     {
     printf("\nOptimum Solution achieved.\nThe solution is:\n");
     for(j=1;j<=n;j++)
       for(i=1;i<=m+1;i++)
             if(j==y[i])
                 printf("X%d=%f\n",y[i],a[i][1]);
     printf("The optimum value of function is:%f\n",zc[1]);
     return(0);
     }
else
     {
         l++;
             gomor(l, m, n, c, x, y, b, a, zc);
             return(0);
     }
  }
else if(y[m+1]>m)
  {
    for(i=1;i<=m;i++)
        if(dv==0 || (1-dv>=0 && 1-dv<=0.0001))
         {
           printf("\nOptimum Solution achieved.\nThe solution is:\n");
            for(j=1;j<=n;j++)
                for(i=1;i<=m;i++)
                          if(j==y[i])
                          printf("X%d=%f\n",y[i],a[i][1]);
           printf("The optimum value of function is:%f\n",zc[1]);
           return(0);
           }
else
     {
             l++;
```

```
                        gomor(l, m, n, c, x, y, b, a, zc);
                        return(0);
               }
        }
}
```

14.4.4   CODES TO DISPLAY GOMORY'S TABLES

```c
int g_table (int l, int m, int n, float c[], int x[],
             int y[], float b[50], float a[50][50], float zc[50])
{
int i,j;
printf("\n");
for(i=1;i<=n+l+1;i++)
printf("---------------");
printf("\n        Cj");
for(i=1;i<=n+l+2;i++)
printf("  %f ",c[i]);
printf("\n");
for(i=1;i<=n+l+1;i++)
printf("---------------");
printf("\nB.V.   Cb   Xb\t\t");
for(i=1;i<=n+l+1;i++)
printf("X%d          ",x[i]);
printf("\n");
for(i=1;i<=n+l+1;i++)
printf("---------------");
for(i=1;i<=m;i++)
   {
     printf("\n");  printf("X%d ",y[i]); printf(" %3.2f",b[i]);
     for(j=1;j<=n+l+2;j++)
     printf("  %f ", a[i][j]);
   }
     printf("\n");
for(i=1;i<=n+l+1;i++)
printf("---------------"); printf("\n        Z=");
     for(j=1;j<=n+l+2;j++)
     printf(" %f  ", zc[j]);
     printf("\n");
     return(0);
}

int gom_t(int l, int m, int n, float c[], int x[],
          int y[], float b[50], float a[50][50], float zc[50])
{
int i,j;
printf("\n");
for(i=1;i<=n+l+1;i++)
```

```
printf("--------------");
printf("\n      Cj");
for(i=1;i<=n+l+2;i++)
printf("  %f ",c[i]);
printf("\n");
for(i=1;i<=n+l+1;i++)
printf("--------------");
printf("\nB.V.   Cb   Xb\t\t");
for(i=1;i<=n+l;i++)
printf("X%d          ",x[i]);
for(i=n+l+1;i<=n+l+1;i++)
printf("G");
printf("\n");
for(i=1;i<=n+l+1;i++)
printf("--------------");
for(i=1;i<=m;i++)
   {
    printf("\n");  printf("X%d ",y[i]); printf(" %3.2f",b[i]);
    for(j=1;j<=n+l+2;j++)
    printf("  %f ", a[i][j]);
   }
   for(i=m+1;i<=m+1;i++)
   {
    printf("\n");  printf(" G "); printf(" %3.2f",b[i]);
    for(j=1;j<=n+l+2;j++)
    printf("  %7.5f ", a[i][j]);
   }
   printf("\n");
for(i=1;i<=n+l+1;i++)
printf("--------------"); printf("\n       Z=");
   for(j=1;j<=n+l+2;j++)
   printf(" %f ", zc[j]);
   printf("\n");
   return(0);
}
```

Example 1. Find the optimum integer solution of the following all integer programming problem:

$$\text{Max. } z = x_1 + 2x_2, \quad \text{subject to the constraints:}$$
$$2x_2 \leq 7, x_1 + x_2 \leq 7, 2x_1 \leq 11, x_1 \geq 0, x_2 \geq 0.$$

Solution:

```
Enter the number of variables in objective function.......:2
Enter the number of constraints in LPP....................:3
Enter the number of inequalities having sign <= ..........:3
Enter the number of inequalities having sign >= ..........:0
Enter the number of inequalities having sign  = ..........:0

This problem is solving by simplex method.
```

```
Welcome by Simplex.
Press 1 for maximization.
Press 2 for minimization.
1
Enter the coefficients of variables in objective function:
C[1]=1
C[2]=2
Constant term:0

Enter the coefficients of constraint having <= sign.....:
Coefficient 1 :0
Coefficient 2 :2
Constant term:7

Coefficient 1 :1
Coefficient 2 :1
Constant term:7

Coefficient 1 :2
Coefficient 2 :0
Constant term:11
```

```
Position of Pivot element is (1, 2)
------------------------------------------------------------------
        Cj  0.000000 1.000000 2.000000 0.000000 0.000000 0.000000
------------------------------------------------------------------
B.V.  Cb    Xb       X1       X2       X3       X4       X5       M.R

X3  0.00  7.000000 0.000000 2.000000 1.000000 0.000000 0.000000  3.5
X4  0.00  7.000000 1.000000 1.000000 0.000000 1.000000 0.000000  7.0
X5  0.00 11.000000 2.000000 0.000000 0.000000 0.000000 1.000000  ---
------------------------------------------------------------------
        Z = 0.000000-1.000000-2.000000 0.000000 0.000000 0.000000

Position of Pivot element is (2, 1)
------------------------------------------------------------------
        Cj 0.000000 1.000000 2.000000 0.000000 0.000000 0.000000
------------------------------------------------------------------
B.V. Cb   Xb        X1       X2        X3       X4       X5       M.R
------------------------------------------------------------------
X2 2.00  3.500000 0.000000 1.000000  0.500000 0.000000 0.000000  ---
X4 0.00  3.500000 1.000000 0.000000 -0.500000 1.000000 0.000000  3.5
X5 0.00 11.000000 2.000000 0.000000  0.000000 0.000000 1.000000  5.5
------------------------------------------------------------------
        Z = 7.000000-1.000000 0.000000  1.000000 0.000000 0.000000


------------------------------------------------------------------
        Cj  0.000000 1.000000 2.000000 0.000000 0.000000 0.000000
```

```
-------------------------------------------------------------------------
B.V. Cb    Xb         X1        X2        X3        X4        X5      M.R
-------------------------------------------------------------------------
X2 2.00 3.500000 0.000000 1.000000  0.500000  0.000000 0.000000   ---
X1 1.00 3.500000 1.000000 0.000000 -0.500000  1.000000 0.000000   3.5
X5 0.00 4.000000 0.000000 0.000000  1.000000 -2.000000 1.000000   ---
-------------------------------------------------------------------------
     Z = 10.500000 0.000000 0.000000  0.500000  1.000000 0.000000
```

Optimum Solution Achieved.
The maximization of the problem is as
Maximum of objective function is=10.500000
X1=3.500000
X2=3.500000

Do you want integer solution of the problem?
Press 1 for yes!
Press 2 for no!
1

Gomory table is:

```
---------------------------------------------------------------------------------
      Cj 0.000000 1.000000 2.000000 0.000000 0.000000 0.000000 0.000000
---------------------------------------------------------------------------------
B.V. Cb    Xb        X1        X2        X3        X4        X5        G
---------------------------------------------------------------------------------
X2  2.00 3.500000 0.000000 1.000000 0.500000 0.000000 0.000000 0.000000
X1  1.00 3.500000 1.000000 0.000000-0.500000 1.000000 0.000000 0.000000
X5  0.00 4.000000 0.000000 0.000000 1.000000-2.000000 1.000000 0.000000
 G  0.00-0.500000 0.000000 0.000000-0.500000 0.000000 0.000000 1.000000
---------------------------------------------------------------------------------
    Z = 10.500000 0.000000 0.000000 0.500000 1.000000 0.000000 0.000000
Position of pivot element is (4, 3)
```

```
---------------------------------------------------------------------------------
    Cj  0.000000 1.000000 2.000000 0.000000 0.000000 0.000000 0.000000
---------------------------------------------------------------------------------
B.V.  Cb  Xb      X1          X2        X3        X4        X5        X6
---------------------------------------------------------------------------------
X2  2.00 3.000000 0.000000 1.000000 0.000000 0.000000 0.000000 1.000000
X1  1.00 4.000000 1.000000 0.000000 0.000000 1.000000 0.000000-1.000000
X5  0.00 3.000000 0.000000 0.000000 0.000000-2.000000 1.000000 2.000000
X3  0.00 1.000000 0.000000 0.000000 1.000000 0.000000 0.000000-2.000000
---------------------------------------------------------------------------------
    Z = 10.000000 0.000000 0.000000 0.000000 1.000000 0.000000 1.000000
```

Optimum Solution achieved.

```
The solution is:
X1=4.000000
X2=3.000000
The optimum value of function is:10.000000
```

Example 2. Find the optimum integer solution of the following integer programming problem:

$$\text{Min. } z = 4x_1 + 3x_2, \quad \text{subject to the constraints:}$$
$$x_1 \le 4, x_2 \le 6, 5x_1 + 3x_2 \ge 30, x_1 \ge 0, x_2 \ge 0.$$

Solution:

```
Enter the number of variables in objective function......:2
Enter the number of constraints in LPP...................:3
Enter the number of inequalities having sign <= .........:2
Enter the number of inequalities having sign >= .........:1
Enter the number of inequalities having sign  = .........:0

This problem is solved by two-phase simplex method.
Welcome by Simplex.
Press 1 for maximization.
Press 2 for minimization.
2
Enter the coefficients of variables in objective function:
C[1]=4
C[2]=3
Constant term:0

Enter the coefficients of constraint having <= sign.....:
Coefficient 1 :1
Coefficient 2 :0
Constant term:4

Coefficient 1 :0
Coefficient 2 :1
Constant term:6


Enter the coefficients of constraint having >= sign.....:
Coefficient 1 :5
Coefficient 2 :3
Constant term:30


Position of Pivot element is (1, 1)
---------------------------------------------------------------------------
     Cj   0.000000 0.000000 0.000000 0.000000 0.000000 0.000000 -1.000000
---------------------------------------------------------------------------
B.V. Cb      Xb        X1       X2       X3       X4       X5       X6    M.R
---------------------------------------------------------------------------
```

```
X3   0.00   4.000000 1.000000 0.000000 1.000000 0.000000  0.000000 0.000000 4.0
X4   0.00   6.000000 0.000000 1.000000 0.000000 1.000000  0.000000 0.000000 ---
X6  -1.00  30.000000 5.000000 3.000000 0.000000 0.000000 -1.000000 1.000000 6.0
-------------------------------------------------------------------------------
     Z = -30.000000-5.000000-3.000000 0.000000 0.000000  1.000000 0.000000
```

Position of Pivot element is (3, 2)
```
-------------------------------------------------------------------------------
     Cj  0.000000 0.000000 0.000000 0.000000 0.000000 0.000000-1.000000
-------------------------------------------------------------------------------
B.V. Cb     Xb        X1       X2       X3       X4       X5       X6     M.R
-------------------------------------------------------------------------------
X1   0.00   4.000000 1.000000 0.000000 1.000000 0.000000 0.000000 0.000000  ---
X4   0.00   6.000000 0.000000 1.000000 0.000000 1.000000 0.000000 0.000000  6.0
X6  -1.00  10.000000 0.000000 3.000000-5.000000 0.000000-1.000000 1.000000  3.3
-------------------------------------------------------------------------------
     Z = -10.000000 0.000000-3.000000 5.000000 0.000000 1.000000 0.000000
```

Pivot element could not found
```
-------------------------------------------------------------------------------
     Cj 0.000000 0.000000 0.000000 0.000000 0.000000 0.000000-1.000000
-------------------------------------------------------------------------------
B.V. Cb     Xb       X1       X2       X3       X4       X5       X6     M.R
-------------------------------------------------------------------------------
X1   0.00 4.000000 1.000000 0.000000 1.000000 0.000000 0.000000 0.000000  ----
X4   0.00 2.666667 0.000000 0.000000 1.666667 1.000000 0.333333-0.333333  ----
X2   0.00 3.333333 0.000000 1.000000-1.666667 0.000000-0.333333 0.333333  ----
-------------------------------------------------------------------------------
     Z= 0.000000 0.000000 0.000000 0.000000 0.000000 0.000000 1.000000
```


Optimization achieved.

PhaseI of simplex method ends here.
All artificial variables have been removed.

Welcome in Phase-II of the Simplex method:-
Pivot element could not found

```
----------------------------------------------------------------------
     Cj-0.000000-4.000000-3.000000 0.000000 0.000000 0.000000
----------------------------------------------------------------------
B.V.  Cb    Xb        X1       X2       X3       X4       X5     M.R
----------------------------------------------------------------------
X1  -4.00 4.000000 1.000000 0.000000 1.000000 0.000000 0.000000  ---
X4   0.00 2.666667 0.000000 0.000000 1.666667 1.000000 0.333333  ---
X2  -3.00 3.333333 0.000000 1.000000-1.666667 0.000000-0.333333  ---
----------------------------------------------------------------------
     Z = -26.000000 0.000000 0.000000 1.000000 0.000000 1.000000
```

The minimization of the problem is as
Minimum of objective function is=26.000000
X1=4.000000

```
X2=3.333333
Do you want integer solution of the problem?
Press 1 for yes!
Press 2 for no!
1

Gomory table is:


------------------------------------------------------------------
      Cj-0.000000-4.000000-3.000000 0.000000 0.000000 0.000000
------------------------------------------------------------------

B.V.  Cb    Xb       X1       X2       X3       X4       G
------------------------------------------------------------------

X1 -4.00 4.000000 1.000000 0.000000 1.000000 0.000000 0.000000
X4  0.00 2.666667 0.000000 0.000000 1.666667 1.000000 0.000000
X2 -3.00 3.333333 0.000000 1.000000-1.666667 0.000000 0.000000
 G  0.00-0.666667 0.000000 0.000000-0.666667 0.000000 1.000000
------------------------------------------------------------------

    Z =-26.000000 0.000000 0.000000 1.000000 0.000000 0.000000
Position of pivot element is (4, 3)


------------------------------------------------------------------
      Cj-0.000000-4.000000-3.000000 0.000000 0.000000 0.000000
------------------------------------------------------------------

B.V.  Cb    Xb       X1       X2       X3       X4       X5
------------------------------------------------------------------

X1 -4.00 3.000000 1.000000 0.000000 0.000000 0.000000 1.500000
X4  0.00 1.000000 0.000000 0.000000 0.000000 1.000000 2.500000
X2 -3.00 5.000000 0.000000 1.000000 0.000000 0.000000-2.500000
X3  0.00 1.000000 0.000000 0.000000 1.000000 0.000000-1.500000
------------------------------------------------------------------

    Z =-27.000000 0.000000 0.000000 0.000000 0.000000 1.500000


An optimum basic solution is attended.

X1=3.000000
X2=5.000000
The optimum value of function is-27.000000
-----------------------------------
Process exited with return value 0
Press any key to continue . . .
```

14.5 DISCUSSION AND SUMMARY

Some methods are involved in the evolution of LP problems. In practice, where typical LP model may involve a large number of variables and constraints, the only feasible way to solve such models is to use the computer. However, if solution is required in integers, LP problems become more complicated. Some reliable softwares such as AIMMS, GAMS, LONGO, MPL, OPL Studio, and X-press model are generally used to solve such type of problems. As the cost of these commercial packages are too high, but they are not efficient to solve LP problems and integer programming problems, if M-method and two-phase method are used to find initial feasible solutions.

The program discussed in this chapter is an efficient way to solve any type of linear as well as integer programming problems. It is more reliable and can be used/understood easily by the individuals. If linear or integer programming models have some $\geq$ or $=$ signs, two-phase algorithm will be executed. In our subsequent study, the scope of the research will focus as the uses of Big-M method in place of traditional two-phase method. It can be observed that the arrays declared in the main function of the program are static and their sizes are fixed as 50-float memory blocks. The sizes can be changed according to our requirement. However, if an integer programming problem required a number of iterations more than the number of declared memory blocks, an abnormal termination of computation will be occurred. In this situation, it will be required to resize the arrays. However, this difficulty can be handled by the concept of dynamic memory allocations or uses of vector classes in place of arrays. This version of the program will be enabling to extend or shrink the memory sizes of declared variables on the time of execution. Suggested amendments by the author are being postponed for his subsequent study. However, interested readers have ample opportunities for further explorations.

ACKNOWLEDGMENT

Authors would also like to express his sincere gratitude to Dr. H.G. Sharma, Former Professor, Department of Mathematics, Indian Institute of Technology, Roorkee, India, for inviting to this work.

Authors are also thankful to Dr. N. K. Baunthiyal and Dr. A. K. Choudhary, Department of Applied Science, Swami Rama Himalayan University, Dehradun, India, for fruitful discussions and suggestions.

REFERENCES

1. K.H. Borgwardt. *The Simplex Method: A Probabilistic Analysis.* Number 1 in Algorithms and Combinatorics. Springer-Verlag, New York, 1980.
2. E. W. Cheney and A. A. Goldstein. Newton's Method for Convex Programming and Tchebycheff Approximation. *Numer. Math.*, 1:253–268, 1959.
3. G. B. Dantzig. *Maximization of a Linear Function of Variables Subject to Linear Inequalities.* Cowles Commission Monograph No. 13. John Wiley & Sons, Inc., New York; Chapman & Hall, Ltd., London, 1951.
4. G. B. Dantzig. *Linear Programming and Extensions.* Princeton University Press, Princeton, NJ, 1963.
5. G. B. Dantzig. Impact of linear programming on computer development. In: D.V. Chudnovsky and R.D. Jenks (Eds) *Computers in Mathematics, Lecture Notes in Pure and Applied Mathematics*, vol. 125, pp. 233–240. Dekker, New York, 1990.

6. M. A. H. Dempster and R. R. Merkovsky. A practical geometrically convergent cutting plane algorithm. *SIAM J. Numer. Anal.*, 32(2):631–644, 1995.

7. C. I. Fábián, G. Mitra, and D. Roman. Processing second-order stochastic dominance models using cutting-plane representations. *Math. Program.*, 130(1, Ser. A):33–57, 2011.

8. P. C. Gilmore and R. E. Gomory. A linear programming approach to the cutting-stock problem. *Oper. Res.*, 9:849–859, 1961.

9. P. C. Gilmore and R. E. Gomory. A linear programming approach to the cutting-stock problem-part ii. *Oper. Res.*, 11:863–888, 1963.

10. A. Hoffman, M. Mannos, D. Sokolowsky, and N. Wiegmann. Computational experience in solving linear programs. *J. Soc. Indust. Appl. Math.*, 1:17–33, 1953.

11. R.S. Hussain. Trim loss minimization and reel cutting at paper mill. *Int. J. Eng. Res. Dev.*, 4(3):13–22, 2012.

12. J. E. Kelley, Jr. The cutting-plane method for solving convex programs. *J. Soc. Indust. Appl. Math.*, 8:703–712, 1960.

13. V. Klee and G. J. Minty. How good is the simplex algorithm? In: O. Shisha (Ed.) *Inequalities III*, pp. 159–175. Academic Press, New York, 1972.

14. W. K. Klein Haneveld and M. H. van der Vlerk. Integrated chance constraints: Reduced forms and an algorithm. *Comput. Manag. Sci.*, 3(4):245–269, 2006.

15. A. Kunzi-Bay and J. Mayer. Computational aspects of minimizing conditional value-at-risk. *Comput. Manag. Sci.*, 3:3–27, 2006.

16. J. Luedtke. New formulations for optimization under stochastic dominance constraints. *SIAM J. Optim.*, 19(3):1433–1450, 2008.

17. I. Maros. *Computational Techniques of the Simplex Method*. Volume 61, International Series in Operations Research & Management Science. Kluwer Academic Publishers, Boston, MA, 2003. With a foreword by András Prékopa.

18. D. K. Nadar. Some applications of simplex method. *Int. J. Eng. Res. Rev.*, 4(1):60–63, 2016.

19. E. Neuman. Linear Programming with MATLAB. Math 472/CS 472.

20. G. Rudolf and A. Ruszczyński. Optimization problems with second order stochastic dominance constraints: duality, compact formulations, and cut generation methods. *SIAM J. Optim.*, 19(3):1326–1343, 2008.

21. D. A. Spielman and S. -H. Teng. Smoothed analysis of algorithms: Why the simplex algorithm usually takes polynomial time. *J. ACM*, 51(3):385–463, 2004.

22. H. A. Taha. *Operations Researchan Introduction*. Macmillan Co., New York, third edition, 1982.

15 Fuzzy EOQ Model with Reliability-Induced Demand and Defuzzification by Graded Mean Integration

Neelanjana Rajput and R.K. Pandey
Hemvati Nandan Bahuguna Garhwal University

Anand Chauhan
Graphic Era (Deemed to be University)

CONTENTS

15.1 INTRODUCTION

There is an imperative role of demand in the economic order quantity (EOQ) models for the production process. As customer satisfaction plays a key role in management of the inventory system, the share in the market, productivity, as well as the total profit of the company increase. We have developed an EOQ model of the inventory system for analyzing the effect of demand function which is dependent on reliability as well as time. The model is developed for perishable products with constant deterioration rate. Due to this, there is a shortage of goods in the system, and there is backlogging which is of partial nature. The classical(crisp) inventory control model with reliability influence demand and partially backlogged items is to be considered. The primary motive of an efficient inventory management is to provide a suitable customer service, thereby keeping a low cost of the inventory system. The cost factor is to consider as a fuzzy random variable. Here, the ordering cost and reliability of production process are the decision variables. Thus, the aim of this chapter is to minimization the cost of the inventory system. This model has been developed for both crisp and fuzzy system. Due to fuzzy parameters, the model become a fuzzy quantity and its defuzzify by sign distance method. The facts used in this research work have been exemplified by the use of appropriate numerical example. Given the current situation, the competition level is obscenely outrageous, so the acceptability of the inventory is prime important in the market. The production and durability of reliable products, which help to create a good reputation in the market, are also importance. Generally, it is observed that a product of high quality or meeting all the guidelines required to be observed a high-grade product is one that is a genuine product, which will firmly ensure an enduring market. The leading objective of reliability is the depletion of failures over a time period considered. Thus, for real circumstances, observing product reliability and time-dependent demand are the major realistic substitute for the complete inventory system. The effect due to deterioration on the inventory system results in a shortage of things. Various researchers have provided their thoughts in this regard. Due to the scarcity of goods on the market, the researchers suggested the concept of totally unsatisfied demand, resulting in a complete backlog of products. Considering the present situation that the customers have numerous choices in hand but they focus primarily on the goods which have greater reliability so as to reduce the expenses on maintenance of goods. A globally recognized brand always earns a higher profit and also becomes a dependable product of repute. The demand of such products increases with an increase in the time span. The sale of the goods with higher reliability factor generally increases as the time passes by because such goods have an impact of positive nature on the prospects. Now, we study an EOQ model in which the demand rate depends on reliability of the product as well as the time. This model has been considered keeping in view the effect of deterioration, and shortages have also been allowed in the proposed model. The nature of shortages is such that they are partially backlogged with lost sale. This model is developed for both crisp and fuzzy environments. Sections 15.5.1 and 15.5.2 give a detailed description of the model in both the environments. Both the models are validated by considering appropriate numerical examples.

15.2 LITERATURE REVIEW

Padmanabhan and Vrat [14] proposed an EOQ model with partial backlogging and demand rate is a function of the order quantity which is already backlogged at that time. Sarkar [22] proposed a model on the concept of demand which is stock dependent. Wee [28] erupted the idea of piece and demand which depends on time. In this sequence, there is a model with generalized demand function was introduced by Hung [9]. Abdulla et al. [1] erupted his research on the inventory model with effect of reliable products. Some research works are done by Krishnamoorthi and Panayappan [11] on imperfect items with shortage in products. Sarkar et al. [23] also proposed an inventory model with reliability factor on imperfect items. Now, we consider the crisp inventory models with the effect of deterioration; in real situation, deterioration has a great effect on the inventory system. Skouri et al. [24] developed an inventory model where he discussed about perishable products and order level. Some research work done by Papachristos and Skouri [17] on inventory system with deterioration as a function of time. In this sequence, Dye et al. [6] introduced a model of inventory system with general type deterioration rate. Panda et al. [16] introduced his model for single-item order level, where the demand rate is ramp-type time-dependent function and shortages are not allowed. Liang and Zhou [13] discussed his research work on inventory model with two separate warehouses, with lower rate of deterioration and a linear demand rate. A single-item single-period EOQ model introduced by Giri et al. [7] where they discussed a model with shortage, ramp-type demand and Weibull deterioration distribution. Later, the fuzzy concepts are applied on classical inventory model with ramp-type demand and Weibull distribution by Pal et al. [15] for finite time horizon. Chu et al. [3] proposed an inventory models with shortages, partially backlog and lost sales. An inventory model for stock-out period proposed by Park [18] is able to reduce the lost sale and backorder. Hsu [8] discussed the supply chain network with a reliability evaluation method to evaluate the role of plants under fuzzy demand. Some customers would not like to wait for backlogging during the shortage period, so Wang et al. [27] studied on an inventory modeling for deteriorating items with shortages and partial backlogging. N. Rajput [20] proposed an inventory model with different types of demand function discussed the importance of fuzzy parameters in healthcare industries. Wu [29] proposed an EOQ model of inventory system with shortages and partial backlogging, and the backlogging rate is variable and is dependent on waiting time for next replenishment. Taleizadeh et al. [25] developed a model with special sale price and partial backlogging, where a constant unit purchase cost is one of the main assumptions in this EOQ model. An EPQ model is developed by P. Anita et al. [2] with Weibull distribution deterioration and stock dependent demand rate in fuzzy environment. In this work machine reliability, flexibility and packing cost are considered and defuzzification with the help of GMI method. N. Rajput [21] in her research work used sign distance method for defuzzification and the optimal inventory cost. B. Khara et al. [10] deal with an inventory model which has an imperfect production process; demand depends on selling price and reliability of the products. They deal with many problems in this model such as long-run process, lack of labor, machinery, and technology. M. Pervin et al. [19] proposed a deterministic control model with deterioration, stochastic deterioration, and time-dependent demand. In this chapter, author investigates the optimal retailer's replenishment decisions to get minimum total inventory cost. A. H. Tai et al. [26] discussed the inventory model with two replenishment policies – (i) quantity-based and (ii) time-based policies; and two inspection positions – (i) one inspection and(ii) continuous monitoring. In the recent research paper work introduced by Guiping Li et al. [12], they studied a joint pricing, replenishment, and preservation technology investment problem for non-instantaneous deteriorating items with price dependent demand and waiting-time-dependent backlog rates.

15.3 PRELIMINARIES

Definition 1 The Fuzzy Number: Let X be a crisp set of objects, whose elements are denoted by x. Membership function in a crisp subset S of X is written as $\mu_{\tilde{S}}(x)$

$$\mu_{\tilde{S}}(x) = \begin{cases} 1, & x \in S \\ 0, & \text{else} \end{cases}$$

A fuzzy set $\tilde{S} \subseteq X$ is said to be normal if $\exists$ at least one $x_0 \in X$ such that $\mu_{\tilde{S}}(x_0) = 1$. A fuzzy set $\tilde{S} \subseteq X$ is said to be convex if $\forall x_1 \in X, \forall x_2 \in X$, and $\lambda \in [0,1]$ such that

$$\mu_{\tilde{S}}(\lambda x_1 + (1 - \lambda)x_2) \geq \min(\mu_{\tilde{S}}(x_1), \mu_{\tilde{S}}(x_2))$$

A covex normalized fuzzy subset $\tilde{x} \in R$ with membership function $\mu_{\tilde{x}} : R \to [0,1]$ is called a fuzzy number (Dubois Prade [5]).

Definition 2 L-R Representation of Fuzzy Numbers: A fuzzy number $\tilde{S} \subseteq R$ is said to be an L-R-type fuzzy number if its membership function is given by Dubois Prade [5].

$$\mu_{\tilde{S}(x)} = \begin{cases} L\left(\frac{m-x}{\alpha}\right), \text{for} & x \leqslant m, \alpha > 0 \\ R\left(\frac{x-m}{\beta}\right), \text{for} & x \geqslant m, \beta > 0 \end{cases}$$

Definition 3 α-Level set: α-Level set(interval of confidence at level α) of a fuzzy set $\tilde{S}$ in X is a crisp subset of X denoted by $S(\alpha)$ and is defined by $S(\alpha) = \{x \in X / \mu_{\tilde{S}(x) \geq \alpha}\} \forall \alpha \in [0,1]$. Let F be the set of all fuzzy numbers.

Then, for any $\tilde{P}, \tilde{Q} \in F$ and for any $\lambda \in R$,

$(\tilde{P} * \tilde{Q})(\alpha) = P(\alpha) * Q(\alpha), (\lambda \tilde{P})(\alpha) = \lambda P(\alpha)$ (Bector Chandre [4])

Definition 4 Trapezoidal Membership Function: A fuzzy number is a convex fuzzy set, defined on given interval of real numbers, each with a grade of membership between 0 and 1. A trapezoidal fuzzy number $\tilde{\eta} = (\eta_1, \eta_2, \eta_3, \eta_4)$. The representation of its membership function is shown in Figure 15.1

$$\mu_{\tilde{\eta}(x)} = \begin{cases} \frac{x - \eta_1}{\eta_2 - \eta_1}, & \text{if } \eta_1 < x < \eta_2 \\ 1, & \text{if } \eta_2 < x < \eta_3 \\ \frac{\eta_4 - x}{\eta_4 - \eta_3}, & \text{if } \eta_3 < x < \eta_4 \\ 0, & \text{otherwise} \end{cases}$$

In the case of trapezoidal fuzzy number, the membership value is obtained for a particular range, which is more realistic with defuzzification method.

Definition 5 Arithmetic Operations: Suppose $\tilde{\tau} = (\tau_1, \tau_2, \tau_3, \tau_4)$ and $\tilde{\varsigma} = (\varsigma_1, \varsigma_2, \varsigma_3, \varsigma_4)$ are two trapezoidal fuzzy numbers, then arithmetical operations are defined as:

1: Addition: $\tilde{\tau} \oplus \tilde{\varsigma} = (\tau_1 + \varsigma_1, \tau_2 + \varsigma_2, \tau_3 + \varsigma_3, \tau_4 + \varsigma_4)$

2: Subtraction: $\tilde{\tau} \ominus \tilde{\varsigma} = (\tau_1 - \varsigma_4, \tau_2 - \varsigma_3, \tau_3 - \varsigma_2, \tau_4 - \varsigma_1)$

3: Multiplication: $\tilde{\tau} \otimes \tilde{\varsigma} = (\Omega_1, \Omega_2, \Omega_3, \Omega_4)$
where $A = \{\tau_1 \varsigma_1, \tau_1 \varsigma_4, \tau_4 \varsigma_1, \tau_4 \varsigma_4\}$ and $B = \{\tau_2 \varsigma_2, \tau_2 \varsigma_3, \tau_3 \varsigma_2, \tau_3 \varsigma_3\}$, $\Omega_1 = \min(A)$, $\Omega_2 = \min(B), \Omega_3 = \max(A), \Omega_4 = \max(B)$;

5: If $\tau_1, \tau_2, \tau_3, \tau_4, \varsigma_1, \varsigma_2, \varsigma_3$, and ς_4 are all nonzero positive real numbers, then division of two trapezoidal fuzzy number is

$$\tilde{\tau} \oslash \tilde{\varsigma} = \left(\frac{\tau_1}{\varsigma_4}, \frac{\tau_2}{\varsigma_3}, \frac{\tau_3}{\varsigma_2}, \frac{\tau_4}{\varsigma_1} \right)$$

Definition 6 Signed Distance Method: If $\eta > 0$, then distance between η and 0 is $d_o(\eta, 0) = \eta$. If $\eta < 0$, then distance between η and 0 is $-d_o(\eta, 0) = -\eta$. $d_o(\eta, 0)$ is signed distance between a and 0. We have that the signed distance of the interval $[S_L(\alpha), S_R(\alpha)]$ measured from 0 is,

$$d_o([S_L(\alpha), S_R(\alpha)], 0) = \frac{1}{2}[[S_L(\alpha), 0] + [S_R(\alpha), 0]] = \frac{1}{2}[S_L(\alpha) + S_R(\alpha)]$$

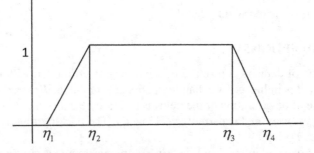

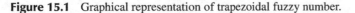

Figure 15.1 Graphical representation of trapezoidal fuzzy number.

Since $\tilde{S}$ is a fuzzy number, $S_L(\alpha)$ and $S_R(\alpha)$ are left and right α-cut of $\tilde{S}$ (Definition 2.2) and they are integrable and exist for $\alpha\varepsilon(0,1)$; then, the distance between $\tilde{S}$ and 0, which is known as signed distance, is given by

$$d(\tilde{S},0) = \frac{1}{2}\int_0^1 [S_L(\alpha) + S_R(\alpha)]d\alpha$$

$$d(\tilde{S},0) = \frac{1}{2}\int_0^1 [s_1 + (s_2 - s_1)\alpha + s_4 - (s_4 - s_3)\alpha]d\alpha = \frac{1}{4}(s_1 + s_2 + s_3 + s_4)$$

Definition 7 Graded Mean Integration Method (GMI): Graded Mean Integration method has been introduced by S.H. Chen and C.H. Hsieh in (1999). Let $\tilde{L} = (s_1, s_2, s_3, s_4)$, be a trapezoidal fuzzy number, then the GMI representation of $\tilde{L}$, is defined as

$$G(\tilde{L}) = \frac{\frac{1}{2}\int_0^1 \alpha[L_L(\alpha) + L_R(\alpha)]d\alpha}{\int_0^1 \alpha\,d\alpha} = \frac{1}{6}(s_1 + 2s_2 + 2s_3 + s_4)$$

15.4 NOTATIONS AND ASSUMPTIONS

15.4.1 NOTATIONS

μ = Demand Parameter.
θ = Rate of deterioration taken as a constant.
ω = Rate of reliability factor.
b = Rate of backlogging.
$I(t)$ = Level of inventory at time t.
c_D = Cost of deterioration.
c_O = Ordering cost for items.
c_H = Cost of holding inventory.
c_S = Shortage cost.
c_L = Cost of lost sales.
t_0 = Time when inventory level becomes zero.
T = One complete cycle under consideration.
TC = Total cost of the inventory system.
$\widetilde{TC}$ = Fuzzy total cost.
$(\sim)$ = tilde symbol is representation of fuzzy.

15.4.2 ASSUMPTIONS

1. The flow of demand is taken to be a non-negative function with respect to time t and is influence by reliability parameter ω that is $d(t) = \mu t \omega^t$.
2. The rate of deterioration of the items is taken to be constant.
3. Shortages occur and is partially backlogged. The backlog function is to be consider $f(x) = e^{-bx}$.
4. Time horizon is infinite and there is a replenishment of good in the next cycle.

15.5 FORMULATION OF MATHEMATICAL MODEL

15.5.1 CRISP MODEL

This model of the inventory system begins with shortage of goods. There is a replenishment of the goods when the time $t = 0$ and the level of the inventory is highest at this time. When the time approaches t_0, there is a reduction in the level of the inventory system. From the Figure (15.2), when time is $t = t_0$, the level of the inventory system becomes equivalent to zero and there is a shortage of goods in the inventory system up to time $t = T$.

Thus, the levels of inventory are as follows:

$$\frac{d}{dt}I(t) + \theta I(t) = -\mu t \omega^t \text{ for } 0 < t < t_0$$

$$\frac{d}{dt}I(t) = -f(T-t)\mu t \omega^t \text{ for } t_0 < t < T$$

After integrating above equations, we get inventory levels:

$$I(t) = \mu e^{-t\theta} \int_t^{t_0} s\omega^s e^{s\theta} ds \text{ for } 0 < t < t_0 \text{ and}$$

$$I(t) = \mu \int_t^{t_0} f(T-s)s\omega^s ds \text{ for } t_0 < t < T$$

Cost of deterioration (DC) in the time interval $[0, t_0]$ is;

$$DC = c_D \times \left[I(0) - \int_0^{t_0} \mu s\omega^s ds \right] = c_D \int_0^{t_0} \mu s\omega^s (e^{s\theta} - 1) ds$$

$$= c_D \mu \left[\frac{(t_0\log\omega - 1)\omega^{t_0}(e^{t_0\theta}(\log\omega)^2 - (\theta + \log\omega)^2)}{(\log\omega)^2(\theta + \log\omega)^2} \right.$$

$$\left. + \frac{\theta t_0 e^{\theta t_0} \omega^{t_0}}{(\theta + \log\omega)^2} + \frac{\theta(2\log\omega + \theta)}{(\log\omega)^2(\theta + \log\omega)^2} \right]$$

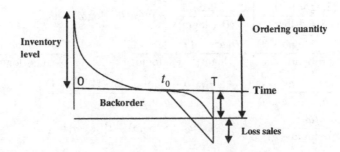

Figure 15.2 EOQ model representation with inventory and time.

Cost of holding inventory (HC) in the time interval $[0, t_0]$ is;

$$HC = c_H \int_0^{t_0} I(t)ds = \mu c_H \int_0^{t_0} e^{-t\theta} \int_t^{t_0} s\omega^s e^{s\theta} \, ds \, dt = \frac{\mu c_H}{(\theta + \log\omega)^2}$$

$$\times \left[\frac{\omega^{t_0}(t_0(\theta + \log\omega) - 1)((e^{t_0\theta} - 1)\log\omega - \theta)}{\theta \log\omega} + \frac{\theta(\omega^{t_0} - 1)}{(\log\omega)^2} + \frac{\omega^{t_0} - 2}{\log\omega} \right]$$

Cost of shortage items (SC) in the time interval $[t_0, T]$ is

$$SC = c_S \int_{t_0}^T -I(t)dt = \mu c_S \int_{t_0}^T \int_{t_0}^t f(t-s)s\omega^s \, ds \, dt$$

$$= \mu c_S \left[\frac{(T\omega^T - t_0\omega^{t_0})}{(b + \log\omega)\log\omega} - \frac{(2\log\omega + b)(\omega^T - \omega^{t_0})}{(\log\omega)^2(b + \log\omega)^2} \right.$$

$$\left. + \omega^{t_0}(t_0(b + \log\omega) - 1)\frac{(e^{b(t_0 - T)} - 1)}{b(b + \log\omega)^2} \right]$$

Cost of lost sales (LS) in the time interval $[t_0, T]$ is

$$LS = \mu c_L \int_{t_0}^T [1 - f(T-t)]t\omega^t \, dt$$

$$= \frac{\mu c_L}{(\log\omega)^2(\log\omega + b)^2} \left[e^{b(t_0 - T)}\omega^{t_0}[t_0(b + \log\omega) - 1](\log\omega)^2 + \omega^{t_0}(1 - t_0\log\omega) \right.$$

$$\left. \times (\log\omega + b)^2 + b\omega^T[T\log\omega(\log\omega + b) - (2\log\omega + b)] \right]$$

Therefore, the total cost for crisp inventory is

$$TC(t_0) = \frac{1}{T}[c_O + DC + HC + SC + LS]$$

$$TC(t_0) = \frac{1}{T}\left[c_O + c_D\mu \left[\frac{(t_0\log\omega - 1)\omega^{t_0}(e^{t_0\theta}(\log\omega)^2 - (\theta + \log\omega)^2)}{(\log\omega)^2(\theta + \log\omega)^2} + \frac{\theta t_0 e^{\theta t_0}\omega^{t_0}}{(\theta + \log\omega)^2} \right. \right.$$

$$+ \frac{\theta(2\log\omega + \theta)}{(\log\omega)^2(\theta + \log\omega)^2} \right] + \frac{\mu c_H}{(\theta + \log\omega)^2}\left[\omega^{t_0}(t_0(\theta + \log\omega) - 1) \right.$$

$$\times \frac{((e^{t_0\theta} - 1)\log\omega - \theta)}{\theta \log\omega} + \frac{\theta(\omega^{t_0} - 1)}{(\log\omega)^2} + \frac{\omega^{t_0} - 2}{\log\omega} \right] + \mu c_S \left[\frac{(T\omega^T - t_0\omega^{t_0})}{(b + \log\omega)\log\omega} \right.$$

$$- \frac{(2\log\omega + b)(\omega^T - \omega^{t_0})}{(\log\omega)^2(b + \log\omega)^2} + \frac{\omega^{t_0}(t_0(b + \log\omega) - 1)(e^{b(t_0 - T)} - 1)}{b(b + \log\omega)^2} \right]$$

$$+ \frac{\mu c_L}{(\log\omega)^2(\log\omega + b)^2}\left[e^{b(t_0 - T)}\omega^{t_0}[t_0(b + \log\omega) - 1](\log\omega)^2 \right.$$

$$\left. \left. + \omega^{t_0}(1 - t_0\log\omega)(\log\omega + b)^2 + b\omega^T[T\log\omega(\log\omega + b) - (2\log\omega + b)] \right] \right] \quad (15.1)$$

15.5.2 FUZZY MODEL

The cost functions of this model have been assumed to be trapezoidal fuzzy numbers, and they are defuzzified by using signed distance method. This model of the inventory system begins with shortage of goods. From Figure (15.2), there is a replenishment of the goods when the time $t = 0$ and the level of the inventory is highest at this time. When the time approaches t_0, there is a reduction in the level of the inventory system. When time is $t = t_0$, the level of the inventory system becomes equivalent to zero, and there is a shortage of goods in the inventory system up to time $t = T$. Therefore, the trapezoidal fuzzy ordering cost, fuzzy deterioration cost, fuzzy holding cost, fuzzy shortage cost, and fuzzy cost for lost sale are, respectively,

$$\tilde{c}_O = (c_{O_1}, c_{O_2}, c_{O_3}, c_{O_4}), \tilde{c}_D = (c_{D_1}, c_{D_2}, c_{D_3}, c_{D_4}),$$
$$\tilde{c}_H = (c_{H_1}, c_{H_2}, c_{H_3}, c_{H_4}), \tilde{c}_S = (c_{S_1}, c_{S_2}, c_{S_3}, c_{S_4}),$$
$$\tilde{c}_L = (c_{L_1}, c_{L_2}, c_{L_3}, c_{L_4}).$$

Fuzzy cost of deterioration $(\widetilde{DC})$ in the time interval $[0, t_0]$ is

$$\widetilde{DC} = \tilde{c}_D[I(0) - \int_0^{t_0} \mu s \omega^s ds] = \tilde{c}_D \int_0^{t_0} \mu s \omega^s (e^{s\theta} - 1) ds$$

$$= \mu(c_{D_1}, c_{D_2}, c_{D_3}, c_{D_4}) \otimes \left[\frac{(t_0 \log \omega - 1) \omega^{t_0} (e^{t_0\theta} (\log \omega)^2 - (\theta + \log \omega)^2)}{(\log \omega)^2 (\theta + \log \omega)^2} \right.$$

$$\left. \oplus \frac{\theta t_0 e^{\theta t_0} \omega^{t_0}}{(\theta + \log \omega)^2} \oplus \frac{\theta(2\log \omega + \theta)}{(\log \omega)^2 (\theta + \log \omega)^2} \right]$$

Fuzzy cost of holding inventory $(\widetilde{HC})$ in the time interval $[0, t_0]$ is

$$\widetilde{HC} = \tilde{c}_H \int_0^{t_0} I(t) dt = \mu \tilde{c}_H \int_0^{t_0} e^{-t\theta} \int_t^{t_0} s \omega^s e^{s\theta} ds dt = \frac{\mu(c_{H_1}, c_{H_2}, c_{H_3}, c_{H_4})}{(\theta + \log \omega)^2}$$

$$\otimes \left[\frac{\omega^{t_0}(t_0(\theta + \log \omega) - 1)((e^{t_0\theta} - 1)\log \omega - \theta)}{\theta \log \omega} \oplus \frac{\theta(\omega^{t_0} - 1)}{(\log \omega)^2} \oplus \frac{\omega^{t_0} - 2}{\log \omega} \right]$$

Fuzzy cost of shortage items $(\widetilde{SC})$ in the time interval $[t_0, T]$ is

$$\widetilde{SC} = \tilde{c}_S \int_{t_0}^T -I(t) dt = \mu \tilde{c}_S \int_{t_0}^T \int_{t_0}^t f(t-s) s \omega^s ds dt$$

$$= \mu(c_{S_1}, c_{S_2}, c_{S_3}, c_{S_4}) \otimes \left[\frac{(T\omega^T - t_0\omega^{t_0})}{(b + \log \omega)\log \omega} \ominus \frac{(2\log \omega + b)(\omega^T - \omega^{t_0})}{(\log \omega)^2 (b + \log \omega)^2} \right.$$

$$\left. \oplus \frac{\omega^{t_0}(t_0(b + \log \omega) - 1)(e^{b(t_0-T)} - 1)}{b(+\log \omega)^2} \right]$$

Fuzzy cost of lost sales $(\widetilde{LS})$ in the time interval $[t_0, T]$ is;

$$\widetilde{LS} = \mu \tilde{c}_L \int_{t_0}^{T} [1 - f(T-t)] t \omega^t dt$$

$$= \frac{\mu(c_{L_1}, c_{L_2}, c_{L_3}, c_{L_4})}{(\log\omega)^2(\log\omega+b)^2} \otimes \left[e^{b(t_0-T)} \omega^{t_0} [t_0(b+\log\omega) - 1](\log\omega)^2 \right.$$

$$\left. \oplus \omega^{t_0}(1 - t_0\log\omega)(\log\omega+b)^2 \oplus b\omega^T [T\log\omega(\log\omega+)b - (2\log\omega+b)] \right]$$

Therefore, total fuzzy cost is

$$\widetilde{TC}(t_0) = \frac{1}{T}\left[\tilde{c}_O \oplus \widetilde{DC} \oplus \widetilde{HC} \oplus \widetilde{SC} \oplus \widetilde{LS} \right]$$

$$= \frac{1}{T}\left[(c_{O_1}, c_{O_2}, c_{O_3}, c_{O_4}) \oplus \widetilde{DC} \oplus \widetilde{HC} \oplus \widetilde{SC} \oplus \widetilde{LS} \right]$$

Using arithmetic operations of fuzzy numbers defined in Definition 5 and with the help of Definition 6, we get fuzzy optimal total cost is

$$\widetilde{TC}(t_0) = \left[\frac{1}{6T}\left((c_{O_1} + 2c_{O_2} + 2c_{O_3} + c_{O_4}) + \frac{(c_{D_1} + 2c_{D_2} + 2c_{D_3} + c_{D_4})\mu}{(\mathrm{Log}[\omega] + \theta)^2(\mathrm{Log}[\theta])^2} \right.\right.$$

$$\times \left(\omega^{t_0}(t_0\mathrm{Log}[\omega] - 1)\left(E^{t_0\theta}(\mathrm{Log}[\omega])^2 - (\mathrm{Log}[\omega] + \theta)^2 \right) + \theta t_0 \omega^{t_0} E^{t_0\theta}(\mathrm{Log}[\omega])^2 \right.$$

$$\left. - \theta(2\mathrm{Log}[\omega] + \theta) \right) + \frac{\mu(c_{H_1} + 2c_{H_2} + 2c_{H_3} + c_{H_4})}{(\mathrm{Log}[\omega] + \theta)^2} \left(\frac{1}{\theta\mathrm{Log}[\omega]} \omega^{t_0}(t_0(\mathrm{Log}[\omega]) \right.$$

$$\left. + \theta) - 1)\left(\left(E^{t_0\theta} - 1 \right)\mathrm{Log}[\omega] - \theta \right) \frac{\theta(\omega^{t_0} - 1)}{(\mathrm{Log}[\omega])^2} + \frac{\omega^{t_0} - 2}{\mathrm{Log}[\omega]} \right)$$

$$+ \mu(c_{S_1} + 2c_{S_2} + 2c_{S_3} + c_{S_4})\left(\frac{T\omega^T - t_0\omega^{t_0}}{(\mathrm{Log}[\omega] + b)\mathrm{Log}[\omega]} - \frac{(2\mathrm{Log}[\omega] + b)(\omega^T - \omega^{t_0})}{(\mathrm{Log}[\omega] + b)^2(\mathrm{Log}[\omega])^2} \right.$$

$$\left. + \omega^{t_0}(t_0(\mathrm{Log}[\omega] + b) - 1)\frac{\left(E^{b(t_0-T)} - 1 \right)}{b(\mathrm{Log}[\omega] + b)^2} \right) + \frac{\mu(c_{L_1} + 2c_{L_2} + 2c_{L_3} + c_{L_4})}{(\mathrm{Log}[\omega] + b)^2(\mathrm{Log}[\omega])^2}$$

$$\times \left(E^{b(t_0-T)}\omega^{t_0}(t_0(\mathrm{Log}[\omega] + b) - 1)(\mathrm{Log}[\omega])^2 + \omega^{t_0}(1 - t_0\mathrm{Log}[\omega])(\mathrm{Log}[\omega] + b)^2 \right.$$

$$\left.\left.\left. + b\omega^T(T\mathrm{Log}[\omega](\mathrm{Log}[\omega] + b) - (2\mathrm{Log}[\omega] + b)) \right)\right) \right] \qquad (15.2)$$

15.6 OPTIMALITY CRITERIA

To get the optimal value of the average total cost $TC(t_0)$ with respect to t_0. The method used to minimize t_0 as follows:

Step 1: First start with $TC(t_0)$.

Step 2: Take the first derivative of $TC(t_0)$ with respect to continuous time variable t_0 and equate the results to zero ($dTC/dt_0 = 0$) and get the critical points (t_0).

Step 3: Evaluate $TC(t_0)$ with the help of t_0, which is found in Step 2.

Step 4: Repeat Steps 2 and 3, until we get $\frac{d^2TC(t_0)}{dt_0^2} > 0$ for t_0. Thus, we get the optimal number t_0^* for which the total cost function $TC(t_0)$ is convex (shown in Appendix)

Step 5: Now, we can find the optimal total cost $TC(t_0)$.

15.7 NUMERICAL EXAMPLE

Suppose we have an example production factory. $\mu = 0.5$, $\omega = 0.4$, $c_O = 300$, $c_H =$ Rs. 100 per unit, $c_D =$ Rs. 75 per unit, $\theta = 0.005$, $c_S =$ Rs. 150, $c_L =$ Rs. 50, $T = 2$ year, and $b = 0.5$. From the equation 15.1 and 15.2, we get optimal total cost in different environments as shown in Table (15.1);

From the Table (15.1), we see that the decrease in total cost will 0.62% from crisp model to fuzzy model. The total cost will decrease and gives more appropriate result with fuzzy parameters(from Figure 15.3).

Table 15.1

Effect on Total Cost in Crisp and Fuzzy Models

t_0	1.1	1.15	1.17731	1.2	1.24
Crisp total cost	161.205	161.148	161.141	161.146	161.184
Fuzzy total cost	160.185	160.133	160.127	160.135	160.175

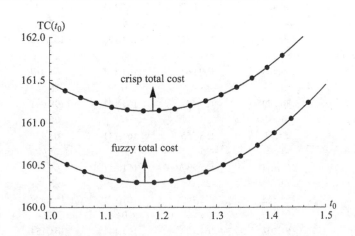

Figure 15.3 Total cost in crisp and fuzzy sense.

15.8   SENSITIVITY ANALYSIS

15.8.1   FOR CRISP MODEL

We have to study that how the parameters affect the optimal solution in crisp environment. Due to some uncertainties in different situations, there are some changes in the value of parameters occurs. To examine the ramification of these changes, the sensitivity analysis will work as a tool in decision-making. Using the numerical example given in the previous Section 15.7, sensitivity analysis with respect to various parameters on total system cost is carried out as follows.

The main conclusion of sensitivity analysis revealed in Table 15.2 as

- There is no effect in total cost due to any change in deterioration cost (c_D) (Figure 15.4) and deterioration rate (θ) (Figure 15.5).
- The effect of changing in the ordering cost (c_O) (Figure 15.6), holding cost (c_H) (Figure 15.7), shortage cost (c_S) (Figure 15.8), and lost sale cost (c_L) (Figure 15.9) is also relatively sensitive to the total cost ($TC(t_0)$).
- There are minor changes in $TC(t_0)$ due to change in rate of reliability (ω)(Figure 15.10), demand parameter (μ) (Figure 15.11), and rate of backlogging (b) (Figure 15.12). Total cost is increasing with respect to these parameters.

Table 15.2
Percentage Effect on Crisp Parameters

% change in c_D	−50	−25	0	25	50
$TC(t_0)$	161.141	161.141	161.141	161.141	161.141
% change in c_H	−50	−25	0	25	50
$TC(t_0)$	156.486	158.037	161.141	162.692	164.244
% change in c_S	−50	−25	0	25	50
$TC(t_0)$	159.026	160.083	161.141	162.198	163.255
% change in c_L	−50	−25	0	25	50
$TC(t_0)$	160.788	160.964	161.141	161.317	161.493
% change in c_O	−50	−25	0	25	50
$TC(t_0)$	86.1405	123.641	161.141	198.641	236.141
% change in μ	−50	−25	0	25	50
$TC(t_0)$	155.57	158.355	161.141	163.926	166.711
% change in θ	−50	−25	0	25	50
$TC(t_0)$	161.134	161.137	161.141	161.144	161.147
% change in ω	−50	−25	0	25	50
$TC(t_0)$	155.363	158.155	161.141	164.309	167.65
% change in b	−50	−25	0	25	50
$TC(t_0)$	161.102	161.123	161.141	161.154	161.165

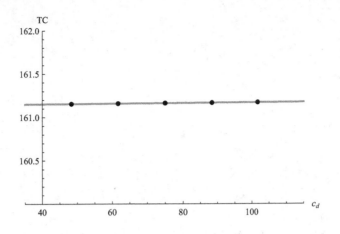

Figure 15.4 Change in total cost due to deterioration cost.

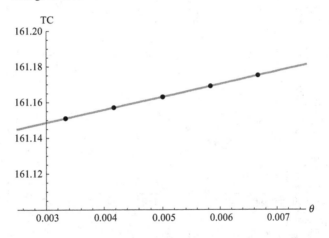

Figure 15.5 Change in total cost due to holding cost.

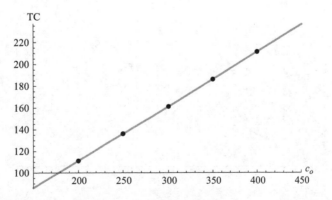

Figure 15.6 Change in total cost due to shortage cost.

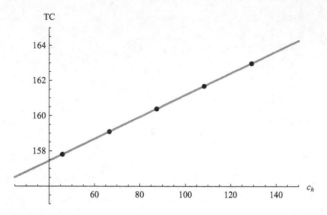

Figure 15.7 Change in total cost due to lost sale cost.

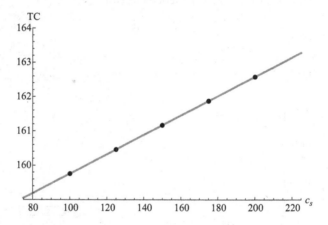

Figure 15.8 Change in total cost due to ordering cost.

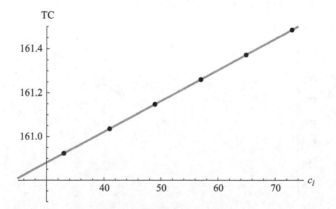

Figure 15.9 Change in total cost due to demand parameter.

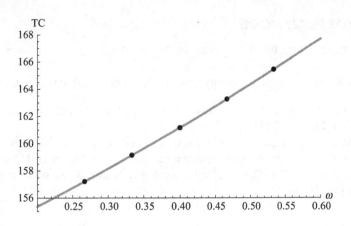

Figure 15.10 Change in total cost due to deterioration rate.

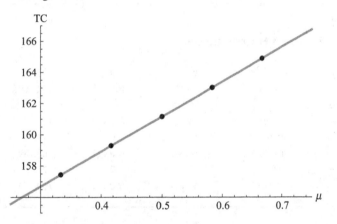

Figure 15.11 Change in total cost due to reliability factor.

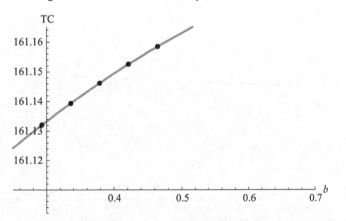

Figure 15.12 Change in total cost due to rate of backlogging.

15.8.2   FOR FUZZY MODEL

Here, we have to see that how fuzzy total cost affected due to change in some parameters.

The main conclusion of sensitivity analysis revealed in Table (15.3) as;

- There are no effect found in fuzzy total cost with the change in deterioration rate (θ) (Figure 15.13).
- There are minor changes in fuzzy total cost $\tilde{T}C(t_0)$ due to change in rate of reliability (ω) (Figure 15.14), demand parameter (μ) (Figure 15.15) and rate of backlogging (b) (Figure 15.16). Total cost is increasing with respect to these parameters.

Table 15.3
Percentage Effect on Fuzzy Parameters

% change in μ	−50	−25	0	25	50
$TC(t_0)$	154.564	157.346	160.127	162.909	165.691
% change in θ	−50	−25	0	25	50
$\tilde{T}C(t_0)$	160.12	160.124	160.127	160.131	160.134
% change in ω	−50	−25	0	25	50
$\tilde{T}C(t_0)$	154.361	157.147	160.127	163.289	166.622
% change in b	−50	−25	0	25	50
$\tilde{T}C(t_0)$	160.082	160.107	160.127	160.144	160.158

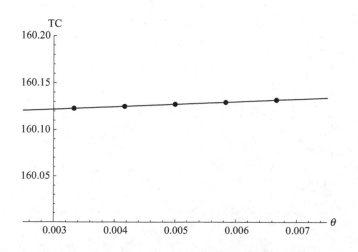

Figure 15.13   Change in total cost due to fuzzy demand parameter.

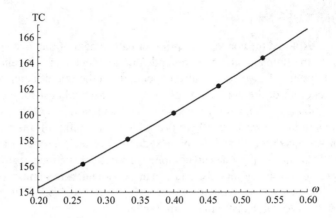

Figure 15.14   Change in total cost due to fuzzy deterioration rate.

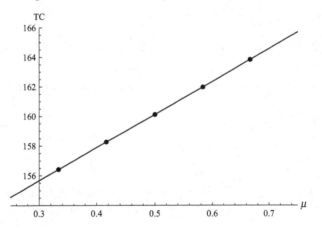

Figure 15.15   Change in total cost due to fuzzy reliability factor.

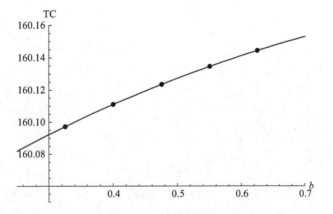

Figure 15.16   Change in total cost due to fuzzy backlogging rate.

15.9 CONCLUSION

Here, we have explored the concept of optimal management of the inventory system with respect to the time and also the demand function dependent on reliability factor. Proposed system of inventory is also dependent on the rate of deterioration and the partial backorder of the items. We conclude that the total cost for the inventory system under fuzzy environment has been reduced comparatively to the crisp environment. Thus in the future scope, there is a wide applicability of this model of the inventory in various domains as it helps in reduction of the total cost of the inventory system. This chapter provides an compulsive topic for the further study of such kind of important inventory models, and it can be extended in numerous ways for future research with exponential demand, ramp-type demand, Verhulst's model-type demand rate, trapezoidal demand rate, inflation etc.

APPENDIX

Lemma 1: Optimality process for total cost $TC(t_0)$ Section (15.1) with respect to t_0 is given below.

Step 1: Take the derivative of total cost TC and put it equals to zero;

$$\frac{\mu \omega^{t_0} \left(b \left(-1 + e^{\theta t_0} \right) \theta c_D + b \left(-1 + e^{\theta t_0} \right) c_H + \left(-1 + e^{b(-T + t_0)} \right) \theta \left(b c_L + c_S \right) \right) t_0}{b T \theta} = 0 \tag{15.3}$$

From the Figures 15.17 and 15.18, $\frac{dTC}{dt_0} < 0$ at $t_0 = 0$ and $\frac{dTC}{dt_0} > 0$ in $[0, T]$. Then there exists a unique critical value t_0^* lies between $[0, T]$ (using Mathematica 7).

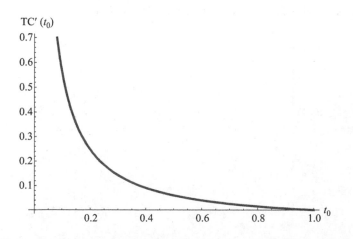

Figure 15.17 First derivative of total cost is negative at some point in the interval $[0, T]$.

Step 2: Now check the value of double derivative $\frac{d^2TC}{dt_0^2}$ at t_0^*;

$$\frac{1}{bT\theta}\mu\omega^{t_0}\left[b\theta c_D\left(-1+e^{\theta t_0}+\left(e^{\theta t_0}\theta+\left(-1+e^{\theta t_0}\right)\mathrm{Log}[\omega]\right)t_0\right)\right.$$
$$+bc_H\left(-1+e^{\theta t_0}+\left(e^{\theta t_0}\theta+\left(-1+e^{\theta t_0}\right)\mathrm{Log}[\omega]\right)t_0\right)+\theta\left(bc_L+c_S\right)$$
$$\left.\times\left(-1+e^{b(-T+t_0)}+\left(be^{b(-T+t_0)}+\left(-1+e^{b(-T+t_0)}\right)\mathrm{Log}[\omega]\right)t_0\right)\right] \quad (15.4)$$

From the Figure 15.19, the slope of double derivative $\frac{d^2TC}{dt_0^2}$ at t_0^* is positive, thus we have $\frac{d^2TC}{dt_0^2}>0$ at t_0^*. Hence we say that t_0^* is the minimum value.

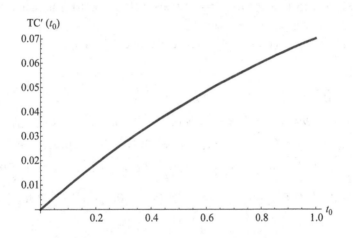

Figure 15.18 First derivative of total cost is positive at some point in the interval [0, T].

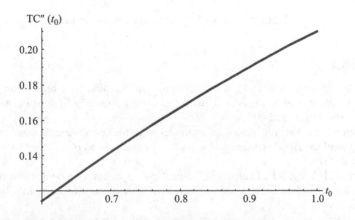

Figure 15.19 Second derivative of total cost is positive in the interval [0, T].

Lemma 2: Optimality process for total cost $\tilde{T}C(t_0)$ Section (15.5.2) with respect to t_0 is given below.

Step 1: Take the derivative of total cost $\tilde{T}C$ and put it equals to zero;

$$
\begin{aligned}
\frac{d\tilde{T}C}{dt_0} = \frac{1}{4bT\theta}\mu\omega^{t_0}\Bigg[& \left(-1+e^{\theta t_0}\right)\theta c_{D_1} + a\left(-1+e^{\theta t_0}\right)\theta c_{D_2} - b\theta c_{D_3} + be^{\theta t_0}\theta c_{D_3}\\
& - b\theta c_{D_4} + be^{\theta t_0}\theta c_{D_4} - bc_{H_1} + be^{\theta t_0}c_{H_1} - bc_{H_2} + be^{\theta t_0}c_{H_2} - bc_{H_3} + be^{\theta t_0}c_{H_3}\\
& - bc_{H_4} + be^{\theta t_0}c_{H_4} - b\theta c_{L_1} + be^{b(-T+t_0)}\theta c_{L_1} - b\theta c_{L_2} + be^{b(-T+t_0)}\theta c_{L_2} - b\theta c_{L_3}\\
& + be^{b(-T+t_0)}\theta c_{L_3} - b\theta c_{L_4} + be^{b(-T+t_0)}\theta c_{L_4} - \theta c_{S_1} + e^{b(-T+t_0)}\theta c_{S_1} - \theta c_{S_2}\\
& + e^{b(-T+t_0)}\theta c_{S_2} - \theta c_{S_3} + e^{b(-T+t_0)}\theta c_{S_3} - \theta c_{S_4} + e^{b(-T+t_0)}\theta c_{S_4}\Bigg] t_0
\end{aligned}
\tag{15.5}
$$

From the Figures 15.17 and 15.18, $\frac{d\tilde{T}C}{dt_0} < 0$ and $\frac{d\tilde{T}C}{dt_0} > 0$ in $[0, T]$. Then there exists a unique critical value t_0^* lies between $[0, T]$.

Step 2: Now check the value of double derivative $\frac{d^2\tilde{T}C}{dt_0^2}$ at t_0^*;

$$
\begin{aligned}
\frac{d^2\tilde{T}C(t_0)}{dt_0^2} = \frac{1}{4bT\theta}\mu\omega^{t_0}\Bigg[& b\left(-1+e^{\theta t_0}\right)\theta c_{D_1} + b\left(-1+e^{\theta t_0}\right)\theta c_{D_2} - b\theta c_{D_3}\\
& + be^{\theta t_0}\theta c_{D_3} - b\theta c_{D_4} + be^{\theta t_0}\theta c_{D_4} - bc_{H_1} + be^{\theta t_0}c_{H_1} - bc_{H_2} + be^{\theta t_0}c_{H_2} - bc_{H_3}\\
& + be^{\theta t_0}c_{H_3} - bc_{H_4} + be^{\theta t_0}c_{H_4} - b\theta c_{L_1} + be^{b(-T+t_0)}\theta c_{L_1} - b\theta c_{L_2} + be^{b(-T+t_0)}\theta c_{L_2}\\
& - b\theta c_{L_3} + be^{b(-T+t_0)}\theta c_{L_3} - b\theta c_{L_4} + be^{b(-T+t_0)}\theta c_{L_4} - \theta c_{S_1} + e^{b(-T+t_0)}\theta c_{S_1}\\
& - \theta c_{S_2} + e^{b(-T+t_0)}\theta c_{S_2} - \theta c_{S_3} + e^{b(-T+t_0)}\theta c_{S_3} - \theta c_{S_4} + e^{b(-T+t_0)}\theta c_{S_4}\Bigg] t_0
\end{aligned}
\tag{15.6}
$$

From the Figure 15.19, the slope of double derivative $\frac{d^2\tilde{T}C}{dt_0^2}$ at t_0^* is positive, thus we have $\frac{d^2\tilde{T}C}{dt_0^2} > 0$ at t_0^*. Hence we say that t_0^* is the minimum value.

REFERENCES

1. A.M.M. Abdulla, S.K. Paul, and A. Azeem .Optimisation of a production inventory model with reliability considerations. *International Journal of Logistics Systems and Management*, 17(1):22–45, 2014.
2. P. Anitha and P. Parvathi. An inventory model with stock dependent demand, two parameter Weibull distribution deterioration in a fuzzy environment. *ICGET-2016*, Coimbatore, pp. 1–8, 2016.
3. P. Chu, K.I. Yang, S.K. Liang, and T. Niu. Note on inventory model with a mixture of back orders and lost sales. *European Journal of Operational Research*, 159(2):470–475, 2004.
4. R.C. Bector and S. Chandra. *Fuzzy Mathematical Programming and Fuzzy Matrix Games*. Springer, New York, 2005.

5. D. Dubois and H. Prade. *Fuzzy Set and System Theory and Applications*. Academic Press, Cambridge, MA, 1980.

6. C.Y. Dye, H.J. Chang, and J.U. Teng. A deteriorating inventory model with time varying demand and shortage-dependent partial backlogging. *European Journal of Operational Research*, 117(2):417–429, 2005.

7. B.C. Giri, A.K. Jalan, and K.S. Chadhuri. EOQ model with Weibull deterioration distribution, shortage and ramp type demand. *International Journal of Systems Science*, 34(4):237–243, 2010.

8. C.I. Hsu and H.C. Li. Reliability evaluation and adjustment of supply chain network design with demand fluctuations. *International Journal of Production Economics*, 132(1):131–145, 2011.

9. K.C. Hung. An inventory model with generalized type demand, deterioration and back order rates. *European Journal of Operational Research*, 208(3):239–242, 2011.

10. B. Khara, J.K. Dey, and S.K. Mondal. An inventory model under development cost-dependent imperfect production and reliability-dependent demand. *Journal of Management Analytics*, 4:258–275, 2017.

11. C. Krishnamoorthi and S. Panayappan. An EPQ model for an imperfect production system with rework and shortages. *International Journal of Operational Research*, 17(1):104–124, 2013.

12. G. Li, X. He, J. Zhou, and H. Wu. Pricing, replenishment and preservation technology investment decision for non-instantaneous deteriorating items. *Omega*, 84:114–126 (2019).

13. Y. Liang and F. Zhou. A two-warehouse inventory model for deteriorating items with linear demand under conditionally permissible delay in payment. *Applied Mathematical Modelling*, 35(5):2221–2231, 2011.

14. G. Padmanabhan and P. Vrat. Inventory model with a mixture of back orders and lost sales. *International Journal of Systems Science*, 21(8):1721–1726, 1990.

15. S. Pal, G.S. Mahapatra, and G.P. Samanta. An EPQ model of ramp type demand with Weibull deterioration under inflation and finite horizon in crisp and fuzzy environment. *International Journal of Production Economics*, 156(C):159–166, 2014.

16. S. Panda, S. Senapati, and M. Basu. Optimal replenishment policy for perishable seasonal products in a season with ramp-type dependent demand. *Computers and Industrial Engineering*, 54(2):301–314, 2008.

17. S. Papachristos and K. Skouri. An optimal replenishment policy for deteriorating items with time varying demand and partial exponential type backlogging. *Operations Research Letters*, 27(4):175–184, 2000.

18. K.S. Park. Inventory model with partial backorders. *International Journal of Systems Sciences*, 13(2):1313–1317, 1982.

19. M. Pervin, S.K. Roy, and G.W. Weber. Analysis of inventory control model with shortage under time-dependent demand and time-varying holding cost including stochastic deterioration. *Annals in Operations Research*, 260(1–2):437-460, 2018.

20. N. Rajput, R.K. Pandey, A.P. Singh, and A. Chauhan. An optimization of fuzzy EOQ model in healthcare industries with three different demand pattern using signed distance technique. *Mathematics in Engineering Science and Aerospace*, 10(2):205–218, 2019.

21. N. Rajput, A.P. Singh, and R.K. Pandey. Optimize the cost of a fuzzy inventory model with shortage using sigh distance method. *International Journal of Research in Advent Technology*, 7(5):198–202, 2019.

22. B. Sarkar and S. Sarkar. An improved inventory model with partial backlogging, time varying deterioration and stock dependent demand. *Economic Modelling*, 30:924–932, 2013.

23. B. Sarkar, S.S. Sana, and K.S. Chaudhuri. Optimal reliability, production lot size and safety stock in an imperfect production system. *International Journal of Mathematics in Operational Research*, 2(4):467–490, 2010.

24. K. Skouri, I. Konstantaras, S. Papachristos, and I. Ganas. Inventory models with ramp type demand rate, partial backlogging and Weibull deterioration rate. *European Journal of Operational Research*, 192(1):79–92, 2009.

25. A.A. Taleizadeh, D.W. Pentico, M. Aryanezhad, and S.M. Ghoreyshi. An economic order quantity model with partial backordering and a special sale price. *European Journal of Operations Research*, 221(3):571–583, 2012.

26. A.H. Tai, Y. Xie, He Wanhua, and W.K. Ching. Joint inspection and inventory control for deteriorating items with random maximum lifetime. *International Journal of Production Economics*, 20:144–162, 2019.

27. S.P. Wang. An inventory replenishment policy for deteriorating items with shortages and partial backlogging. *Computers and Operations Research*, 29(14):2043–2051, 2002.

28. H.M. Wee. Joint pricing and replenishment policy for deteriorating inventory with declining market. *International Journal of Production Economics*, 40(2–3):163–171, 1995.

29. K.S. Wu. An EOQ inventory model for items with Weibull distribution deterioration, ramp type demand rate and partial backlogging. *Production Planning and Control*, 12(8):787–793, 2001.

16 Inventory Model for Decaying Products with Shortages and Inflation under Trade Credit in Two Warehouses

Deo Datta Aarya
Acharya Narendra Dev College (Delhi University)

Mukesh Kumar and S. J. Singh
Graphic Era (Deemed to be University)

CONTENTS

16.1   INTRODUCTION

We normally observe that improved inventory levels offer the purchaser an extensive collection of good quality goods and boost the possibility of production. In this case, the demand rate is internally associated with the seller and is a functional level of inventory. Therefore, the retailer has reason to preserve higher levels of inventory and inventory level-dependent demand function which is increased. Mostly, models are expanded with the particular storage space facility. On the other hand, in the field of supply chain system the seller can purchase a bulky collection of items when the supplier offers the cut rate on bulkiness purchases. Therefore, the retailer purchased the items in surplus amount. Surplus number of products cannot be accumulated in on-hand own warehouse (OW) due to limited space. So, it is always essential to maintain additional stock of items in a new warehouse, i.e., called rented warehouse (RW) of unlimited capability which is near to OW. In addition an extensive, cost of inventory of RW is larger than that of OW. Therefore, retailer must initiate putting up their sale from RW in place of OW to keep away from the situation of stock out. Mostly, researchers do not consider simultaneously the observable fact to develop model with two warehouses. Because these phenomena are not special in actual time, we include them in our model.

    In the existing models, two warehouses are considered one is OW, and the other is RW— having limited capacity and involving the present value of total cost for the whole planning horizon with different cases.

Sarma [1] developed a deterministic inventory model (Economic Order Quantity) with two levels of storage and an optimum release rule under OW and RW. Yang H.L. [2] considered a two-warehouse inventory model for deteriorating items with shortages under inflation by assuming the demand rate is deterministic and is completely backlogged.

Singh and Malik [4] presented an inventory model with inflationary environment on two-warehouse production inventory organization with exponential demand and changeable deterioration rate. Kumar et al. [3] proposed a deterministic inventory control model for fading items with price-dependent demand and time-varying carrying cost under permissible delay in payment. Kumar et al. [6] developed an optimal payment policy with price-dependent demand and three parameters dependent deterioration rate under the influence trade credit. Huang [7] described the best possible cycle time and best possible payment time under the permissible delay in payment and cash discount strategies. Kumar et al. [8] proposed a deterministic inventory control model for fading items with price-dependent demand and time-varying carrying cost under permissible delay in payment. Liao et al. [9] investigated the effects of inflation and time value of money on a fading inventory when the supplier permitted allowable delay in payments, with a fix demand rate. Inventory models for deteriorating items with stock-dependent selling price were formulated by Padmanabhan and Vrat [10].

16.2 ASSUMPTIONS AND NOTATIONS

16.2.1 ASSUMPTIONS

The demand rate $D(t)$ at time t is

$$D(t) = \begin{cases} \alpha + \beta I(t) - \eta p, & I(t) > 0, \\ \alpha - \eta p, & I(t) < 0, \end{cases}$$

where α and β are positive constants, $\alpha \geq \beta$, p is selling price, and $I(t)$ denotes the inventory at time t.

- Consider the shortages are permitted to arise, and assume a fraction of demand is backlogged. Moreover, the higher the waiting time, the lesser the backlogging rate. Let $B(t)$ represent fraction where t is the waiting time up to the next replenishment. We consider $B(t) = \frac{1}{(1+\delta t)}$, where $\delta > 0$ is the backlogging parameter.
- The deterioration time (product life) t has a probability density function $f(t) = \theta e^{-\theta(t-t_d)}$ for $t > t_d$, where the duration of time (t_d) in which we consider fresh products which has no deterioration rate. At the end of this stage (t_d), a constant fraction $0 < \theta < 1$ of the on-hand inventory deteriorates and there is no replacement or repair of the deteriorated products. The c. d. f. of time t is known as

$$F(t) = \int_{t_d}^{t} f(x)dx = 1 - e^{-\theta(t-t_d)}; \qquad t > t_d.$$

- t_d is the duration time in which there is no deterioration of the products.
- In the permissible delay time M, the account is not developed. A sales profit is generated and submitted interest in account. At the end of the time, the retailer pays off all bought units and starts to pay the capital opportunity cost (OC) for the items in stock.
- The model is considered for infinite planning horizon.
- A borrowed warehouse is used to build up the too much units over the fixed capacity W of the OW, and we assumed that the deterioration rate in RW is the same as that in OW. The holding cost in RW > OW. The transportation cost is not considered when stocks of RW are transported to OW in continuous release pattern.
- The items in RW are consumed first and then the items in OW for economical purpose.
- The OW has restricted capacity of W units, and the RW has unlimited capacity.

16.2.2 NOTATIONS

The following are the notations used:

Q: Denotes order quantity.
T: Denotes length of order.
A: Denotes ordering cost.
W: Denotes capacity of the owned warehouse (OW).
k: Denotes unit stock holding cost for items in RW.
h: Denotes unit stock holding cost (HC) for items in OW.
c: Denotes purchase cost.
p: Denotes selling price.
μ: Denotes inflation rate.
s: Denotes shortage cost (SC) for backlogged items per unit per year.
π: Denotes unit cost of lost sales per unit.
I_p: Denotes capital opportunity cost (OC) in stock per dollar per year.
I_e: Denotes interest earned (IE) per dollar per year.
t_w: Denotes the period in which the inventory becomes zero in RW and no deterioration rate.
t_d: Denotes the period at which the inventory stage in OW and no deterioration rate.
t_1: Denotes the time at which the inventory level reaches zero in OW and length of time with no shortage.
$I_r(t)$: Denotes the level of positive inventory in RW at time t ($0 \leq t \leq t_w$) in which products are not deteriorated.
$I_{01}(t)$, $I_{02}(t)$: Denotes the level of positive inventory in OW at time t ($0 \leq t \leq t_w$ and $t_w \leq t \leq t_d$) in which product are not deteriorated.
$I_{03}(t)$: Denotes the level of positive inventory in OW at time ($t_d \leq t \leq t_1$) in which the products are deteriorated.

$I_S(t)$: Denotes the level of negative inventory at time t ($t_1 \leq t \leq T$) at this stage shortage occurred.

M: Denotes the permissible delay period.

TC (t_1, T): Denotes the total yearly inventory cost.

16.3 MATHEMATICAL FORMULATION AND SOLUTION OF THE MODEL

In this study, we discussed the two-warehouse inventory system. Considering replenishment problem of a single non-instantaneous deteriorating item with partial backlogging rate, W units of goods are stored in OW and the remaining is dispatched to the RW. The RW is utilized only after OW is filled, but stocks in RW are dispatched first. During the time period [0, t_w], the inventory level is decreasing only remaining to stock-dependent demand rate from RW. At the same time, the inventory level is equal to the W in OW. At the period [t_w, t_d], the inventory level is decreasing only remaining to stock-dependent demand rate from OW. The inventory level becomes to zero due to the combined effect of order, and items deteriorate in the interval [t_d, t_1]. Then, the shortage continues till the last of the present-order cycle, and the complete process is repeated (Figure 16.1).

Differential equations for the inventory conditions at RW and OW are as follows:

$$I'_r(t) = -[\alpha + \beta I_r(t) - \eta p], \quad 0 \leq t \leq t_w \tag{16.1}$$

$$I_{01}(t) = W, \quad 0 \leq t \leq t_w \tag{16.2}$$

$$I'_{02}(t) = -[\alpha + \beta I_{02}(t) - \eta p], \quad t_w \leq t \leq t_d \tag{16.3}$$

$$I'_{03}(t) + \theta I_{03}(t) = -[\alpha + \beta I_{03}(t) - \eta p], \quad t_d \leq t \leq t_1 \tag{16.4}$$

with the boundary conditions $I_r(t_w) = 0$, $I_{02}(t_w) = w$, $I_{03}(t_1) = 0$.

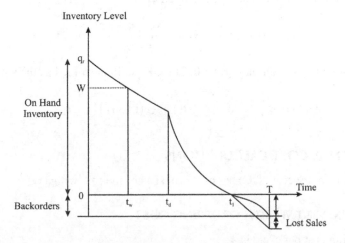

Figure 16.1 Two warehouse graphical representation of inventory system.

Solving Equations (16.1), (16.3), and (16.4), the inventory stage is as follows:

$$I_r(t) = \frac{(\alpha - \eta p)}{\beta}\left[e^{\beta(t_w - t)} - 1\right], \quad 0 \leq t \leq t_w \tag{16.5}$$

$$I_{01}(t) = W, \quad 0 \leq t \leq t_w \tag{16.6}$$

$$I_{02}(t) = We^{\beta(t_w - t)} + \frac{(\alpha - \eta p)}{\beta}\left[e^{\beta(t_w - t)} - 1\right], \quad t_w \leq t \leq t_d \tag{16.7}$$

$$I_{03}(t) = \frac{(\alpha - \eta p)}{(\theta + \beta)}\left[e^{(\theta + \beta)(t_1 - t)} - 1\right], \quad t_d \leq t \leq t_1 \tag{16.8}$$

Due to continuity of $I(t)$ at $t = t_d$, from Equations (16.7) and (16.8), it can be written as

$$I_{02}(t_d) = I_{03}(t_d)$$

$$We^{\beta(t_w - t_d)} + \frac{(\alpha - \eta p)}{\beta}\left[e^{\beta(t_w - t_d)} - 1\right] = \frac{(\alpha - \eta p)}{(\theta + \beta)}\left[e^{(\theta + \beta)(t_1 - t_d)} - 1\right]$$

This implies that

$$t_w = \frac{(\alpha - \eta p)t_1 - W(1 - \beta t_d)}{(\alpha - \eta p + \beta W)} \tag{16.9}$$

Here, t_w is a function of t_1.

Differentiating Equation (16.9) with respect to t_1, we obtain

$$\frac{dt_w}{dt_1} = \frac{(\alpha - \eta p)}{(\alpha - \eta p + \beta W)} \tag{16.10}$$

For the duration of the shortage period $[t_1, T]$, at time t the demand rate is partially backlogged at fraction $B(T - t)$.

Thus, the inventory at time t is developed as

$$I'_s(t) = -(\alpha - \eta p)B(T - t) = -\frac{(\alpha - \eta p)}{1 + \delta(T - t)}, \quad t_1 \leq t \leq T \tag{16.11}$$

with the Boundary condition $I_s(t_1) = 0$. Thus, Equation (16.11) has the solution:

$$I_s(t) = -(\alpha - \eta p)\left[(1 - \delta T)(t - t_1) + \frac{\delta}{2}\left(t^2 - t_1^2\right)\right], \quad t_1 \leq t \leq T \tag{16.12}$$

16.4   TOTAL COST CALCULATIONS

The total inventory cost (TIC) per cycle consists of the following costs:

16.4.1   PRESENT WORTH ORDERING COST

The ordering cost per cycle is A.

$$OC = A \tag{16.13}$$

16.4.2 PRESENT WORTH HOLDING COST FOR RW

Inventory holding cost (IHC) in RW is obtained in the following two cases:

16.4.2.1 Case 1: When $Q \leq W$

For the above case, there is no requirement for RW. Thus, the holding cost for items in RW is zero.

16.4.2.2 Case 2: When $Q > W$

The IHC in RW per cycle is shown in Figure 16.2.

$$(HC_{RW}) = k \int_0^{t_w} I_r(t) e^{-\mu t} dt$$

$$= \frac{k(\alpha - \eta p)}{\beta} \left[\frac{1}{(\beta + \mu)} \left(e^{\beta t_w} - e^{-\mu t_w} \right) + \frac{1}{\mu} \left(e^{-\mu t_w} - 1 \right) \right] \quad (16.14)$$

16.4.3 PRESENT WORTH HOLDING COST FOR OW

The following two cases define the IHC in OW:

16.4.3.1 Case 1: When $Q \leq W$

IHC in OW per cycle is shown in Figure 16.3.

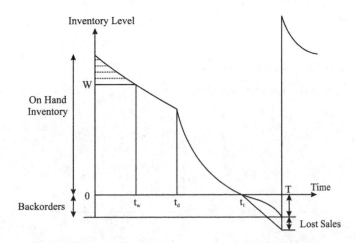

Figure 16.2 The holding cost in RW for $Q > W$ during the period $(0, T)$.

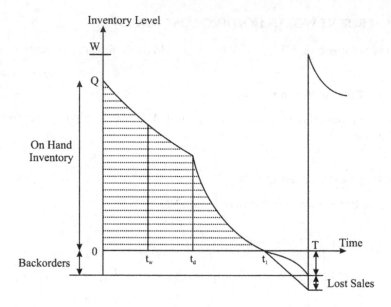

Figure 16.3   The total holding cost in OW for $Q \leq W$ during the period $(0, T)$.

$$(\text{HC}_{\text{OW}}) = h\left[\int_0^{t_d} I(t)e^{-\mu t}dt + \int_{t_d}^{t_1} I_{03}(t)e^{-\mu t}dt\right] \quad (\text{See Appendix I})$$

$$= h\left[\frac{(\alpha - \eta p + \beta Q)}{(\beta + \mu)\beta}\left(1 - e^{-(\beta + \mu)t_d}\right) + \frac{(\alpha - \eta p)}{\beta\mu}\left(e^{-\mu t_d} - 1\right) + \frac{(\alpha - \eta p)}{(\theta + \beta)}\right.$$

$$\left. \times \left\{\frac{1}{(\beta + \theta + \mu)}\left(e^{(\theta + \beta)(t_1 - t_d) - \mu t_d} - e^{-\mu t_1}\right) + \frac{1}{\mu}\left(e^{-\mu t_1} - e^{-\mu t_d}\right)\right\}\right]$$

$$(16.15)$$

16.4.3.2   Case 2: When $Q > W$

IHC in OW per cycle is shown in Figure 16.4.

$$(\text{HC}_{\text{OW}}) = h\left[\int_0^{t_w} We^{-\mu t}dt + \int_{t_w}^{t_d} I_{02}(t)e^{-\mu t}dt + \int_{t_d}^{t_1} I_{03}(t)e^{-\mu t}dt\right]$$

$$= h\left[\frac{W}{\mu}\left(1 - e^{-\mu t_w}\right) - \frac{(\alpha - \eta p + \beta W)}{\beta(\beta + \mu)}\left(e^{\beta(t_w - t_d) - \mu t_d} - e^{-\mu t_w}\right)\right.$$

$$+ \frac{(\alpha - \eta p)}{\beta\mu}\left(e^{-\mu t_d} - e^{-\mu t_w}\right) + \frac{(\alpha - \eta p)}{(\theta + \beta)(\beta + \theta + \mu)}$$

$$\left. \times \left(e^{(\theta + \beta)(t_1 - t_d) - \mu t_d} - e^{-\mu t_1}\right) + \frac{(\alpha - \eta p)}{(\theta + \beta)\mu}\left(e^{-\mu t_1} - e^{-\mu t_d}\right)\right] \quad (16.16)$$

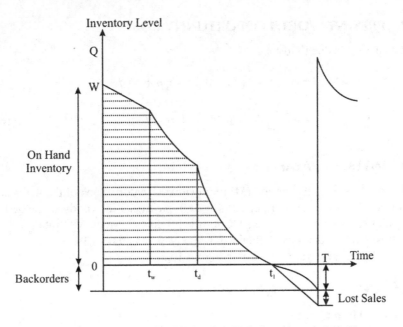

Figure 16.4   The total holding cost in OW for $Q > W$ during the period $(0, T)$.

16.5   PRESENT WORTH DETERIORATION COST

The deterioration cost (DC) per cycle is

$$
\begin{aligned}
DC &= c\theta \int_{t_d}^{t_1} I_{03}(t) e^{-\mu t}\, dt \\
&= \frac{c\theta\,(\alpha - \eta p)}{(\theta + \beta)} \left[\frac{1}{(\theta + \beta + \mu)} \left(e^{(\theta+\beta)(t_1 - t_d) - \mu t_d} - e^{-\mu t_1} \right) + \frac{1}{\mu} \left(e^{-\mu t_1} - e^{-\mu t_d} \right) \right]
\end{aligned}
$$

$$(16.17)$$

16.6   PRESENT WORTH SHORTAGE COST

The SC per cycle due to backlog is

$$
\begin{aligned}
SC &= s \int_{t_1}^{T} -\{I_s(t)\} e^{-\mu t}\, dt \\
&= s\,(\alpha - \eta p) \left[\frac{(1 - \delta T)}{\mu^2} \left\{ ((t_1 - T)\mu - 1) e^{-\mu T} + e^{-\mu t_1} \right\} \right. \\
&\quad \left. + \frac{\delta}{2\mu^3} \left\{ ((t_1^2 - T^2)\mu^2 - 2T\mu - 2) e^{-\mu T} + 2(t_1\mu + 1) e^{-\mu t_1} \right\} \right]
\end{aligned}
$$

$$(16.18)$$

16.7   PRESENT WORTH OPPORTUNITY COST

The OC per cycle due to lost sale is

$$
\begin{aligned}
\text{OC} &= \pi \int_{t_1}^{T} [1 - B(T-t)](\alpha - \eta p)\,e^{-\mu t}\,dt \\
&= \frac{\pi(\alpha - \eta p)\delta}{\mu^2}\left[e^{-\mu T} - \{(t_1 - T)\mu + 1\}e^{-\mu t_1}\right]
\end{aligned} \tag{16.19}
$$

16.8   INTEREST PAYABLE

Finally, permissible delay period (M) for positive inventory stock of the items is less than or equal to the length of period ($M \le t_1$), payments for goods are settled, and the retailer starts paying the assets OC for the products in stock with rate I_p.

Interest payable per year is obtained, when $Q \le W$ and $Q > W$.

Thus, the OC per cycle is shown in Figure 16.5.

16.8.1   CASE 1: $0 \le M \le t_d$

16.8.1.1   When $Q \le W$

$$
\begin{aligned}
\text{IP}_1 &= cI_p \left[\int_{M}^{t_d} I(t)e^{-\mu t}\,dt + \int_{t_d}^{t_1} I_{03}(t)e^{-\mu t}\,dt \right] \quad (\text{see Appendix I}) \\
&= cI_p \left[\frac{(\alpha - \eta p + \beta Q)}{(\beta + \mu)\beta}\left(e^{-(\beta+\mu)M} - e^{-(\beta+\mu)t_d}\right) + \frac{(\alpha - \eta p)}{\beta\mu}\left(e^{-\mu t_d} - e^{-\mu M}\right) \right. \\
&\quad \left. + \frac{(\alpha - \eta p)}{(\theta + \beta)}\left\{ \frac{1}{(\beta + \theta + \mu)}\left(e^{(\theta+\beta)(t_1-t_d)-\mu t_d} - e^{-\mu t_1}\right) + \frac{1}{\mu}\left(e^{-\mu t_1} - e^{-\mu t_d}\right)\right\} \right]
\end{aligned} \tag{16.20}
$$

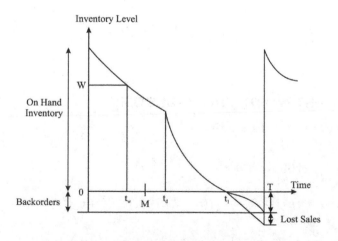

Figure 16.5   Graphical representation for case 1 ($0 < M \le t_d$).

16.8.1.2 When $Q > W$

$$\text{IP}_1 = cI_p \left[\int_M^{t_d} I_{02}(t)e^{-\mu t}\,dt + \int_{t_d}^{t_1} I_{03}(t)e^{-\mu t}\,dt \right]$$

$$= cI_p \left[\frac{(\alpha - \eta p + \beta W)e^{-\beta t_w}}{(\beta + \mu)\beta} \left(e^{-(\beta+\mu)M} - e^{-(\beta+\mu)t_d} \right) + \frac{(\alpha - \eta p)}{\beta\mu} \left(e^{-\mu t_d} - e^{-\mu M} \right) \right.$$

$$\left. + \frac{(\alpha - \eta p)}{(\theta + \beta)} \left\{ \frac{1}{(\beta + \theta + \mu)} \left(e^{(\theta+\beta)(t_1-t_d)-\mu t_d} - e^{-\mu t_1} \right) + \frac{1}{\mu} \left(e^{-\mu t_1} - e^{-\mu t_d} \right) \right\} \right] \tag{16.21}$$

16.8.2 CASE 2: $t_d \leq M \leq t_1$ (SHOWN IN FIGURE 16.6)

16.8.2.1 When $Q \leq W$ and $Q > W$

$$\text{IP}_2 = cI_p \int_M^{t_1} I_{03}(t)e^{-\mu t}\,dt$$

$$= \frac{cI_p(\alpha - \eta p)}{(\theta + \beta)} \left[\frac{1}{(\theta + \beta + \mu)} \left\{ e^{(\theta+\beta)(t_1-M)-\mu M} - e^{-\mu t_1} \right\} + \frac{1}{\mu} \left(e^{-\mu t_1} - e^{-\mu M} \right) \right] \tag{16.22}$$

16.8.3 CASE 3: WHEN $M \geq t_1$

There is no OC for $Q \leq W$ and $Q > W$ (shown in Figure 16.6)

$$\text{Therefore,} \quad \text{IP}_3 = 0 \tag{16.23}$$

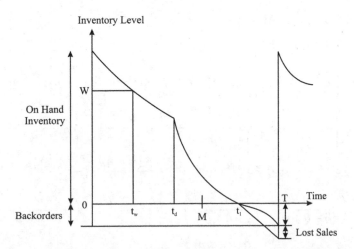

Figure 16.6 Graphical representation for the Case 2 ($t_d \leq M \leq t_1$) and graphical representation for the Case 3 ($M > t_1$).

16.9 INTEREST EARNED (I_e) FROM SALES REVENUE

Following special cases are given to determine the interest earned, and at this time, we assume that the account is not established, the retailer sells the products, collects sales revenue, and earns the interest with rate (I_e).

Thus, the interest earned (I_e) per cycle is given for the following three cases:

16.9.1 CASE 1: $0 < M \leq t_d$

16.9.1.1 When $Q \leq W$

$$IE_1 = pI_e \left[\int_0^M \{\alpha + \beta I(t) - \eta p\} t e^{-\mu t} dt \right]$$

$$= pI_e (\alpha - \eta p + Q\beta) \left(\frac{\left(1 - e^{-(\beta+\mu)M} \right)}{(\beta+\mu)^2} - \frac{M e^{-(\beta+\mu)M}}{(\beta+\mu)} \right) \tag{16.24}$$

16.9.1.2 When $Q > W$

$$IE_1 = pI_e \left[\int_0^{t_w} \{\alpha + \beta I_r(t) - \eta p\} t e^{-\mu t} dt + \int_{t_w}^M \{\alpha + \beta I_{02}(t) - \eta p\} t e^{-\mu t} dt \right]$$

$$= pI_e \left[(\alpha - \eta p) \left\{ \frac{(e^{\beta t_w} - e^{-\mu t_w})}{(\beta+\mu)^2} - \frac{t_w e^{-\mu t_w}}{(\beta+\mu)} \right\} + (\alpha - \eta p + \beta W) \right.$$

$$\left. \times \left\{ \frac{\left(e^{\beta t_w - (\beta+\mu)M} - e^{-\mu t_w} \right)}{(\beta+\mu)^2} + \frac{\left(M e^{\beta t_w - (\beta+\mu)M} - t_w e^{-\mu t_w} \right)}{(\beta+\mu)} \right\} \right] \tag{16.25}$$

16.9.2 CASE 2: $t_d < M \leq t_1$

16.9.2.1 When $Q \leq W$

$$IE_2 = pI_e \left[\int_0^{t_d} \{\alpha + \beta I(t) - \eta p\} t e^{-\mu t} dt + \int_{t_d}^M \{\alpha + \beta I_{03}(t) - \eta p\} t e^{-\mu t} dt \right]$$

$$= pI_e \left[(\alpha - \eta p + \beta Q) \left\{ \frac{\left(1 - e^{-(\beta+\mu)t_d} \right)}{(\beta+\mu)^2} - \frac{t_d e^{-(\beta+\mu)t_d}}{(\beta+\mu)} \right\} + \frac{(\alpha - \eta p)}{\mu^2} \right.$$

$$\times \left\{ e^{-\mu t_d} (1 + \mu t_d) - e^{-\mu M} (1 + \mu M) \right\} + \frac{\beta (\alpha - \eta p)}{(\theta+\beta)} \left\{ -e^{(\beta+\theta)(t_1-t_d)-\mu t_d} \right.$$

$$\times \left(\frac{t_d}{(\beta+\theta+\mu)} + \frac{1}{(\beta+\theta+\mu)^2} \right) - e^{(\beta+\theta)(t_1-M)-\mu M} \left(\frac{M}{(\beta+\theta+\mu)} \right.$$

$$\left. + \frac{1}{(\beta+\theta+\mu)^2} \right) - \frac{\beta (\alpha - \eta p)}{(\theta+\beta)\mu^2} \left\{ e^{-\mu t_d} (1 + \mu t_d) - e^{-\mu M} (1 + \mu M) \right\} \right] \tag{16.26}$$

16.9.2.2 When $Q > W$

$$IE_2 = pI_e \left[\int_0^{t_w} \{\alpha + \beta I_r(t) - \eta p\} t e^{-\mu t} dt + \int_{t_w}^{t_d} \{\alpha + \beta I_{02}(t) - \eta p\} t e^{-\mu t} dt \right.$$

$$\left. + \int_{t_d}^{M} \{\alpha + \beta I_{03}(t) - \eta p\} t e^{-\mu t} dt \right]$$

$$= pI_e \left[(\alpha - \eta p) \left\{ \frac{e^{\beta t_w}}{(\beta + \mu)^2} - e^{-\mu t_w} \left(\frac{t_w}{(\beta + \mu)} + \frac{1}{(\beta + \mu)^2} \right) \right\} + (\alpha - \eta p + \beta W) \right.$$

$$\times \left\{ e^{\beta(t_w - t_d) - \mu t_d} \left(\frac{t_d}{(\beta + \mu)} + \frac{1}{(\beta + \mu)^2} \right) - e^{-\mu t_w} \left(\frac{t_w}{(\beta + \mu)} + \frac{1}{(\beta + \mu)^2} \right) \right\}$$

$$+ \frac{(\alpha - \eta p) \theta}{\mu^2} \left\{ e^{-\mu t_d} (1 + \mu t_d) - e^{-\mu M} (1 + \mu M) \right\} + \frac{\beta (\alpha - \eta p)}{(\theta + \beta)}$$

$$\times \left\{ e^{(\beta + \theta)(t_1 - t_d) - \mu t_d} \left(\frac{t_d}{(\beta + \theta + \mu)} + \frac{1}{(\beta + \theta + \mu)^2} \right) \right.$$

$$\left. \left. - e^{(\beta + \theta)(t_1 - M) - \mu M} \left(\frac{M}{(\beta + \theta + \mu)} + \frac{1}{(\beta + \theta + \mu)^2} \right) \right\} \right] \qquad (16.27)$$

16.9.3   CASE 3: $M > t_1$

16.9.3.1   When $Q \leq W$

$$IE_3 = pI_e \left[\left\{ \int_0^{t_d} \{\alpha + \beta I(t) - \eta p\} t e^{-\mu t} dt + \int_{t_d}^{t_1} \{\alpha + \beta I_{03}(t) - \eta p\} t e^{-\mu t} dt \right\} \right.$$

$$\left. + (M - t_1) e^{-\mu t_1} \left\{ \int_0^{t_d} \{\alpha + \beta I(t) - \eta p\} dt + \int_{t_d}^{t_1} \{\alpha + \beta I_{03}(t) - \eta p\} dt \right\} \right]$$

$$= pI_e \left[(\alpha - \eta p + \beta Q) \left\{ \frac{1}{(\beta + \mu)^2} - \frac{e^{-(\beta + \mu)t_d}}{(\beta + \mu)^2} \left(\frac{t_d}{(\beta + \mu)} + \frac{1}{(\beta + \mu)^2} \right) \right\} \right.$$

$$+ \frac{(\alpha - \eta p) \theta}{\mu^2} \left\{ e^{-\mu t_d} (1 + \mu t_d) - e^{-\mu t_1} (1 + \mu t_1) \right\} + \frac{\beta (\alpha - \eta p)}{(\theta + \beta)}$$

$$\times \left\{ e^{(\beta + \theta)(t_1 - t_d) - \mu t_d} \left(\frac{t_d}{(\beta + \theta + \mu)} + \frac{1}{(\beta + \theta + \mu)^2} \right) \right.$$

$$\left. - e^{-\mu t_1} \left(\frac{t_1}{(\beta + \theta + \mu)} + \frac{1}{(\beta + \theta + \mu)^2} \right) \right\} + e^{-\mu t_1} (M - t_1) \left\{ \frac{(\alpha - \eta p + \beta Q)}{\beta} \right.$$

$$\left. \left. \times \left(1 - e^{-\beta t_d} \right) + \frac{\theta (\alpha - \eta p)(t_1 - t_d)}{(\theta + \beta)} + \frac{\beta (\alpha - \eta p)}{(\theta + \beta)^2} \left(e^{(\beta + \theta)(t_1 - t_d)} - 1 \right) \right\} \right] \qquad (16.28)$$

16.9.3.2 When $Q > W$

$$IE_3 = pI_e \left[\left\{ \int_0^{t_w} \{\alpha + \beta I_r(t) - \eta p\} t e^{-\mu t} dt + \int_{t_w}^{t_d} \{\alpha + \beta I_{02}(t) - \eta p\} t e^{-\mu t} dt \right. \right.$$

$$\left. + \int_{t_d}^{t_1} \{\alpha + \beta I_{03}(t) - \eta p\} t e^{-\mu t} dt \right\} + (M - t_1) e^{-\mu t_1} \left\{ \int_0^{t_w} \{\alpha + \beta I_r(t)\} dt \right.$$

$$\left. \left. + \int_{t_w}^{t_d} \{\alpha + \beta I_{02}(t)\} dt + \int_{t_d}^{t_1} \{\alpha + \beta I_{03}(t)\} dt \right\} \right]$$

$$= pI_e \left[(\alpha - \eta p) \left\{ \frac{e^{\beta t_w}}{(\beta + \mu)^2} - e^{-\mu t_w} \left(\frac{t_w}{(\beta + \mu)} + \frac{1}{(\beta + \mu)^2} \right) \right\} + (\alpha - \eta p + \beta W) \right.$$

$$\times \left\{ e^{\beta(t_w - t_d) - \mu t_d} \left(\frac{t_d}{(\beta + \mu)} + \frac{1}{(\beta + \mu)^2} \right) - e^{-\mu t_w} \left(\frac{t_w}{(\beta + \mu)} + \frac{1}{(\beta + \mu)^2} \right) \right\}$$

$$+ \frac{(\alpha - \eta p)}{\mu^2} \left\{ e^{-\mu t_d} (1 + \mu t_d) - e^{-\mu M} (1 + \mu M) \right\} + \frac{\beta(\alpha - \eta p)}{(\theta + \beta)}$$

$$\times \left\{ e^{(\beta + \theta)(t_1 - t_d) - \mu t_d} \left(\frac{t_d}{(\beta + \theta + \mu)} + \frac{1}{(\beta + \theta + \mu)^2} \right) \right.$$

$$\left. - e^{-\mu t_1} \left(\frac{t_1}{(\beta + \theta + \mu)} + \frac{1}{(\beta + \theta + \mu)^2} \right) \right\} + e^{-\mu t_1} (M - t_1)$$

$$\times \left\{ \frac{(\alpha - \eta p)}{\beta} \left(e^{\beta t_w} - 1 \right) - W e^{\beta(t_w - t_d)} + W + \frac{(\alpha - \eta p)}{\beta} \left(e^{\beta(t_w - t_d)} - 1 \right) \right.$$

$$\left. \left. + \frac{\theta(\alpha - \eta p)(t_1 - t_d)}{(\theta + \beta)} + \frac{\beta(\alpha - \eta p)}{(\theta + \beta)^2} \left(e^{(\beta + \theta)(t_1 - t_d)} - 1 \right) \right\} \right] \tag{16.29}$$

16.10   PRESENT WORTH TOTAL COST

Total inventory cost (TIC) of the retailer for entire planning horizon is the summation of the individual costs given by Equations (16.13) to (16.29) according to different cases (Figures 16.7 and 16.8).

Thus, TIC per unit time is

$$\text{TC}(t_1, T) = 1/T \, [\text{Inventory OC} + \text{Inventory HC per cycle in RW}$$

$$+ \text{Inventory HC per cycle in OW} + \text{DC per cycle}$$

$$+ \text{SC per cycle due to backlog} + \text{OC per cycle due to lost sales}$$

$$+ \text{Interest earned} - \text{Interest payable}]$$

The total cost when $Q \le W$, which depends on t_1, and T is given by

$$\text{TC}(t_1, T) = \begin{cases} \text{TC}_1(t_1, T) & \text{if} \quad 0 < M \le t_d \\ \text{TC}_2(t_1, T) & \text{if} \quad t_d < M \le t_1 \\ \text{TC}_3(t_1, T) & \text{if} \quad M > t_1 \end{cases}$$

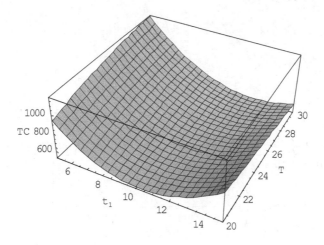

Figure 16.7 Convexity of the TC for proposed model when $Q \leq W$.

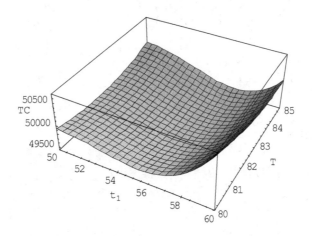

Figure 16.8 Convexity of the TC for proposed model when $Q > W$.

where

$$
\begin{aligned}
\text{TC}_1(t_1, T) = \frac{1}{T} &\left[A + h \left[\frac{(\alpha - \eta p + \beta Q)}{(\beta + \mu)\beta} \left(1 - e^{-(\beta + \mu)t_d} \right) + \frac{(\alpha - \eta p)}{\beta \mu} \left(e^{-\mu t_d} - 1 \right) \right.\right. \\
&+ \left.\frac{(\alpha - \eta p)}{(\theta + \beta)} \left\{ \frac{1}{(\beta + \theta + \mu)} \left(e^{(\theta + \beta)(t_1 - t_d) - \mu t_d} - e^{-\mu t_1} \right) + \frac{1}{\mu} \left(e^{-\mu t_1} - e^{-\mu t_d} \right) \right\} \right] \\
&+ \frac{c\theta(\alpha - \eta p)}{(\theta + \beta)} \left[\frac{1}{(\theta + \beta + \mu)} \left(e^{(\theta + \beta)(t_1 - t_d) - \mu t_d} - e^{-\mu t_1} \right) + \frac{1}{\mu} \left(e^{-\mu t_1} - e^{-\mu t_d} \right) \right] \\
&+ \left. s(\alpha - \eta p) \left[\frac{(1 - \delta T)}{\mu^2} \left\{ ((t_1 - T)\mu - 1) e^{-\mu T} + e^{-\mu t_1} \right\} \right. \right.
\end{aligned}
$$

$$+ \frac{\delta}{2\mu^3} \left\{ \left((t_1^2 - T^2)\mu^2 - 2T\mu - 2 \right) e^{-\mu T} + 2(t_1\mu + 1) e^{-\mu t_1} \right\} \bigg]$$

$$+ \frac{\pi(\alpha - \eta p)\delta}{\mu^2} \left[e^{-\mu T} - \{(t_1 - T)\mu + 1\} e^{-\mu t_1} \right]$$

$$+ cI_p \left[\frac{(\alpha - \eta p + \beta Q)}{(\beta + \mu)\beta} \left(e^{-(\beta+\mu)M} - e^{-(\beta+\mu)t_d} \right) + \frac{(\alpha - \eta p)}{\beta\mu} \left(e^{-\mu t_d} - e^{-\mu M} \right) \right.$$

$$+ \frac{(\alpha - \eta p)}{(\theta + \beta)} \left\{ \frac{1}{(\beta + \theta + \mu)} \left(e^{(\theta+\beta)(t_1 - t_d) - \mu t_d} - e^{-\mu t_1} \right) + \frac{1}{\mu} \left(e^{-\mu t_1} - e^{-\mu t_d} \right) \right\} \bigg]$$

$$- pI_e (\alpha - \eta p + Q\beta) \left(\frac{\left(1 - e^{-(\beta+\mu)M}\right)}{(\beta + \mu)^2} - \frac{M e^{-(\beta+\mu)M}}{(\beta + \mu)} \right) \bigg] \tag{16.30}$$

$$TC_2(t_1, T) = \frac{1}{T} \left[A + h \left[\frac{(\alpha - \eta p + \beta Q)}{(\beta + \mu)\beta} \left(1 - e^{-(\beta+\mu)t_d} \right) + \frac{(\alpha - \eta p)}{\beta\mu} \left(e^{-\mu t_d} - 1 \right) \right.\right.$$

$$+ \frac{(\alpha - \eta p)}{(\theta + \beta)} \left\{ \frac{1}{(\beta + \theta + \mu)} \left(e^{(\theta+\beta)(t_1 - t_d) - \mu t_d} - e^{-\mu t_1} \right) + \frac{1}{\mu} \left(e^{-\mu t_1} - e^{-\mu t_d} \right) \right\} \bigg]$$

$$+ \frac{c\theta(\alpha - \eta p)}{(\theta + \beta)} \left[\frac{1}{(\theta + \beta + \mu)} \left(e^{(\theta+\beta)(t_1 - t_d) - \mu t_d} - e^{-\mu t_1} \right) + \frac{1}{\mu} \left(e^{-\mu t_1} - e^{-\mu t_d} \right) \right]$$

$$+ s(\alpha - \eta p) \left[\frac{(1 - \delta T)}{\mu^2} \left\{ ((t_1 - T)\mu - 1) e^{-\mu T} + e^{-\mu t_1} \right\} \right.$$

$$+ \frac{\delta}{2\mu^3} \left\{ \left((t_1^2 - T^2)\mu^2 - 2T\mu - 2 \right) e^{-\mu T} + 2(t_1\mu + 1) e^{-\mu t_1} \right\} \bigg]$$

$$+ \frac{\pi(\alpha - \eta p)\delta}{\mu^2} \left[e^{-\mu T} - \{(t_1 - T)\mu + 1\} e^{-\mu t_1} \right]$$

$$+ \frac{cI_p(\alpha - \eta p)}{(\theta + \beta)} \left[\frac{1}{(\theta + \beta + \mu)} \left\{ e^{(\theta+\beta)(t_1 - M) - \mu M} - e^{-\mu t_1} \right\} + \frac{1}{\mu} \left(e^{-\mu t_1} - e^{-\mu M} \right) \right]$$

$$- pI_e \left[(\alpha - \eta p + \beta Q) \left\{ \frac{\left(1 - e^{-(\beta+\mu)t_d}\right)}{(\beta + \mu)^2} - \frac{t_d e^{-(\beta+\mu)t_d}}{(\beta + \mu)} \right\} + \frac{(\alpha - \eta p)}{\mu^2} \right.$$

$$\times \left\{ e^{-\mu t_d}(1 + \mu t_d) - e^{-\mu M}(1 + \mu M) \right\} + \frac{\beta(\alpha - \eta p)}{(\theta + \beta)} \left\{ -e^{(\beta+\theta)(t_1 - t_d) - \mu t_d} \right.$$

$$\times \left(\frac{t_d}{(\beta + \theta + \mu)} + \frac{1}{(\beta + \theta + \mu)^2} \right) - e^{(\beta+\theta)(t_1 - M) - \mu M} \left(\frac{M}{(\beta + \theta + \mu)} \right.$$

$$+ \frac{1}{(\beta + \theta + \mu)^2} \right) - \frac{\beta(\alpha - \eta p)}{(\theta + \beta)\mu^2} \left\{ e^{-\mu t_d}(1 + \mu t_d) - e^{-\mu M}(1 + \mu M) \right\} \bigg] \bigg]$$

$$\tag{16.31}$$

$$
TC_3(t_1, T) = \frac{1}{T}\left[A + h\left[\frac{(\alpha - \eta p + \beta Q)}{(\beta + \mu)\beta}\left(1 - e^{-(\beta+\mu)t_d}\right) + \frac{(\alpha - \eta p)}{\beta\mu}\left(e^{-\mu t_d} - 1\right)\right.\right.
$$

$$
+ \frac{(\alpha - \eta p)}{(\theta + \beta)}\left\{\frac{1}{(\beta + \theta + \mu)}\left(e^{(\theta+\beta)(t_1 - t_d) - \mu t_d} - e^{-\mu t_1}\right) + \frac{1}{\mu}\left(e^{-\mu t_1} - e^{-\mu t_d}\right)\right\}\right]
$$

$$
+ \frac{c\theta(\alpha - \eta p)}{(\theta + \beta)}\left[\frac{1}{(\theta + \beta + \mu)}\left(e^{(\theta+\beta)(t_1 - t_d) - \mu t_d} - e^{-\mu t_1}\right) + \frac{1}{\mu}\left(e^{-\mu t_1} - e^{-\mu t_d}\right)\right]
$$

$$
+ s(\alpha - \eta p)\left[\frac{(1 - \delta T)}{\mu^2}\left\{((t_1 - T)\mu - 1)e^{-\mu T} + e^{-\mu t_1}\right\}\right.
$$

$$
+ \frac{\delta}{2\mu^3}\left\{((t_1^2 - T^2)\mu^2 - 2T\mu - 2)e^{-\mu T} + 2(t_1\mu + 1)e^{-\mu t_1}\right\}\right]
$$

$$
+ \frac{\pi(\alpha - \eta p)\delta}{\mu^2}\left[e^{-\mu T} - \{(t_1 - T)\mu + 1\}e^{-\mu t_1}\right]
$$

$$
+ cI_p\left[\frac{(\alpha - \eta p + \beta Q)}{(\beta + \mu)\beta}\left(e^{-(\beta+\mu)M} - e^{-(\beta+\mu)t_d}\right) + \frac{(\alpha - \eta p)}{\beta\mu}\left(e^{-\mu t_d} - e^{-\mu M}\right)\right.
$$

$$
+ \frac{(\alpha - \eta p)}{(\theta + \beta)}\left\{\frac{1}{(\beta + \theta + \mu)}\left(e^{(\theta+\beta)(t_1 - t_d) - \mu t_d} - e^{-\mu t_1}\right) + \frac{1}{\mu}\left(e^{-\mu t_1} - e^{-\mu t_d}\right)\right\}\right]
$$

$$
- pI_e\left[(\alpha - \eta p + \beta Q)\left\{\frac{1}{(\beta + \mu)^2} - \frac{e^{-(\beta+\mu)t_d}}{(\beta + \mu)^2}\left(\frac{t_d}{(\beta + \mu)} + \frac{1}{(\beta + \mu)^2}\right)\right\}\right.
$$

$$
+ \frac{(\alpha - \eta p)\theta}{\mu^2}\left\{e^{-\mu t_d}(1 + \mu t_d) - e^{-\mu t_1}(1 + \mu t_1)\right\} + \frac{\beta(\alpha - \eta p)}{(\theta + \beta)}
$$

$$
\times \left\{e^{(\beta+\theta)(t_1 - t_d) - \mu t_d}\left(\frac{t_d}{(\beta + \theta + \mu)} + \frac{1}{(\beta + \theta + \mu)^2}\right)\right.
$$

$$
\left. - e^{-\mu t_1}\left(\frac{t_1}{(\beta + \theta + \mu)} + \frac{1}{(\beta + \theta + \mu)^2}\right)\right\} + e^{-\mu t_1}(M - t_1)\left\{\frac{(\alpha - \eta p + \beta Q)}{\beta}\right.
$$

$$
\left.\left.\left.\times\left(1 - e^{-\beta t_d}\right) + \frac{\theta(\alpha - \eta p)(t_1 - t_d)}{(\theta + \beta)} + \frac{\beta(\alpha - \eta p)}{(\theta + \beta)^2}\left(e^{(\beta+\theta)(t_1 - t_d)} - 1\right)\right\}\right]\right]
$$

$$
\tag{16.32}
$$

The total annual cost when $Q > W$, which is a function of t_1 and T, is given by

$$
TC(t_1, T) = \begin{cases} TC_4(t_1, T) & \text{if } 0 < M \leq t_d \\ TC_5(t_1, T) & \text{if } t_d < M \leq t_1 \\ TC_6(t_1, T) & \text{if } M > t_1 \end{cases}
$$

where

$$
TC_4(t_1, T) = \frac{1}{T}\left[A + \frac{k(\alpha - \eta p)}{\beta}\left[\frac{1}{(\beta + \mu)}\left(e^{\beta t_w} - e^{-\mu t_w}\right) + \frac{1}{\mu}\left(e^{-\mu t_w} - 1\right)\right]\right.
$$

$$
+ h\left[\frac{W}{\mu}\left(1 - e^{-\mu t_w}\right) - \frac{(\alpha - \eta p + \beta W)}{\beta(\beta + \mu)}\left(e^{\beta(t_w - t_d) - \mu t_d} - e^{-\mu t_w}\right)\right.
$$

$$+ \frac{(\alpha - \eta p)}{\beta \mu} \left(e^{-\mu t_d} - e^{-\mu t_w} \right) + \frac{(\alpha - \eta p)}{(\theta + \beta)(\beta + \theta + \mu)} \left(e^{(\beta + \theta)(t_1 - t_d) - \mu t_d} - e^{-\mu t_1} \right)$$

$$+ \frac{(\alpha - \eta p)}{(\theta + \beta) \mu} \left(e^{-\mu t_1} - e^{-\mu t_d} \right) \bigg] + \frac{c\theta(\alpha - \eta p)}{(\theta + \beta)}$$

$$\times \left[\frac{1}{(\theta + \beta + \mu)} \left(e^{(\theta + \beta)(t_1 - t_d) - \mu t_d} - e^{-\mu t_1} \right) + \frac{1}{\mu} \left(e^{-\mu t_1} - e^{-\mu t_d} \right) \right]$$

$$+ s(\alpha - \eta p) \left[\frac{(1 - \delta T)}{\mu^2} \left\{ ((t_1 - T)\mu - 1) e^{-\mu T} + e^{-\mu t_1} \right\} \right.$$

$$\left. + \frac{\delta}{2\mu^3} \left\{ ((t_1^2 - T^2)\mu^2 - 2T\mu - 2) e^{-\mu T} + 2(t_1 \mu + 1) e^{-\mu t_1} \right\} \right]$$

$$+ \frac{\pi(\alpha - \eta p)\delta}{\mu^2} \left[e^{-\mu T} - \left\{ (t_1 - T)\mu + 1 \right\} e^{-\mu t_1} \right]$$

$$+ cI_p \left[\frac{(\alpha - \eta p + \beta W) e^{-\beta t_w}}{(\beta + \mu)\beta} \left(e^{-(\beta + \mu)M} - e^{-(\beta + \mu)t_d} \right) + \frac{(\alpha - \eta p)}{\beta \mu} \left(e^{-\mu t_d} - e^{-\mu M} \right) \right.$$

$$\left. + \frac{(\alpha - \eta p)}{(\theta + \beta)} \left\{ \frac{1}{(\beta + \theta + \mu)} \left(e^{(\theta + \beta)(t_1 - t_d) - \mu t_d} - e^{-\mu t_1} \right) + \frac{1}{\mu} \left(e^{-\mu t_1} - e^{-\mu t_d} \right) \right\} \right]$$

$$- pI_e \left[(\alpha - \eta p) \left\{ \frac{(e^{\beta t_w} - e^{-\mu t_w})}{(\beta + \mu)^2} - \frac{t_w e^{-\mu t_w}}{(\beta + \mu)} \right\} + (\alpha - \eta p + \beta W) \right.$$

$$\left. \times \left\{ \frac{\left(e^{\beta t_w - (\beta + \mu)M} - e^{-\mu t_w} \right)}{(\beta + \mu)^2} + \frac{\left(Me^{\beta t_w - (\beta + \mu)M} - t_w e^{-\mu t_w} \right)}{(\beta + \mu)} \right\} \right] \right] \quad (16.33)$$

$$TC_5(t_1, T) = \frac{1}{T} \left[A + \frac{k(\alpha - \eta p)}{\beta} \left[\frac{1}{(\beta + \mu)} \left(e^{\beta t_w} - e^{-\mu t_w} \right) + \frac{1}{\mu} \left(e^{-\mu t_w} - 1 \right) \right] \right.$$

$$+ h \left[\frac{W}{\mu} \left(1 - e^{-\mu t_w} \right) - \frac{(\alpha - \eta p + \beta W)}{\beta(\beta + \mu)} \left(e^{\beta(t_w - t_d) - \mu t_d} - e^{-\mu t_w} \right) \right.$$

$$+ \frac{(\alpha - \eta p)}{\beta \mu} \left(e^{-\mu t_d} - e^{-\mu t_w} \right) + \frac{(\alpha - \eta p)}{(\theta + \beta)(\beta + \theta + \mu)} \left(e^{(\beta + \theta)(t_1 - t_d) - \mu t_d} - e^{-\mu t_1} \right)$$

$$+ \frac{(\alpha - \eta p)}{(\theta + \beta) \mu} \left(e^{-\mu t_1} - e^{-\mu t_d} \right) \bigg]$$

$$+ \frac{c\theta(\alpha - \eta p)}{(\theta + \beta)} \left[\frac{1}{(\theta + \beta + \mu)} \left(e^{(\theta + \beta)(t_1 - t_d) - \mu t_d} - e^{-\mu t_1} \right) + \frac{1}{\mu} \left(e^{-\mu t_1} - e^{-\mu t_d} \right) \right]$$

$$+ s(\alpha - \eta p) \left[\frac{(1 - \delta T)}{\mu^2} \left\{ ((t_1 - T)\mu - 1) e^{-\mu T} + e^{-\mu t_1} \right\} \right.$$

$$\left. + \frac{\delta}{2\mu^3} \left\{ ((t_1^2 - T^2)\mu^2 - 2T\mu - 2) e^{-\mu T} + 2(t_1 \mu + 1) e^{-\mu t_1} \right\} \right]$$

$$+ \frac{\pi(\alpha - \eta p)\delta}{\mu^2} \left[e^{-\mu T} - \{(t_1 - T)\mu + 1\} e^{-\mu t_1} \right]$$

$$+ \frac{c I_p (\alpha - \eta p)}{(\theta + \beta)} \left[\frac{1}{(\theta + \beta + \mu)} \left\{ e^{(\theta + \beta)(t_1 - M) - \mu M} - e^{-\mu t_1} \right\} + \frac{1}{\mu} \left(e^{-\mu t_1} - e^{-\mu M} \right) \right]$$

$$- p I_e \left[(\alpha - \eta p) \left\{ \frac{e^{\beta t_w}}{(\beta + \mu)^2} - e^{-\mu t_w} \left(\frac{t_w}{(\beta + \mu)} + \frac{1}{(\beta + \mu)^2} \right) \right\} + (\alpha - \eta p + \beta W) \right.$$

$$\times \left\{ e^{\beta(t_w - t_d) - \mu t_d} \left(\frac{t_d}{(\beta + \mu)} + \frac{1}{(\beta + \mu)^2} \right) - e^{-\mu t_w} \left(\frac{t_w}{(\beta + \mu)} + \frac{1}{(\beta + \mu)^2} \right) \right\}$$

$$+ \frac{(\alpha - \eta p)\theta}{\mu^2} \left\{ e^{-\mu t_d} (1 + \mu t_d) - e^{-\mu M} (1 + \mu M) \right\} + \frac{\beta(\alpha - \eta p)}{(\theta + \beta)}$$

$$\times \left\{ e^{(\beta + \theta)(t_1 - t_d) - \mu t_d} \left(\frac{t_d}{(\beta + \theta + \mu)} + \frac{1}{(\beta + \theta + \mu)^2} \right) \right.$$

$$\left. \left. - e^{(\beta + \theta)(t_1 - M) - \mu M} \left(\frac{M}{(\beta + \theta + \mu)} + \frac{1}{(\beta + \theta + \mu)^2} \right) \right\} \right] \right] \qquad (16.34)$$

$$\mathrm{TC}_6(t_1, T) = \frac{1}{T} \left[A + \frac{k(\alpha - \eta p)}{\beta} \left[\frac{1}{(\beta + \mu)} \left(e^{\beta t_w} - e^{-\mu t_w} \right) + \frac{1}{\mu} \left(e^{-\mu t_w} - 1 \right) \right] \right.$$

$$+ h \left[\frac{W}{\mu} \left(1 - e^{-\mu t_w} \right) - \frac{(\alpha - \eta p + \beta W)}{\beta(\beta + \mu)} \left(e^{\beta(t_w - t_d) - \mu t_d} - e^{-\mu t_w} \right) \right.$$

$$+ \frac{(\alpha - \eta p)}{\beta \mu} \left(e^{-\mu t_d} - e^{-\mu t_w} \right) + \frac{(\alpha - \eta p)}{(\theta + \beta)(\beta + \theta + \mu)} \left(e^{(\beta + \theta)(t_1 - t_d) - \mu t_d} - e^{-\mu t_1} \right)$$

$$\left. + \frac{(\alpha - \eta p)}{(\theta + \beta)\mu} \left(e^{-\mu t_1} - e^{-\mu t_d} \right) \right]$$

$$+ \frac{c\theta(\alpha - \eta p)}{(\theta + \beta)} \left[\frac{1}{(\theta + \beta + \mu)} \left(e^{(\theta + \beta)(t_1 - t_d) - \mu t_d} - e^{-\mu t_1} \right) + \frac{1}{\mu} \left(e^{-\mu t_1} - e^{-\mu t_d} \right) \right]$$

$$+ s(\alpha - \eta p) \left[\frac{(1 - \delta T)}{\mu^2} \left\{ ((t_1 - T)\mu - 1) e^{-\mu T} + e^{-\mu t_1} \right\} \right.$$

$$\left. + \frac{\delta}{2\mu^3} \left\{ ((t_1^2 - T^2)\mu^2 - 2T\mu - 2) e^{-\mu T} + 2(t_1 \mu + 1) e^{-\mu t_1} \right\} \right]$$

$$+ \frac{\pi(\alpha - \eta p)\delta}{\mu^2} \left[e^{-\mu T} - \{(t_1 - T)\mu + 1\} e^{-\mu t_1} \right]$$

$$- p I_e \left[(\alpha - \eta p + \beta Q) \left\{ \frac{1}{(\beta + \mu)^2} - \frac{e^{-(\beta + \mu)t_d}}{(\beta + \mu)^2} \left(\frac{t_d}{(\beta + \mu)} + \frac{1}{(\beta + \mu)^2} \right) \right\} \right.$$

$$+ \frac{(\alpha - \eta p)\theta}{\mu^2} \left\{ e^{-\mu t_d} (1 + \mu t_d) - e^{-\mu t_1} (1 + \mu t_1) \right\} + \frac{\beta(\alpha - \eta p)}{(\theta + \beta)}$$

$$
\times \left\{ e^{(\beta+\theta)(t_1-t_d)-\mu t_d} \left(\frac{t_d}{(\beta+\theta+\mu)} + \frac{1}{(\beta+\theta+\mu)^2} \right) \right.
$$

$$
\left. - e^{-\mu t_1} \left(\frac{t_1}{(\beta+\theta+\mu)} + \frac{1}{(\beta+\theta+\mu)^2} \right) \right\} + e^{-\mu t_1} (M-t_1) \left\{ \frac{(\alpha-\eta p + \beta Q)}{\beta} \right.
$$

$$
\left. \left. \times \left(1 - e^{-\beta t_d} \right) + \frac{\theta(\alpha-\eta p)(t_1-t_d)}{(\theta+\beta)} + \frac{\beta(\alpha-\eta p)}{(\theta+\beta)^2} \left(e^{(\beta+\theta)(t_1-t_d)} - 1 \right) \right\} \right]
$$

$$
\tag{16.35}
$$

16.11   NUMERICAL ILLUSTRATIONS AND ANALYSIS

On the basis of the previous studies, the following data in proper units have been measured for solving the equations of both the OW and the RW. We discuss two examples based on trade credits.

16.11.1   EXAMPLE

When $M \le t_d$: $A = 250$; $W = 300$; $\alpha = 800$; $\beta = 1.5$; $\delta = 0.58$; $h = 15$; $k = 18$; $s = 30$; $p = 75$; $c = 60$; $\eta = 0.1$; $\mu = 0.09$; $\pi = 25$; $I_p = 0.15$; $I_e = 0.12$; $M = 0.1856$; $t_d = 0.2064$ in proper units.

When $M > t_d$: $A = 250$; $W = 300$; $\alpha = 800$; $\beta = 1.5$; $\delta = 0.58$; $h = 15$; $k = 18$; $s = 30$; $p = 75$; $c = 60$; $\eta = 0.1$; $\mu = 0.09$; $\pi = 25$; $I_p = 0.15$; $I_e = 0.12$; $M = 0.2432$; $t_d = 0.2094$ in proper units.

With these values, we find different solutions of the system for the two cases: $M \le t_d$ and $M > t_d$.

Optimal values of t_1, T, and TC have been computed for Example 16.11. Evaluated outcomes are tabulated in Tables 16.1 and 16.2:

16.12   SENSITIVITY ANALYSIS

With the provided data, we also perform a sensitivity analysis of most favorable system cost with respect to the various system parameters for Example 16.11 (Case $Q > W$). The results obtained are tabulated in Tables 16.3 and 16.4

Table 16.1

Optimal Solution for the Model When $Q \le W$

Cases	t_1	T	Total Cost (TC)
$M \le t_d$	1.1835	3.3118	13206.9
	1.0989	2.9032	13982.7
$M > t_d$	1.7638	1.8797	8631.13
	1.3229	1.6523	8124.43

Table 16.2
Optimal Solution for the Model When $Q > W$

Cases	t_1	T	Total Cost (TC)
$M \leq t_d$	0.2269	1.2785	10618.2
	0.2649	1.8519	10958.0
$M > t_d$	0.2186	0.8582	4476.48
	0.2293	0.8746	4329.59

Table 16.3
Effects of Changes in Parameters Ordering Cost (A), Selling Price (c), Shortage Cost (s), and Capacity of OW (W) of the Inventory Model

Parameters	% Change in Parameters	$M \leq t_d$ % Change in Total Cost	$M > t_d$ % Change in Total Cost
A	−50	−0.9235	−3.2449
	−25	−0.4622	−1.6224
	+25	+0.4622	+1.6222
	+50	+0.9244	+3.2446
P	−50	+2.4663	+17.8597
	−25	+1.2332	+8.9299
	25	−1.2322	−8.9301
	50	−2.4654	−17.8599
S	−50	−48.2945	−63.7561
	−25	−24.1471	−31.8781
	25	+24.1480	+31.8778
	50	+48.2952	+63.7559
W	−50	+4.2219	+23.8285
	−25	+2.3749	+12.7261
	25	−2.7732	−13.9478
	50	−5.8532	−28.8378

16.13 OBSERVATIONS

16.13.1: From Table 16.3, it is clear that the % change in the total cost is decreasing for the parameters p and W, whereas it is increasing for the parameters A and s, when the trade credit period is $\leq t_d$.

16.13.2: From Table 16.3, it is observed that the % change in the total cost is highly sensitive with respect to parameters p and W, whereas it is increasing for the parameters A and s, when the trade credit period is $> t_d$.

16.13.3: From Table 16.4, it is clear that the % change in the total cost is decreasing for the parameter β, whereas it is increasing for the parameters c, α, δ, h, and I_p, when the trade credit period is $\leq t_d$.

Table 16.4

Effects of Changes in Purchasing Cost (p), Demand Parameter (α), Backlogging Rate (δ), Stock Parameter in Demand Rate (β) and Holding Cost of the OW (h), Capital Opportunity Cost (I_p), and Interest Earned (I_e) of the Inventory Model

Parameters	% Change in Parameters	$M \le t_d$ % Change in Total Cost	$M > t_d$ % Change in Total Cost
c	-50	-0.2071	$+3.1293$
	-25	-0.1035	$+1.5647$
	25	$+0.1045$	-1.5648
	50	$+0.2080$	-3.1295
A	-50	-55.4209	-77.1029
	-25	-27.3647	-37.4909
	25	$+26.9598$	$+36.2472$
	50	$+53.6618$	$+71.7030$
Δ	-50	-2.9332	-3.8724
	-25	-1.4666	-1.9362
	25	$+1.4666$	$+1.9362$
	50	$+2.9332$	$+3.8724$
B	-50	$+13.2182$	$+51.7885$
	-25	$+3.4547$	$+13.6720$
	25	-1.2134	-4.9619
	50	-1.4713	-6.1868
H	-50	-0.1054	-0.1160
	-25	-0.0537	-0.0580
	25	$+0.0537$	$+0.0578$
	50	$+0.1064$	$+0.1158$
I_p	-50	-0.2071	$+3.1293$
	-25	-0.1035	$+1.5647$
	25	$+0.1045$	-1.5649
	50	$+0.2080$	-3.1295
I_e	-50	$+2.4663$	17.8619
	-25	$+1.2331$	8.9299
	25	-1.2322	-8.9301
	50	-2.4654	-17.8599

16.13.4: From Table 16.4, it is clear that the percentage change in the total cost is decreasing for the parameters c, β, and I_p, whereas it is increasing for the parameters α, δ, and h, when the trade credit period is $>t_d$.

16.14   CONCLUDING REMARKS

This inventory model comprises the some practical features that are feasible to be related to some types of inventory models. We developed a model, which will help the retailer find out the optimal replenishment policy for non-instantaneous deteriorating products with restricted storage capacity. In this model, shortages are allowed and partially backlogged. Mostly, few customers will wait for backorders, and a few others will satisfy their needs from other sellers. The whole study is prepared in inflationary atmosphere. This model is very useful for an inventory control of constructive non-instantaneous deteriorating products, such as ration, electronic machinery, and designer products. In the future study, this model will incorporate some other assumptions, such as probabilistic order and a restricted replenishment.

APPENDIX I

From Figure 16.3, in the time period $[0, t_d]$, at time t in OW, the stage of inventory is governed by the following differential equation:

$$\frac{dI(t)}{dt} = -[\alpha + \beta I(t)] \quad 0 \le t \le t_d$$

under the boundary condition $I(0) = Q$. The solution of the above differential equation is

$$I(t) = Qe^{-\beta t} + \frac{\alpha}{\beta}\left(e^{-\beta t} - 1\right), \quad 0 \le t \le t_d$$

APPENDIX II

Necessary conditions for the TIC per unit time of Equations (16.30)–(16.32) for the case when $Q \le W$ and Equations (16.33)–(16.35) for the case when $Q > W$ to be minimum, the optimal values ($t_1 = t_1^*$ and $T = T^*$) can be evaluated using successive equations,

$$\frac{\partial \text{TC}_i(t_1, T)}{\partial t_1} = 0 \qquad (16.36)$$

$$\text{and } \frac{\partial \text{TC}_i(t_1, T)}{\partial T} = 0, \qquad (16.37)$$

$i = 1, 2, \ldots, 5, 6$ for both cases when $Q \le W$ and $Q > W$, provided

$$\left.\frac{\partial^2 \text{TC}_i(t_1, T)}{\partial t_1^2}\right|_{(t_1^*, T^*)} > 0, \left.\frac{\partial^2 \text{TC}_i(t_1, T)}{\partial T^2}\right|_{(t_1^*, T^*)} > 0 \text{ and}$$

$$\left.\left(\frac{\partial^2 \text{TC}_i(t_1, T)}{\partial t_1^2}\right)\left(\frac{\partial^2 \text{TC}_i(t_1, T)}{\partial T^2}\right) - \left(\frac{\partial^2 \text{TC}_i(t_1, T)}{\partial t_1 \partial T}\right)^2\right|_{(t_1^*, T^*)} > 0,$$

REFERENCES

1. Sarma K.V.S., A deterministic inventory model with two levels of storage and an optimum release rule. *Opsearch*, 1983, 20(3):174–180.
2. Yang H.L., Two-warehouse inventory models for deteriorating items with shortages under inflation. *European Journal of Operational Research*, 2004, 157: 344–356.
3. Kumar M., Chauhan A., and Kumar R., A deterministic inventory model for deteriorating items with price dependent demand and time varying cost under trade credit. *International Journal of Soft Computing and Engineering*, 2012, 2(1): 99–105.
4. Singh S.R., Malik A.K. Effect of inflation on two warehouse production inventory systems with exponential demand and variable deterioration. *International Journal of Mathematical and Applications*, 2009, 4(1): 2–12.
5. Zhou Y.W. and Yang S.L., A two-warehouse inventory model for items with stock-level-dependent demand rate. *International Journal of Production Economics*, 2005, 95: 215–228.
6. Kumar M. and Aarya D.D., Optimal payment policy with price-dependent demand and three parameters dependent deterioration rate under the influence trade credit. *International Journal of Operational Research*, 2016, 26(2): 236–251.
7. Huang Y.-F., Buyer's optimal ordering policy and payment policy under supplier credit. *International Journal of Systems Science*, 2005, 36(13): 801–807.
8. Kumar M., Singh S.R., and Pandey R.K., An inventory model with quadratic demand rate for decaying items with trade credits and inflation. *Journal of Interdisciplinary Mathematics*, 2009 12(3): 331–343.
9. Liao H.C., Tsai C.H., and Su C.T., An inventory model for deteriorating items under inflation when a delay in payment is permissible. *International Journal of Production Economics*, 2000, 63: 207–214.
10. Padmanabhan G. and Vrat P., EOQ models for perishable items under stock dependent selling rate. *European Journal of Operational Research*, 1995, 86: 281–292.

Index